Multimedia Services and Streaming for Mobile Devices:

Challenges and Innovations

Alvaro Suarez Sarmiento
Las Palmas de Gran Canaria University, Spain

Elsa Macias Lopez
Las Palmas de Gran Canaria University, Spain

Information Science
REFERENCE

Senior Editorial Director:	Kristin Klinger
Director of Book Publications:	Julia Mosemann
Editorial Director:	Lindsay Johnston
Acquisitions Editor:	Erika Carter
Development Editor:	Myla Harty
Production Editor:	Sean Woznicki
Typesetters:	Adrienne Freeland, Mackenzie Snader, Jennifer Romanchak
Print Coordinator:	Jamie Snavely
Cover Design:	Nick Newcomer

Published in the United States of America by
Information Science Reference (an imprint of IGI Global)
701 E. Chocolate Avenue
Hershey PA 17033
Tel: 717-533-8845
Fax: 717-533-8661
E-mail: cust@igi-global.com
Web site: http://www.igi-global.com

Library of Congress Cataloging-in-Publication Data

Multimedia services and streaming for mobile devices: challenges and innovation / Alvaro Suarez Sarmiento and Elsa Macias Lopez, editors.
p. cm.
Includes bibliographical references and index.
ISBN 978-1-61350-144-3 (hbk.) -- ISBN 978-1-61350-145-0 (ebook) -- ISBN 978-1-61350-146-7 (print & perpetual access) 1. Multimedia communications. 2. Wireless communication systems. 3. Multimedia systems. 4. Internet industry. I. Suarez Sarmiento, Alvaro, 1963- II. Macias Lopez, Elsa, 1971-
TK5105.15.M85 2012
006.7--dc23

2011024237

British Cataloguing in Publication Data
A Cataloguing in Publication record for this book is available from the British Library.

All work contributed to this book is new, previously-unpublished material. The views expressed in this book are those of the authors, but not necessarily of the publisher.

To our daughter Alba Suárez Macías for all the time we stole her for editing this book. To our families. To the Editorial Advisory Board for their work, and to other people that knows they have been actively implicated in this book.

Table of Contents

Foreword

MEDIA STREAMING, MOBILE DEVICES AND RELATED SERVICES

The media streaming and the mobile devices are two research and development areas that are in constant change. The link between them is the mobile and wireless networks that are used to communicate the mobile devices naturally. The combination of mobile devices and media streaming leads to very interesting mobile multimedia services and new business models.

In Spanish starting a sentence with the expression "Three words ..." means the writer wants to present only the main ideas and starting points to readers that allow them to seek more information. In this sense we have titled our first sections starting by this Spanish expression. The objective is to write a short foreword about the things the reader can find in this book.

"THREE WORDS" ABOUT MEDIA STREAMING

Streaming. It is difficult to understand technical concept for novice researchers and under-graduate students. *Media Streaming*. It is a difficult to understand technique that is not always understood in their precise sense. Let us write "three words" about this.

Streaming is related to the theoretic concept of System Theory named *overlapping* of concurrent atomic actions. An atomic action is the one that starts, executes and finishes with no interruption of other action. That is, it is observable and is the one that occurs in a given interval of time. Concurrent atomic actions can occur in different parts of a system at the same time and independently. For example, one shop assistant is shelling tomatoes (action one) while another one is reading the bill of a buyer (action two). The system is the shop where the shop assistants are and a person outside the shop can see these two concurrent actions. The overlapping means the instants of time when the actions start (execute and finish) is not the same. But after their starting instant of time there is an interval of time when they can be simultaneously observed. It is easy to realize that a network of devices is a system in which these overlapping occur for every instance of time. Readers interested in full understanding of Streaming technique can read several works. As an example we mention the following: For instance, Rodriguez and Campelo (2003) hid the memory latency in front of the processor speed, Chan and Wong (2005) overlapped actions in the Job-Shop Scheduling problem, Kang and Zhou (2007) hid the latency of Web page's objects communication, Padmanabhan and Lily (2009) hid the latency of communication among parallel processors and Sharanya and Matthew (2009) efficiently used bandwidth in wireless sensor networks.

Now is the turn to explain Media Streaming. In few words, Media streaming is the application of Streaming technique to *Streams*. But what is a Stream? Brownlee and Claffy (2002) defined a Stream as an "individual *Internet Protocol* (*IP*) session between ports on pairs of hosts". English dictionary defines Stream as "*narrow river*". Informally, Stream can be defined as a set of data packets that flow in the path between an origin node and a final node (extreme to extreme). The overlapping of communication actions (in the links) and processing actions (in the intermediate devices) is clear and natural. Thus, *Media Streaming* is the overlapping of the processing of the information pieces in the hosts (terminals) and the communication of their associated streams in the network. Austerberry (2005) presented an understandable introduction to the typical architecture of video and audio streaming system. Rao and Bojkovic (2006) did a high quality review of media streaming, discussed a very interesting wide vision of current research topics in media streaming and presented different network architectures, protocols, coding and *Quality of Service* (*QoS*) mechanisms. Media Streaming is the key technique to implement multimedia services in actual networks.

Several indicators have shown that video streaming is well implanted in Information Society: First Erman and Gerber (2009) showed that the usage of the *HiperText Transfer Protocol* (*HTTP*) to distribute multimedia content exhibited an 83% annualized growth rate. Second, recent statistics showed streaming traffic surpasses *Peer-to-Peer* (*P2P*) traffic in Internet. Third, streaming was the key technique to implement multimedia services on broadband Internet (Esaki & Sunahara, 2008). Lastly, a *Publish/Subscribe* architecture of Internet is proposed by Katsaros and Fotiou (2009). Furthermore, video streaming has had important consequences in Enterprise 2.0 (Tuten 2008). These consequences have forced video streaming to meet some technical and QoS requirements (Xipeng, 2008). Zhu et al. (2011) have recently edited a book on media streaming.

The future of multimedia services is linked to streaming technique. Moreover, the implementation of streaming technique in wireless networks is challenging. This book contributes to this research topic reviewing and presenting research and telco companies' works.

OTHER "THREE WORDS" ABOUT MOBILE DEVICES

Wireless communications and mobile devices are also well implanted technologies in Information Society (Pagani, 2005). Currently, mobile manufacturers offer high-end mobile telephones with on chip *Wireless Network Interface Cards* (*WNIC*) like Bluetooth, *Wireless Fidelity* (*WiFi*), *Global Positioning System* (*GPS*) and *Third Generation* (*3G*) Wireless networks. Mobile operators have improved their networks to facilitate the access to *Video on Demand* (*VoD*) servers (Qadeer & Ahmad, 2009). Their objectives have been to stay competitive, to offer better and more cost-efficient provisioning of old services as well as new services (Dahlman & Parkval, 2007), and to satisfy the user requirements (Pagani & Schipani, 2005). Savo (2006) identified some general challenges that must be taken on to satisfy those user requirements. On the one hand, the dynamic adaptation of the wireless channel and *Medium Access Control (MAC)* protocol to the number of active users in a wireless *Access Point* (*AP*). On the other hand, the design of routing algorithms, transport protocols, security procedures and mobility management. Apostolopoulos and Trott (2005) remarked that the efficient implementation of wireless video streaming service presents some important differences as compared to other services. Verkasalo (2009) studied the access pattern of European users to multimedia services and concluded that video streaming technique implementation in mobile networks, in the years to come, must be improved to provide a good user experience.

Last years have seen an explosion of wireless mobile devices available in the market. All of them are allowed to communicate using the above wireless technologies. Among the recent devices are: e-readers, video consoles, intelligent cars, intelligent cameras, ultra small portable computers and the smart phones. The evolution in the hardware of smart phones (also other mobile widgets) include high performance processors, memory and codec support. Several companies have launched their own version of the desktops operating systems for these devices (*Windows Phone, iOS, Linux* (*Android*)). The success of these operating systems is high and is the responsible of selling a big amount of devices in the last year. *Android* devices (netbooks, smart phones and tablets) have a high volume market. This heterogeneity in the implementation of mobile devices has provoked a problem with the design of multiplatform software (the iOS versions of software must be adapted to Android devices). This heterogeneity also influences the design and implementation of streaming protocols appropriated for these devices. Main companies that dominate the market with their proprietary versions of Streaming servers (*Windows Media Server* (generally included in the *Windows Server* operating system), the *RealNetworks Media Server*, the *Darwin Streaming Server* (*Apple*) and the *Adobe Flash Media Server* (Sanders, 2008)), have serious competitors: free software and open source projects like *VideoLAN* and *Openflash*. Also the above heterogeneity is a problem for the implementation of a standard protocol for streaming. Traditional version, *Real Time Streaming Protocol* (*RTSP*) (and its variation *Real Time Multimedia Protocol* (*RTMP*)), have a serious competitor: the *HiperText Transfer Protocol* (*HTTP*) that is adapted to issue the streaming to mobile devices. An additional remark is that the success of media streaming is due in part to the increase usage of Internet videos and well known Web sites. That could provoke the massive use of adapted versions of HTTP in the next recent future.

This book contributes to this state of the art presenting some works in this direction in order to analyze the challenges and solutions of several authors.

MOBILE MULTIMEDIA SERVICES

Some recent works include improved mechanisms for mobile multimedia services. We will briefly comment on some of them.

Frossard and Martin (2008) presented multiple paths routing algorithms to take advantage of the structure and organization of the network to efficiently deliver video packets from the VoD Server. Buyya and Pathan (2008) argued in favor of proxy caching, *Content Distribution Network* (*CDN*) and *P2P* paradigms to alleviate two bottlenecks: the VoD Server scheduling and load balancing. Bruneo and Iellamo (2009) argued that Grid Technology reports better results than CDN and P2P. Man-Fung and Chan (2007) considered mobile telephones to access a broadcast TV Server using 3G. These mobile telephones re-sent the TV signal to other mobile telephones using ad hoc WiFi or Bluetooth WNIC. But they neither considered the power consumption nor possible service disruptions in the ad hoc network. Shorfuzzaman and Eskicioglu (2009) analyzed efficient multicast of video streaming for heterogeneous devices (portable computers and Personal Digital Assistants (PDA)) using in-network trans-coding in the network edges and network processors in the routers' WNIC. But current and upcoming high-end mobile telephones and portable computers have powerful processing hardware to efficiently implement transcoding. Bolla and Mangialardi (2009) proposed the adaptation of *Real Time Streaming Protocol* (*RTSP*) to solve access problems of nomadic users. For instance, if the mobile device of the nomadic user changes its IP address, the VoD Server will terminate the streaming session, thus the mobile device starts a new session from scratch with its new IP address.

At the beginning of the twenty-first century the three dominant multimedia services were on line learning (Kurbel, 2003), telemedicine (Murthy & Krishnamurthy, 2004) and m-commerce (Ghinea & Angelides, 2004). Ten years later there is a constellation of services based on video streaming. Examples of very recent research efforts are mobile interactive *Television* (*TV*), wireless video sensors, multiparty audio conferencing (Zhang & Mao, 2008), multicast mobile services in 3G (Rümmler & Gluhak, 2009) and others reviewed in Section 6.9 of (Sauter, 2009). The following is a brief description of recent services:

- *Mobile e-learning*. Learning administrative tools like *moodle* and others more specialized: *e-aula*, *Claroline* and *Manhattan* allow VoD from mobile telephones. Some universities expose the lessons of their star professors in *youtube*. Learning enhanced by technology (Cycon & Schmidt, 2009) is named *Mobile learning* or *seamless learning* or *ubiquitous learning* (Druin, 2009). It allows students to collaborate, on learning topics and tasks, in any time and place using video based dialogs.
- *Media streaming in e-commerce*. Information transmission of digital goods and decision-making (video meeting, real-time surveillance, information services …) can be improved by video streaming technique as indicated by Hou and Xia (2008).
- *Video and music sharing applications*. A user of a *Mobile Social Network* (*MSN*) interacts with each other exclusively using a mobile telephone (Dong & Song, 2009). Silva and Salgado (2006) explained that a *WirelessSpontaneous Network* (WSN) allows the members of MSN to share music and video when they physically meet. Examples of MSN are *JuiceCaster* and *MocoSpace*. A step ahead is the demonstration that a mobile user can cooperate using a WSN to form cooperative groups (Grob & Kuhn, 2009).
- *Podcasting, vodcasting and broadcatching*. A mobile streaming Client can download an audio stored file (*podcast*) or a video stored file (*vodcast*) using a Web interface and *Really Simple Syndication* (*RSS*). Less common for mobile telephones (also MP3 or MP4 gadgets) is to use RSS and P2P technology (*Broadcatching*) to download music or video stored files.
- *Mobile P2P TV and Mobile Interactive TV*. Mobile telephones use 3G or WiFi to access the Internet P2P mobile TV service (Bruns, 2009). But this service is not exempt of inefficiencies (Wu & Li, 2009). The *Justin.tv* and *current.tv* companies follow the *youtube* model. The *Mobile Streaming Live Broadcasting Service* (*Mobile Interactive TV*) is still in its infancy; however some initiatives have appeared recently. For instance, the *Qik*, *Kyte*, *Upstream.tv* and *FlixWagon* enterprises implement a *publish/view* model of mobile video. It exhibits very short times between the publication and the viewing (quasi-interactive service).
- *Multimedia Collaborative Web Servers Applications*. Yankelovich and Horan (2009) thought *Voice Over IP* (*VoIP*) and video streaming should augment the quality of the cooperation among users of a Collaborative Web Server Applications. Current high-end mobile telephones use such applications naturally.
- *Universal Plug and Play Media Server*. Examples of this kind of Server are *TVMOBiLi*, *FreeNAS* and *FUPPES*. Moreover, there is a galaxy of free media servers adapted to specific home gadgets and gaming devices. An additional requirement is they must be *Digital Living Network Alliance (DLNA)* compliant. The mobile telephone can act as a *Control Point* (Client) for these media servers. For instance, *Windows Mobile*, *Symbian*, *iPhone* and *Android* have compatible clients for some of these servers.

Foreseeable niches for streaming technique based multimedia services could be wireless gaming, medical tele-care, on line learning and home networking (Dixit & Prasad, 2008; Panton, 2008).

"THREE FINAL WORDS" ABOUT TECHNOLOGY INNOVATION SYSTEMS

Sometimes the word innovation is confusing. We have heard people confusing innovation and research on one hand and innovation and creativity on the other hand. But innovation is not research and also innovation is not creativity.

Let us write some things about innovation and its relation with research and creativity.

Creativity is the motor of Research. That is, a researcher could be involved in the solution of a particular challenge during long time. That researcher could consult a lot of solutions and understand them perfectly. Probably, they could be put into practice. But a new solution needs necessary the researcher has certain grade of creativity. Without creativity no new solution can arise. The process the researcher has done is named Research. That it, the researcher discovered a problem (sometimes the discovery also needs creativity), then the researcher studies its importance and solutions and finally he/she will present the best solution. It is important to remark that usually, the researcher seeks for the best solution (in Basic Research, the amount of time employed in finding the best solution is not the critical factor to have success in the research process). A lot of research results are written and never will be implemented and transferred to the technological market. Innovation means that exactly. Innovation is the process by which the research results are implemented in the Industry.

The development of an Innovation system in technological industries is very important. That is, a well-tuned innovation system could allow a particular company to improve its product and gain a competitive advantage (over other companies). That is the reason why a lot of big companies have their own research and innovation departments. In those departments there are experts in research and expert in innovation (a better option is to involve all the employees in innovation as important and famous enterprises do currently). An innovation system does not seek for the best improvement of products: it seeks economic productivity. It has a certain amount of time to put the improved product in the Market. We think it is very important to review research works done in the University or research centers but also to present works from the Industry. We think this because the well-tuned research (University) and innovation (Industry) will produce the best for the Information Society in which we are living.

After reading the above paragraphs, the reader perhaps could understand better why we have included in this book some works from the Industry and other works from research center (located at universities mainly). In our case it is important to understand both types of works in order to generate new multimedia services and mobile devices.

Alvaro Suarez Sarmiento
University of Las Palmas de Gran, Spain

Elsa Macias Lopez
University of Las Palmas de Gran, Spain

REFERENCES

Apostolopoulos, J., Trott, M., Kalker, T., & Wai-Tian, T. (2005). Enterprise Streaming: Different Challenges from Internet Streaming. In *IEEE International Conference on Multimedia and Expo.* (pp. 1386-1391). Amsterdam, The Netherlands: IEEE Press.

Austerberry, D. (2005). *The Technology of Video and Audio Streaming*. Oxford, UK: Focal Press.

Bolla, R., Mangialardi, S., Rapuzzi, R., & Repetto, M. (2009). Streaming Multimedia Contents to Nomadic Users in Ubiquitous Computing Environments. In *Proceedings of the 28th Conference on Computer Communications (Mobile Video Delivery Workshop), INFOCOM2009,* (pp. 1-6). IEEE Press.

Brownlee, N. & Claffy, K.C. (2002). Understanding Internet Traffic Streams: Dragonflies and Tortoises. *IEEE Communications Magazine, 40*(10), 110:117.

Bruneo, D., Iellamo, G., Minutoli, G. & Puliafito, A. (2009). GridVideo: A Practical Example of Nonscientific Application on the Grid. *IEEE Transactions on Knowledge and Data Engineering, 21*(5), 666:680.

Bruns, A. (2009). The User-Led Disruption: Self-(Re)broadcasting at Justin.tv and Elsewhere. In *ACM Seventh European Conference on European Interactive Television Conference.* (pp. 87-90). Leuven, Belgium: ACM Press.

Chan, F., Wong, T., & Chan, P. (2005). Lot Streaming Technique in Job- shop Environment, *IEEE Mediterranean Conference on Control and Automation: Vol. 1,* (pp. 364-369). IEEE Press.

(2008). Content Distribution Networks . InBuyya, R., Pathan, M., & Vakali, A. (Eds.), *Lecture Notes in Electrical Engineering* (*Vol. 9*). Berling, Germany: Springer Verlag.

Cycon, H. L., Schmidt, T. C., Hege, G., Wählisch, M., & Palkow, M. (2009). Let's Meet at the Mobile – Learning Dialogs with a Video Conferencing Software for Mobile Devices. *International Journal of Interactive Mobile Technologies*, *3*(3), 21–25.

Dahlman, E., Parkval, S., Skol, J., & Beming, P. (2007). *3G Evolution: HSPA and LTE for Mobile Broadband.* London, UK: Academic Press.

Dixit, S., & Prasad, R. (Eds.). (2008). *Technologies for Home Networking*. Hoboken, New Jersey: John Wiley & Sons, Ltd.doi:10.1002/9780470196533

Dong, Z., Song, G., Xien, K., & Wang, J. (2009). An Experimental Study of Large-Scale Mobile Social Network. In 18th *ACM World Wide Web Conference.* (pp. 1175-1176). Madrid, Spain: ACM Press.

Druin, A. (Ed.). (2009). *Mobile Technology for Children. Designing for Interaction and Learning*. Burglington, MA, USA: Morgan Kauffman Publishers.

Erman, J., Gerber, A., Hajiaghayi, M. T., Pei, D., & Spatscheck, O. (2009). Network-Aware Forward Caching. In 18th *ACM World Wide Web Conference.* (pp. 291-300). Madrid, Spain: ACM Press.

Esaki, H., Sunahara, H., & Sunahara, J. (Eds.). (2008). *Broadband Internet Deployment in Japan*. Tokyo, Japan: IOS Press.

Frossard, P., Martin, J. C., & Civanlar, M. R. (2008). Media Streaming With Network Diversity. *Proceedings of the EEE, 96*(1), 39–53.

Ghinea, G., & Angelides, M. C. (2004). Quality of Perception in M- Commerce . In Shi, N. (Ed.), *Mobile Commerce Applications* (pp. 284–302). Hershey, PA: IRM Press. doi:10.4018/9781591401827.ch013

Grob, R., Kuhn, M., Wattenhofer, R., & Wirz, M. (2009). Cluestr: Mobile Social Networking for Enhanced Group Communication. In *ACM GROUP '09:* (pp. 81-90). Sanibel Island (FL), USA: ACM Press.

Hou, D., & Xia, H. (2008). Application of Streaming Media Technology in Electronic Commerce. In *Proceedings of the IEEE 2008 International Symposium on Electronic Commerce and Security,* (pp. 850-853). IEEE Computer Society.

Kang, J., Zhou, H., & Yang, H. (2007). Task Decomposition for Communication Computation Overlap to Reengineer a Web-Based System. In *Proceedings of the 11th IEEE International Workshop on Future Trends of Distributed Computing Systems,* (pp. 205-213). IEEE Press.

Katsaros, K., Fotiou, N., Polyzos, G. C., & Xylomenos, G. (2009). Supporting Mobile streaming services in future publish/subscribe networks, *IEEE International Wireless Telecommunications Symposium.* (pp. 1-7). IEEE Press.

Kurbel, K. (2003). *Video Streaming Solutions for Web-Based E-Learning Courses*; in: Khosrow- Pour, M. (Ed.): Information Technology and Organizations: Trends, Issues, Challenges, Solutions, Proceedings of the *2003 IRMA International Conference*, Philadelphia, PA, pp. 874- 876.

Man-Fung, L. & Chan, S.-H.G. (2007), Broadcast-Based Peer-to-Peer Collaborative Video Streaming Among Mobiles. *IEEE Transactions on Broadcasting, 53*(1), 350:361.

Murthy, V. K., & Krishnamurthy, E. V. (2004). Multimedia Computing Environment for Telemedical Applications . In Shi, N. (Ed.), *Mobile Commerce Applications* (pp. 95–115). Hershey, PA: IRM Press. doi:10.4018/9781591401827.ch005

Padmanabhan, S. P., Lily, B. M., Steven, W. S., Rahul, S., & Casey, J. H. (2009). SLIPstream: Scalable Low-latency Interactive Perception on Streaming Data, *Proceedings of the ACM 18th International Workshop on Network and Operating Systems Support for Digital Audio and Video.* (pp. 43-48). IEEE Press.

Pagani, M. (Ed.). (2005). *Mobile and Wireless Systems Beyond 3G: Managing New Business Opportunities.* Hershey, PA: IRM Press. doi:10.4018/978-1-59140-570-2

Pagani, M., & Schipani, D. (2005). Motivations and Barriers to the Adoption of 3G Mobile Multimedia Services: An End User Perspective in the Italian Market . In Deans, P. (Ed.), *E-commerce and M-commerce Technologies* (pp. 80–95). PA: IGI Global.

Panton, M. (2008). *DLNA for Media Streamers--What Does It All Mean?* Crave - CNET. Retrieved January 30, 2009, from http://news.cnet.com/ 8301-17938_105-10007069-1.html.

Qadeer, M. A., Ahmad, R., Khan, M. S., & Ahmad, T. (2009). Real time video streaming over heterogeneous networks, *IEEE 11th International Conference on Advanced Communication Technology: Vol. 2.* (pp. 1117 - 1122). IEEE Press.

Rao, K., Bojkovic, Z., & Milovanovic, D. (2006). *Introduction to Multimedia Communications: Applications, Middleware, Networking*. West Sussex, England: John Wiley & Sons, Ltd.

Rodriguez, F., Campelo, J. C. & Serrano, Juan J. (2003). Improving the Interleaved Signature Instruction Stream Technique, *IEEE Canadian Conference on Electrical and Computer Engineering: Vol. 1*. (pp. 93-96). IEEE Press.

Rümmler, R., Gluhak, A., & Aghvami, A. H. (2009). *Multicast in Third-Generation Mobile Networks Services, Mechanisms and Performance*. West Sussex, England: John Wiley & Sons, Ltd. doi:10.1002/9780470742549

Sanders, W. B. (2008). *Learning Flash Media Server 3*. Sebastopol, CA (USA): O' Reily Media, Inc.

Sauter, M. (2009). *Beyond 3G – Bringing Networks, Terminals and the Web Together. LTE, WiMAX, IMS, 4G Devices and the Mobile Web 2.0*. West Sussex, England: John Wiley & Sons, Ltd.

Savo, G. G. (2006). *Advanced Wireless Networks: 4G Technologies*. West Sussex, England: John Wiley & Sons, Ltd.

Sharanya, E., Matthew, P. J., Archan, M. & Thomas, L. (2009). Distributed Utility-based Rate Adaptation Protocols for Prioritized, Quasi-Elastic Flows. *ACM Computing and Communications Review, 13*(1), 2:13.

Shorfuzzaman, M., Eskicioglu, R., & Graham, P. (2009). In-Network Adaptation of Video Streams Using Network Processors. *Hindawi Publishing Corporation Advances in Multimedia, 2009*, 1–20.

Silva, V. H., Salgado, E. I., & Quintana, F. R. (2006). AWISPA: An Awareness Framework for Collaborative Spontaneous Networks. In *ASEE/IEEE Frontiers in Education Conference*. (pp. 1-6). San Diego, CA (USA): IEEE Press.

Tuten, T. L. (2008). *Advertising 2.0, Social Media Marketing in a Web 2.0 World*. West Port, USA: Praeger Publishers.

Verkasalo, H. (2009). Contextual Patterns in Mobile Service Usage. *Spinger Verlag Personal Ubiquotous Computing, 13*, 331–342. doi:10.1007/s00779-008-0197-0

Xipeng, X. (2008). *Technical, Commercial, and Regulatory Challenges of QoS*. Burglington, MA, USA: Morgan Kauffman Publishers.

Yankelovich, N., Horan, B., Kaplan, J., Simpson, N., & Slott, J. (2009). Collaboration Tools for Distributed Work - projects include Project Wonderland, MPK20, jVoiceBridge, Porta-Person, Conference Manager, and the Sun Labs Meeting Suite. *Collaborative Environments (Sun Labs)*. Retrieved Jannuary 30, 2009, from http://research.sun.com/projects /dashboard.php?id=85.

Zhu, C., Li, Y., & Niu, X. (Eds.). (2011). *Streaming Media Architectures, Techniques, and Applications: Recent Advances*. Hershey, PA: IGI Global.

Preface

Multimedia services include the processing and communication of image, audio and text in real time. The convergence of computer communication and telecommunication has re-defined current multimedia and coming services. Internet has come to be the new backbone for convergent multimedia services. Examples of convergent services over Internet are *Voice over Internet Protocol* (*VoIP*), *Television over IP* (IPTV), IP Telephony, Real time video streaming, *Video on Demand* (*VoD*) … In the last recent years, a lot of services like these have appear in the Market and new kind of services are expected in the next recent future (in fact every day appear new forms of multimedia services over Internet).

It is well known the high evolution of wireless networks in the recent past. But it is also true that this evolution is being increased every day at present. Current and coming mobile devices (sometimes defined as ultra-small-mobile-computers) are being defined every day: From smart mobile telephones to new multimedia tablets. The topology of interconnection of these wireless mobile devices varies from a *Wireless Personal Area Network* (*WPAN*) to a *Wireless Metropolitan Area Network* (*WMAN*). These networks and mobile devices allow their users to use Internet multimedia services at any time and at any place where there is wireless coverage.

Taking into account the joint work of multimedia services and mobile devices, it is easy to realize a scenario of future impact of coming mobile multimedia services to be implemented in coming mobile devices. The research area of mobile multimedia is active nowadays and some commercial interests are also very important. In this research area there are a lot of challenges and innovations. Examples of challenges are the efficient communication of media streaming in different wireless topologies and technologies, the definition of efficient application level mobile streaming protocols, the codec adaptation to efficiently stream multimedia information, the content adaptation for mobile devices and the efficient distribution of streaming services to mobile users defining a good *Quality of Experience* (*QoE*).

The book editors have a long experience in mobile multimedia streaming over mobile devices. They have organized this book for young and experimented researchers in mobile multimedia, graduate and under-graduate students, university professors (exposition of relevant description cases), people interested in mobile multimedia, network engineers and designers, leaders of emerging technical companies, multimedia mobile applications designers...

The editors (guided by the reviewers) selected 49% of the received chapter proposals. The selection was guided in order to show the reader some challenges and innovations in multimedia services for mobile devices. They selected research works and reviews of current state of the art challenges. The contributors are currently working in research centers at the University and in the department of research and innovation of Telco companies. The idea is to provide not only a vision of current challenges and research solution, but also actual works in the Technical companies. These final works will aid researchers in defining better solutions to the problems they are studying because they will have information about which works companies have interests in (favoring in this way the innovation).

The first challenge is reviewing the identification of Real time constraints for actual mobile telephones and wireless networks. Once these constraints are identified some challenges in streaming video are also identified and specifically those challenges that are influenced by the limited computational resources of mobile devices and bandwidth of mobile networks.

In order to be specific and clear, it is analyzed the specific challenges for different kinds of mobile network topologies. First we analyze the problems of Multimedia in a mobile *Worldwide Interoperability for Microwave Access* (*WiMAX*) network for VoD as well as broadcast TV services. Additional challenges like rate and congestion control are analyzed. And particular solutions: multi-connection streaming and data-partitioning are proposed. The main idea is to show the problems of Streaming in this kind of network. Networks in the short range (*Wireless Personal Area Networks* (*WPAN*) and *Wireless Local Area Networks* (*WLAN*)) are also excellent candidates for implementing future multimedia services based on local streaming servers. It is clear that in the next years wireless technologies for short distances could change the current cable communications for domestic mobile devices. In this area there are several kinds of multimedia streaming based services that could use new standards of *Institute of Electrical and Electronic Engineers* (*IEEE*) and *European Computer Manufacturers Association* (*ECMA*). It is also reviewed those standards to inform the reader of possible new improvements in this kind of communications related to multimedia services. Some challenges and solutions are also presented in order to discover the key features of these new standards favoring innovation in multimedia and mobile devices design.

The Streaming is hard to implement in wireless networks, in part due to time and space varying nature of wireless channels. Thus, to maintain the quality of wireless multimedia streaming throughout the transmission is a challenging task. The adaptation of the codec in real time to the wireless channel conditions using information provided by the *Received Signal Strength Indication* (*RSSI*) and *Medium Access Control* (*MAC*) is a good solution. This topic is also reviewed to show solutions to provide high quality video to mobile users mitigating the adverse effects of wireless channels.

The only way to improve the quality of video reception in mobile devices is not to adapt the codec. There are recent proposals to modify the streaming protocols or propose new ones. Traditional Streaming protocols must be adapted to efficiently issue multimedia information to mobile devices. Recently the old project *hotmedia* have received renewed verve. That is, *HyperText Transfer Protocol* (*HTTP*) is the new candidate to be used for mobile media streaming. Two well known and well commercially established Media platforms mark the forefront in this new tendency. An interesting overview of these two platforms is presented. The utility of this overview is high due to the researchers could analyze the state of the commercial business of media streaming for mobile devices. In this manner they can propose practical and challenging ideas in their streaming research works. Also the graduate students, network engineers, designer … can observe the tendencies in media streaming and mobile devices business. The convergence process between computer communications and traditional telecommunication business is sometimes guided by commercial initiatives. For this reason, an interesting initiative named WIMS 2.0 (born in a Telco company) is presented. Some barriers to converge between Internet multimedia services and traditional Telco services are presented and also some challenges and a new business model.

Nowadays there is an intense diversification among the devices that can display multimedia content to the user. This does not always respond to marketing strategy for selling more devices but in other cases that diversification is necessary in order to attend the different user profiles. In any case, this represents a problem from the point of view of multimedia content visualization in the mobile devices. Media Resource Adaptation for Multimedia Services and Streaming Media for Mobile Telephones are important topics in actual mobile devices. Some devices are unable to receive large audio-visual data volumes in the original quality that they were stored in on the VoD server. The original multimedia content must be issued to different mobile devices and this means the content must be adapted to different formats in real time. This adaptation must also take into account current mobile device capabilities, network characteristics, server capacities and user preferences. To illustrate this challenge several works are presented. The first one is a survey on media resource adaptation in which related standards and specifications are presented. The aim of this survey is to introduce this topic to young researchers on and to update the knowledge of experienced researchers. The second one is a work centered on the design and implementation of remote visualization applications that presents solutions for streaming interactive and customizable (adaptation) multimedia contents to mobile devices. This work introduces the design of applications that treats large multimedia data sets in the mobile clients. Most of today's smart-phones are geared towards a single user experience: reading a book, watching a movie, playing a game or listening to music. Smart mobile phones offer the access to Social networks to share information and access to shared documents and videos. Some timid effort has been done in the area of real time video sharing in mobile devices. The need of a general programming platform to develop applications that help users to connect and share information with each other is discovered in another chapter. It is presented to the Misco framework in order to solve this challenge and treat large data sets in mobile devices. Another very interesting kind of application is m-learning, one that automatically adapts to the current situation and intention of the user. It is explained as the technical architecture behind these kind of applications. The aim is to show recent and future trends of mobile applications for mobile devices that could take into account streaming technique to share large data sets.

In the last few years a lot of work has been done to obtain good metrics for *Quality of Experience* (*QoE*). To cope with this topic an interesting survey is presented aimed at young researchers to introduce in this research area. It also overviews the most recent advances and challenges in assessment and traffic conditioner procedures for wireless multimedia streaming systems. It is presented as the use of well-known objective and subjective QoE metrics, namely, Peak Signal-to-Noise Ratio, Video Quality Metric, Structural Similarity Index and Mean Option Score in a case study.

At this point we turned back to the adaptation of the design of mobile multimedia applications for wireless network topologies. To do this, the book presents two examples of how to optimize Quality of Service (*QoS*) and *QoE* in two interesting wireless network topologies: mesh connected wireless networks and Vehicular *Ad-hoc NETwork* (*VANET*). These two topologies share the ad hoc interconnection of mobile devices. In the first case, an additional backbone of fixed network routers is in charge to efficiently manage the routing of multimedia content. In the second case is a pure ad hoc network in which the mobile devices (we can consider the vehicles as the mobile terminals in this case) circulate at a certain velocity and have several movement restrictions. In both kinds of networks is a large challenge to provide QoE and QoS. Two simulated case studies are shown for both types of topologies. In the first one, results show that Multi-description Coding (not an adaptation of multimedia content technique) achieves better quality on video transmissions when nodes have medium or high mobility; especially when using multipoint-to-point transmission or disjoint paths in a hierarchical structure. In the second one, it is proposed the concept of resilient multi-source streaming using a peer-to-peer overlay network over an urban VANET and the examination of the impact of differing traffic densities and road layouts upon a peer-to-peer overlay network's performance when streaming video. These two updated works are representative of the research that has been done in streaming over new and future kinds of wireless networks and mobile devices. The aim is to show recent and possible future works in this research area.

The editors end the book with a chapter that introduces interesting actual mobile multimedia services. Challenges of mobile video streaming are also presented as well as characteristics of the actual mobile telephones, operating systems, programming environments and commercial video server. The aim is to resume and present new challenges in mobile video streaming. Also the editors present a new manner to solve the analyzed problems using a software agent based proposal to cope with the problem of multiplatform solutions to mobile video streaming. The editors think that in the next years it would be very interesting to cope with efficient multiplatform solutions in mobile multimedia and streaming software.

The different works presented in this book present different faces of the mobile streaming and multimedia services in actual mobile devices. On one hand, an intention of the editors was that the experimenting researcher could find new material in this book (research and commercial proposals). Another intention of the editors was that the interested public could find an interesting introduction to the challenges and innovations in this research area.

Acknowledgment

The editors of this book would like to thank The Editorial Advisory Members for their continuous support and encouragement during the preparation of this book.

Our gratitude to IGI Global staff for the preparation of the book, especially to David DeRicco and Myla Harty, who followed the first and second phases of the edition of this book. We also are grateful to Jan Travers for their support.

We would also like to point out work done by the reviewers. In this case, the reviewers have done key work commenting on the chapters and doing professional work.

Last and mostly important, we are indebted to our daughter Alba, and our families. Their invaluable and relentless support, encouragement, and love are without a doubt the most important reasons behind our successes.

Alvaro Suarez Sarmiento
University of Las Palmas de Gran, Spain

Elsa Macias Lopez
University of Las Palmas de Gran, Spain

Chapter 1
The Challenges of Compressing and Streaming Real Time Video Originating from Mobile Devices

Ngozi V. Uti
University of Cincinnati, USA

Richard Fox
Northern Kentucky University, USA

ABSTRACT

In recent years, mobile phones have become the de facto system of communication across the planet. Mobile phones have helped increase economic growth and critical response in many parts of the world. Mobile phones are even being used for data transmission. However, little academic research has been done on the specific problem of streaming real time video originating from the cameras of mobile devices over cell phone networks. There are many factors that complicate this problem including the limited computational resources of mobile phones, the low and variable bandwidth of cell phone networks, and the need for video compression and streaming algorithms that can be supported by both the mobile phones and cell phone networks. This chapter examines the problems involved and discusses on-going research on the topic. The main goal of this chapter is to identify the real time constraints and challenges of compressing and streaming video from mobile devices for the purpose of designing efficient video compression and streaming techniques that are able to work within the constraints of the limited computational resources and bandwidth available to mobile devices.

INTRODUCTION

Due to the proliferation of mobile communication devices, mobile phones and cell phone networks are rapidly becoming the de facto system of communication and even data transmission in many parts of the world. Just a few years ago in many developing regions of the world, access to a telephone was considered a luxury, but today mobile phones are a common aspect of everyday life. Privatized cell phone networks have succeeded in extending international communications access to regions where state phone monopolies had failed

DOI: 10.4018/978-1-61350-144-3.ch001

to provide even basic service. As an extreme example, in Somalia cell phone companies have extended communications services to the populace over the past decade, despite the collapse of the national government in 1992 (Leeson, 2007).

Contemporary mobile phones have the potential to do much more than old style telephones since they are in reality light computing devices. Aside from the obvious use as telephones, they offer a number of other computational and telecommunication capabilities. In support of communication, for instance, people store address and phone books, calendars and notes on their mobile phones. Thousands of other applications are available, including the ability to access the World Wide Web. Additionally, many mobile phones are capable of producing and viewing high resolution photos and videos. It is only a matter of time before mobile phone users come to expect the ability to stream real time video from mobile phones.

The advantages of streaming real time video from mobile phones are explored in the next section of this chapter. The timeliness of this topic is illustrated by the newest Apple iPhone (Sutter, 2010), which can capture video and transmit it in real time to other iPhones. Most mobile phones today only capture video, store the video in a data file, and then require that the file be transmitted over the cell phone network to a recipient to be played once the entire transmission has been completed, as opposed to a live video stream. The iPhone is able to accomplish the live streaming video only if it is transmitted over *Wireless Fidelity* (*WiFi*).

There are significant challenges that must be dealt with to realize the capability of transmitting live streaming video over cell phone networks. First is the inherently low and variable bandwidth of cell phone networks. Cell phone network performance is impacted by a number of independent problems with the potential for dropping numerous packets. Because the size of a raw video stream greatly exceeds the average bandwidth capacity of cell phone networks, the video must undergo some form of compression prior to transmission. The second challenge is that mobile phones and other mobile devices lack computing resources to compress raw video frames efficiently in real time. These devices have processors that compare with desktop/laptop processors of 5-10 years ago, limited memories, and battery management systems which further reduce computing power as the battery drains. The necessity of compressing raw video in real time for streaming requires video compression algorithms that can execute quickly and yet the limited and variable bandwidth of the cell phone networks require that each video frame be reduced in data size to something that the cell phone network will not drop. And so the ultimate challenge is to identify video compression and streaming techniques that can accomplish the task amid these difficult challenges while still delivering a video of watchable quality.

There are several video compression and streaming algorithms. These algorithms, designed for use on desktop/laptop computers, have computational demands that are not suited to the processors of mobile phones. Therefore, these algorithms cannot simply be ported to mobile phones. These ideas will be explored in the chapter.

The remainder of this chapter addresses the challenges of compressing and streaming real time video from mobile phones over cell phone networks and examines existing algorithms and those now being researched. The second section of this chapter contains an analysis of the significant impact that real time streaming video is envisioned to have on society from the point of view of business, emergency response, law enforcement, and education. The third section addresses the notion of "Real Time". The fourth and fifth sections examine the limitations imposed on this problem by the bandwidth of cell phone networks and the computational resources of mobile devices respectively. Continuing with an examination of mobile devices, the sixth section considers the nature of the video that mobile phones can produce. In the seventh section, video compression algorithms are examined. The eighth section combines all of the concepts from this chapter into a complete strategy to solve the problem, and also provides a discussion of on-going research. The ninth section describes future research directions. Finally, the last section offers some concluding thoughts.

THE SIGNIFICANCE OF REAL TIME MOBILE STREAMING VIDEO

The potential for using mobile devices and cell phone networks to send and receive real time video is significant. Applications can extend well beyond the video phone idea of seeing who you are talking to. The ability of individuals to stream video in real time will open up new opportunities for improved emergency response, document human rights atrocities, aid law enforcement, provide a new tool for journalistic blogging, and improve education in areas that have poor educational capabilities, as well as the more obvious uses as a form of entertainment.

For medical applications, the availability of real time video will allow first-responders (or even bystanders) to send detailed visual information regarding patient trauma to emergency physicians. This can greatly reduce the potential for miscommunication and misinformation when time is at a premium. Physicians could then provide critical feedback and guidance to first-responders in order to save lives. There are also significant applications in terms of data collection for the support of tracking epidemic disease, particularly in places where computers are scarce. Similarly, emergency response for fire and utility interruptions can be greatly enhanced. Witnesses at the scene would be able to stream live video to responders in order to help guarantee appropriate response to developing situations.

A case in point is the Open Data Kit project implemented by Anokwa et al., wherein mobile phones were used in the *Academic Model for the Prevention and Treatment of HIV program* to collect data, including stored video, critical to the monitoring, prevention, and treatment of HIV, Malaria, and Tuberculosis in Kenya (Anokwa, Hartung, Brunette, & Lerer, 2009). An example of using live streaming video is that of ReMoteCare, a system that combines streaming video with a wireless sensor network of pulse oximeters. This combined technology provides health monitoring for an aged care facility with the video streamed directly to a local computer rather than over a network (Fischer, Lim, Lawrence, Ganguli, & Kargl, 2008). The ability to efficiently stream real-time video from mobile devices would significantly enhance programs like these.

There are numerous possible law-enforcement applications. The delivery of real time video already provides police and other security professionals with the ability to perform remote monitoring of a variety of settings. These include routine traffic (for instance, the monitoring of drivers who run red lights or for traffic control), monitoring of people in airports and train stations, crowd control, and even home owner monitoring. While these forms of streaming video are already in use, they either use a private network or the Internet. By streaming directly from a mobile device, surveillance can be performed more cheaply to enhance criminal investigations.

There are a number of economic and development applications as well. Mobile phones have already helped overcome the "information famine" faced by many small business operators in developing countries and rural areas. For example in (Molony, 2009), Tanzanian farmers use mobile phone communications to exchange important information about the supply and demand of products between rural-based farmers and wholesalers in the commercial capital. The ability of these small entrepreneurs to send real time visual information regarding products and services will help to further advance economic development.

Mobile phones are often used to capture events as they unfold. In the youth protests in Iran in

2009, much of the violence from government forces was captured and shared. However, without streaming video, the person capturing the video must make sure that there is time to transmit the file after it has been saved. If the person is caught and the mobile phone confiscated, the video can be erased without having ever been seen by others. Therefore, streaming live video can support journalists who are creating video blogs in real time as well as the ordinary citizen who wants to immediately share their video with others.

Finally, streaming real time video over mobile wireless networks can have a broad impact on education, especially to people with limited literacy skills and limited or no computer Internet access. A survey by The Sloan Consortium found that in the fall of 2008, 4.6 million students in the United States had signed up for at least one Online course (Allen & Seaman, 2010). Online courses may be offered asynchronously or synchronously. However, the expense for synchronous distance education is typically far greater than asynchronous techniques due to the requirement of real time audio/video delivery. The ability to stream video over mobile wireless networks can reduce the cost of synchronous distance education dramatically. Further, in developing countries where state resources are limited, there are great challenges in the ability to provide *any* educational services to rural populations. Real time mobile video streaming offers a cheap mechanism by which to deliver a portable yet rich educational environment to end users.

THE CONCEPT OF "REAL TIME" FOR VIDEO STREAMING

The concept of "Real Time" for streaming video is a complex and at times ambiguous issue. There is neither a single clear definition of the term nor a scholarly consensus on what does or does not qualify as real time. Kehtarnavaz and Gamadia (2006) have examined the literature on this subject, and have provided a very useful three-part categorization of the different views of what denotes real time as tied to different applications and disciplines. In so doing, they present three different interpretations of what constitutes real time video. These are real time in a *perceptual sense*, real time in a *software engineering sense*, and real time in a *signal processing sense*. Each of these interpretations is based upon a certain degree of tolerance for computational delay in the processing and streaming of video. This section discusses and provides examples of each of these three definitions of real time. This section will also provide an explanation for the utilization of real time from a signal processing perspective as used in the remainder of this chapter.

Real time in the perceptual sense relies first and foremost on the human perception of the video. If there is no discernable lag in the reception of the video as interpreted by the user, then the video qualifies as real time. Crucial here is the ability of the human mind to string together images into a perceived smooth motion. Thus, even if frame rates are lower than optimal – research shows that humans can detect flickers at video frame rates below 48 *frames per second* (*fps*) (Seeling, Fitzek, & Reisslein, 2007) – users may still interpret the video as adequate at frame rates as low as 7 fps. This is especially true given the expectations of lower-quality video available in resource-constrained environments such as mobile devices and low-bandwidth Internet connections. Nonetheless, there is a limit to what humans will tolerate, depending on the amount of motion in the video, the smoothness of the frame rates, and whether or not there are gaps evident in the reception of the video. Some researchers, including (Bovik, 2005), have even described the delivery of slightly delayed video as real time in this perceptual sense, so long as the recipient does not detect the delay resulting from processing a few future video frame samples. Thus, even if the future frames are slightly delayed while the current frame is being processed, the result to the user may still appear to be real time video.

Real time in the software engineering sense is concerned with how quickly an accurate output of a process can be made available in an expected amount of time. That is, real time is achieved when a successful and predictable performance is made within a time frame that is required by an application (Stankovic & Ramamritham, 1990). Notably, from the software engineering point of view, video processing need not be completed as quickly or often as possible, but rather the output of the software process simply needs to be available on time, perhaps for input into another dependent software process. As an example, in (Boccignone, Esposito, & Marcelli, 1993), a quality control vision inspection tool using a camera to inspect products in an industrial environment at one second intervals could easily achieve real time processing given that the inspection process is completed in 0.1 s, allowing ample time before the next input requiring inspection occurs. The camera might be capable of far more fps, but this would be unnecessary given that the application parameters only demand product inspection once per second. There is certain flexibility in this definition of real time: as long as video processing is able to complete the particular task of video analysis in a predictable time, the video can be considered to be real time.

In signal processing, video processing is considered to be real time if each video frame is able to be compressed and streamed within the time frame between the production of two consecutive video frames. For example, for a 15 fps video camera, assuming the time to produce the video frame is negligible, there is 0.067 s between consecutive frames. A real time video compression algorithm must be able to completely compress, package and stream each video frame into the network within 0.067 s, in time before the video camera generates the next video frame. As such, according to Kehtarnavaz (2005), to be real time, the time to execute the number of instructions in the video compression algorithm and video streaming process must not exceed the time between video frame samples. The definition of real time in the signal processing sense is strict in that the mobile device must implement an efficient video compression algorithm that is able to utilize the limited computing resources on the device to compress the video frame fast enough to have it streamed without the computing resources on the device becoming a bottleneck. There are other factors that impact this restriction. The computing power available to a mobile device varies according to the battery resources available to the device at that time (Al-Turkistany, Helal, & Schmalz, 2009). The size of the video frame will vary based on the complexity of its picture content. Larger sized video frames (in terms of data) may take longer to encode. If the bandwidth of the cell phone network is low, the compressed video frame will take longer to be streamed. The bandwidth of cell phone networks is highly variable. Therefore, to commit to real time in the signal processing sense requires that all of these factors be taken into account.

For the remainder of this chapter, the definition of real time in the signal processing sense will be used when describing streaming video in real time. This is because mobile devices typically lack the computing resources necessary for the accumulation and processing of several video frames in memory so as to have the video still appear to be real time. Given these limited computing resources, it is necessary for the mobile device to process each video frame as quickly as possible to free up the memory and *Central Processing Unit (CPU)* for use by the video camera to capture the next video frame to be compressed and streamed. In addition, the specific purpose of the video may not be known in advance. With a clear understanding of the impact of real time requirements in the compression and streaming of live video from a signal processing sense, the following sections discus the bandwidth of cell phone networks, the computing resources available to mobile devices, and the nature of the video to be compressed and streamed in real time.

THE BANDWIDTH OF CELL PHONE NETWORKS

Although *Third Generation* (*3G*) cell phone networks are rated to support upwards of 15 Mbps, the typical bandwidth available can be much less, for instance below 300 Kbps (Wisely, 2009). Among the more commonly used networks in the United States are the *Enhanced Data rates for GSM Evolution* (*EDGE*), *High Speed Data Packet Access* (*HSDPA*) and *Evolution-Data Optimized* (*EVDO*) networks. The 2.5G EDGE network has a maximum rated bandwidth of 120 Kbps, but typical rates are around 30 Kbps (Wisely, 2009). The HSPA and EVDO are 3G networks with maximum rated bandwidth of up to 14 Mbps and 3.1 Mbps respectively (Wisely, 2009). 3G network coverage and bandwidth can be inconsistent even in major cities, with actual measured throughput typically ranging between 50 Kbps and 300 Kbps for HSPA, and about 90 Kbps for EVDO (Wisely, 2009).

The causes of the degradation in performance from the rated bandwidth is manifold but include network congestion, bit errors in wireless links, and mobility handoff (Kurose & Ross, 2009). In addition, the bandwidths of wireless links fluctuate significantly due to channel fading and interference (Hassan & Krunz, 2007). The user's physical location, mobility and environmental conditions all play a role in the available bandwidth of wireless networks (Lenders & Martonosi, 2009). Together, these various factors can result in highly variable and unpredictable bandwidth. It is critical that real time video be adapted to meet the actual bandwidth available at the time of the video streaming.

The choice of a transport protocol is also necessary to insure that video is delivered in real time. There are two transport protocols available, *User Datagram Protocol* (*UDP*) and *Transmission Control Protocol* (*TCP*). TCP is typically utilized for services which require a guarantee of delivery. Examples include *File Transfer Protocol* (*FTP*), e-mails, and Internet applications that are not time or delay sensitive. However, the time sensitive nature of real time video requires the use of UDP in order to avoid delays in the delivery of packets. The unlimited retransmission of packets in TCP renders this protocol unacceptable for real time video streaming. UDP does not guarantee delivery, but can deliver packets with minimal delay to meet the needs of real time video streaming.

To illustrate the bandwidth variability in a real 3G HSPA cell phone network, experiments were conducted using UDP to stream 5 minutes of fixed size video frames from a mobile device over this 3G network. While the data sizes of actual video frames vary significantly, the use of fixed size video frames in this experiment highlights the influence of bandwidth on video delivery. Six different streams of video frames were sent, with data sizes ranging from 3 KB to 8 KB. The video frames were transmitted at a rate of 15 fps. Figure 1 presents a graph of the delivered fps rate of each video frame data size. Notice the variability in the delivery rate. Smaller size video frames were delivered far more consistently and with far fewer dropped video frames than larger size video frames. For example, at the 3 KB size an average of 14.2 fps were delivered, while at 8 KB only an average of 5.25 fps were delivered.

Figure 2 shows the percentage of delivery of the video frames data sizes from the same experiment as Figure 1. The percentage of delivery drops with increasing frame data size. At the time of the experiments, video frames of 3 KB had a 94.6% delivery and video frames of 8 KB were delivered at only 35%, dropping nearly two of every three video frames. Clearly, at the time of the experiments, the 3G network could not handle streaming video of 4KB video frames or larger.

There is a greater problem than the low delivery rate of large video frames. Some video compression algorithms use inter-frame compression techniques that create dependencies between multiple frames (inter-frame compression is covered later in this chapter). With such an algorithm,

Figure 1. Delivery rate of fixed size video frames of 3 KB, 4 KB, 5 KB, 7 KB, and 8 KB across a 3G HSPA cell phone network. All frames sizes were transmitted at a rate of 15 fps

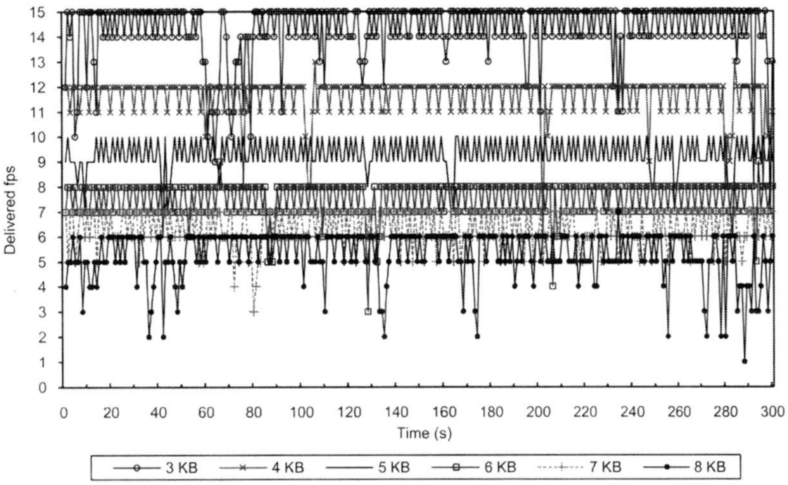

Figure 2. Percentage of delivery for various sizes of video frames

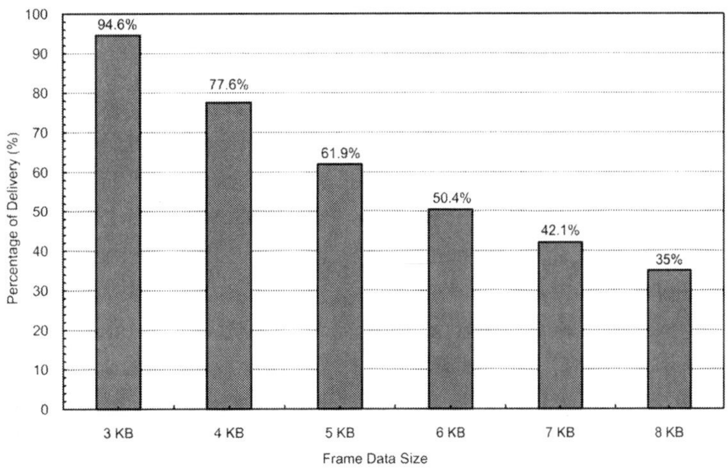

a lost reference frame can exacerbate the problem of limited bandwidth leading to errors propagating across all dependent frames. In (Hassan & Krunz, 2007) an MPEG-4 (MPEG-4, 2005) video experiment utilized inter-frame compression with *Intra* (*I*) frames of approximately 10 KB and *Bi-predicted* (*B*) and *Predicted* (*P*) frames of approximately 4 KB. Based upon the above 3G bandwidth experiment, if I frames of size 10 KB were transmitted at the time of the experiment, their percentage of delivery would likely be less than 35% since frames of 8 KB were delivered at only 35%. These losses would propagate across all dependent B and P frames, which would render these follow-up frames erroneous because they were received without their reference I frame. Indeed in that MPEG-4 experiment, each I frame had 15 dependent B and P frames in the same *Group of Pictures* (*GoP*), all of which would be lost along with their reference I frame.

It should be noted that some mobile phone users are now using their phones to connect to WiFi hot spots. In such cases, their bandwidth is greatly improved as the hot spots connect to the Internet typically through broadband allowing for a much greater and more reliable bandwidth than the cell phone networks. While it may be simpler to stream video from a mobile phone via WiFi, this is presently an incomplete solution to the general problem discussed in this chapter because WiFi has greatly reduced coverage areas compared to that of cell phone network coverage. Using WiFi also restricts the degree of roaming. A truly mobile person using their mobile phone will quickly go beyond the reception area for the WiFi that they are using and will have to locate a new hot spot. Since finding and establishing a new WiFi connection takes some time (possibly several seconds or more), any live streaming video will be interrupted.

There is progress being made in the availability of WiFi, particularly in metropolitan areas that either have poor cell phone network connectivity or extremely high traffic. As noted in (Gass & Diot, 2010), experiments in the Pittsburgh metropolitan area indicate that WiFi access is nearly as prevalent as 3G cell phone network coverage and that both offer roughly equivalent bandwidth when roaming is performed *slowly*. However, at present because WiFi coverage is less than cell phone network coverage, any streaming video technology that relies on WiFi will not provide the flexibility needed for the majority of mobile phone users. This is particularly true in developing regions of the world where WiFi coverage is very limited or non-existent.

THE COMPUTING RESOURCES AVAILABLE TO MOBILE DEVICES FOR VIDEO STREAMING

Mobile phones will become predominant and overtake PCs by 2013 (Albright, 2010). Yet the computer architectures of mobile phones lag behind commercial desktop and laptop units by many years. This section examines the processing and memory layouts of many popular mobile phones and describes how these resources greatly limit processing power.

The GSM 9000-based Bold Blackberry mobile phone uses the *Reduce Instruction Set Computer (RISC) Machine* (*ARM*) 9 processor with up to 128 MB of on board memory and up to 1 GB of flash memory. The Apple iPhone 4 uses the Apple A4 microprocessor (manufactured by Samsung), which is based on the ARM Cortex-A8 processor. The iPhone has up to 512 MB of on board memory, and the ability to store up to 32 GB via flash memory. The Motorola Droid uses the ARM 8 processor with up to 512 MB of on board memory and 16 GB of removable memory (expandable up to 32 GB). The Nokia N9x series of mobile phones use the ARM 11 processor with up to 128 MB of on board memory and up to 16 GB of flash memory. The HTC Touch Pro 2 uses the Qualcomm MSM7201A processor, which is also based on the ARM 11 family of processors, and has up to 288 MB of on board memory and up to 32 GB of flash memory. The Samsung Omnia mobile phones use the Marvel PXA312 with up to 256 MB of on board memory and up to 16GB of flash memory. Caches for these processors vary greatly but are consistently smaller than typical desktop or laptop caches (for instance, the Apple iPhone processor has a comparable L1 cache of 64 KB but a lesser L2 cache of 640 KB whereas modern processor L2 caches are at least 1 MB and as much as 8 MB).

The ARM processors are part of a family of Acorn 32-bit RISC processors, originally developed for the now defunct Acorn Computers. The ARM 8 and ARM 9 processors are centered around a 5-stage pipeline. The ARM 11 was expanded to a 9-stage pipeline. The Cortex-A8 was expanded to a 13-stage superscalar pipeline. Each of these is a typical RISC architecture with a hardwired control unit and a load-store instruction set with a fixed length 32-bit instruction. Most, but not all, instructions require a single cycle for execution (the exceptions being floating point operations). With the exception of the Cortex-A8, operations taking more than a single cycle of execution time will cause pipeline stalls. Being a superscalar, the Cortex-A8 is able to potentially reduce the impact of stalls. This is significant because floating point operations will reduce performance (Uti & Fox, 2010), particularly on the ARM 8 through ARM 11 processors. The instruction set of ARM processors include conditional execution (whereby an instruction can begin executing but is aborted if the condition is computed to be false) and thus reduce branching and branch penalties. The ARM processors each offer 16 32-bit general purpose registers, and condition codes which are set automatically by arithmetic operations but only if explicitly requested (that is, condition codes are not set in most cases).

The ARM processors include the "Thumb" feature, which allows the processor to execute 16-bit instructions by making some of the operands implicit and which thereby can be dropped from the 32-bit operation format. This creates a space savings in terms of the executable code and therefore, permits more applications in memory at one time. It does not however impact the processor's performance when executing any single process.

The Marvel PXA312 processor is itself based on the Intel XScale core, a RISC-based architecture consisting of a seven-stage integer pipeline and an eight-stage memory pipeline. Like the Intel Pentium processor, the pipeline executes instructions at the microinstruction level (instead of machine level). More information about the architectural overview of the XScale processors can be found in (Srinivasan & Velev, 2003).

The processor speeds greatly vary on these mobile phones ranging from 312 MHz to 1 GHz. However, these speeds are not important by themselves because the processors' performances are impacted by many other factors such as number of pipeline stalls, size of caches, speed of memory, and whether the power management is slowing down the clock speed to preserve battery power. Many of these mobile phones have separate memory management units (for instance, the Marvel PXA312 uses an 8-stage pipelined memory management unit), graphics accelerators, and on-chip data and instruction caches. As noted above, only the Apple iPhone's processor is of a superscalar pipeline and therefore floating point operations will have less of an impact on the pipeline's performance. In comparison, all commercial processors for desktop and laptop computers incorporated superscalar pipelines by 1998. The significance here is that some compression algorithms rely heavily on floating point operations and therefore most mobile phone processors will perform far more slowly than their desktop and laptop counterparts due to the stalls that arise because of the singular pipeline used.

In order to stream video in real time, a mobile phone should be able to generate a video frame, compress and stream it over the cell phone network before the next video frame becomes available. Most cameras are set to generate 15 fps. Given this amount, it means that the generation, compression and streaming of each frame should occur within $1/15^{th}$ of a second, or 0.067 s. Further, the size of each compressed video frame will vary based on the scene being captured. As previously discussed, the cell phone network's bandwidth varies and attempting to transmit large quantities of data in 0.067 s is impractical. A target size should be much less than 8 KB per 0.067 s, perhaps as little as 3 KB.

Another limiting factor for mobile phones is the available battery power. As the battery becomes depleted, the mobile phone's operating system uses a power management routine to reduce the drain on the battery by reducing processing power. Thus, the already limited processing power is further reduced. The specifics of power management are proprietary and thus not readily discussed, as such the actual performance of applications are dependent on the particular power management routine on the mobile device.

In summary, mobile phones lag behind desktop and laptop computers by one generation or more. Table 1 outlines the differences between the processors of desktop/laptop computers and the processors of mobile phones. Most mobile phone processor architectures are not multi-core and do not support fast floating point operations. Although most use RISC architectures and pipelines, many fail to have the elaborate pipelines that have been developed over the past 10 years to support superscalar issuing of parallel instructions. The limited memories further restrict the size of the data (e.g., multiple video frames) being processed. Finally, power management can further reduce the processing power as the battery begins to drain from the 100% capacity.

THE NATURE OF RAW VIDEO FRAMES FROM THE CAMERAS OF MOBILE DEVICES

The video software of mobile phones produce already compressed video which cannot be streamed from the device in real time. Further, the video software on these devices may not provide complete information regarding the resolution of video which can be obtained from the camera. This chapter deals with issues related to the compressing of *raw* video directly from the camera for the purpose of streaming it in real time. The information in this section, describing the nature of the raw video frames from mobile devices, is based upon the authors' direct examination of video cameras on a variety of mobile phones. The raw video frame formats and video frame resolutions described in this section cannot be obtained by simply viewing the mobile devices or their installed video software. They were obtained by programmatic communication with the video cameras in the mobile phones.

The nature of the raw video produced by mobile device cameras presents challenges for real time compression and streaming. The video camera on the mobile device produces enormous amounts of data in each raw video frame. This is especially important given that these devices have limited memory, battery life, and CPU power, and yet the video must be compressed significantly. From programmatic access to the cameras of mobile devices, the authors observed that these devices can produce raw video frames in one of two color spaces, RGB and YUV. In the RGB color space, each pixel in a video frame is represented with three components red, green and blue. In the YUV color space each pixel is represented with a luminance Y, which is the grayscale component, and two chrominance components, U for hue and V for intensity.

For the RGB color space, the R, G, and B components of a pixel are typically stored in 24 bits, 8 bits per component. The mobile device cameras that produce video in the RGB color space can usually also produce sub-sampled RGB video where each component is stored with less than 8 bits per component. One such format stores each of the R, G and B components in 5 bits for a total of 15 bits per pixel. A slight variant uses 6 bits for the G component, thus storing each pixel in 16 bits. However, due to fewer bits being used to represent each component, these sub-sampled

Table 1. Comparison between the processors of desktop/laptop computers and the processors of mobile phones

Desktop/laptop computers	**Most mobile phones**
Superscalar pipelines	Single pipelines
Multiple cores	Single cores
1 GHz+ processor speeds	No more than 1 GHz processor speed, usually less than 512 MHz
On-chip caches of up to 64 KB and off-chip caches of up to 8 MB	On-chip caches of 16 KB to 32 KB and off-chip caches of up to 1 MB
Hundreds of registers	32 registers

RGB formats typically produce video frames that are of poorer visual quality (even before video compression is applied). As a result, it is better to use the 24-bit RGB format and then apply forms of data reduction to best retain video quality before compressing the video.

YUV video can be sampled at 8-bits per Y, U, V component for a total of 24 bits per pixel. This is known as the YUV 4:4:4 format. Like RGB, the YUV video format also permits various sub-sampling formats. For mobile devices, the prevalent YUV format is YUV 4:2:0 where the Y component is stored in 8-bits, and the U and V components are sub-sampled and shared across 4 pixels in a block of two pixels high and two pixels wide (2x2 pixels). The rationale behind this approach is that the human eye is more sensitive to information loss in the luminance component and less sensitive to loss of information in the chrominance components. Since the U and V chrominance components are shared across 4 pixels, only 16 bits are used to represent the U and V component for each set of 4 pixels in the YUV 4:2:0 format. As such, each pixel is stored in only 12 bits, compared to the 24 bits required to store a single pixel in either of the RGB 24 or YUV 4:4:4 formats. Such a significant reduction in data size is very important when streaming video from mobile devices where computational and memory capacity are limited. Additionally, the YUV 4:2:0 format does not introduce a visible loss of quality unlike sub-sampled RGB formats.

YUV components may also be stored in either planar or packed methods. In planar storage, the Y components are stored in the first part of an array, and are followed by the U components, and the V components, while in packed storage the YUV components are stored pixel-wise. The planar method is commonly used for the storage of the YUV 4:2:0 raw video frames from the cameras of mobile devices. Figure 3 illustrates an example of an 8x4 video frame with YUV 4:2:0 components stored in the planar method. Each 2x2 block of pixels has individual Y component values, but share the same U and V components.

The RGB color space is used primarily for representing pixels for display on standard desktop and laptop computers screens. However, most video compression algorithms have been optimized to exploit the spatial redundancy of video frames in the YUV color space. It is therefore worthwhile to convert from an RGB to YUV video format prior to compression in order to decorrelate the RGB components and take advantage of the superior compression that can be achieved in the YUV color space. A conversion from 24-bit RGB to 12-bit YUV 4:2:0 video can be done with integer computations and shifts (YUV, 2010). Thus, the conversion is quick and there is a 50% reduction in data size without vis-

Figure 3. YUV 4:2:0 in planar storage method. Each pixel in a block of 2x2 pixels in a video frame has an individual Y component but shares U and V components with other pixels in the block. A video frame of resolution 8x4 will require 48 bytes. The Y, U, and V components use 8 bits each

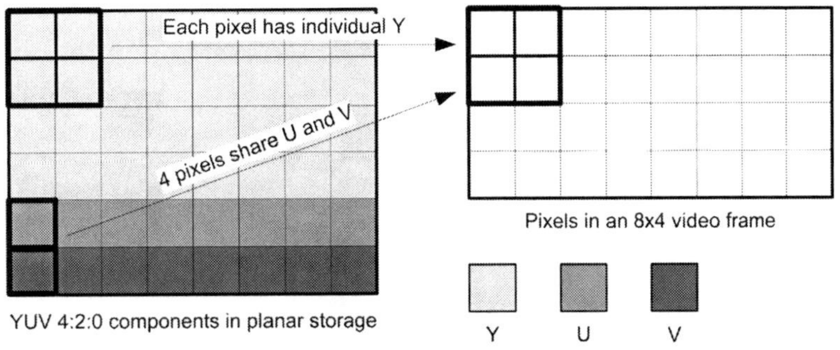

ible loss of quality. Video compression can now be applied to further reduce the data size.

Aside from the format of the video frame, each video frame has a resolution measured in the number of horizontal pixels by the number of vertical pixels. Mobile phones can produce video frames in several different resolutions. The most common resolutions include: 120x120, 176x144, 176x176, 240x320, 320x240, 400x240, 352x288, and 640x480. Higher frame resolutions, however, come at a cost. As the resolution of a video frame increases, the camera on the mobile device cannot produce as many fps. For example, a video camera on a mobile device that is capable of producing video frames of 176x144 resolution at 15 fps may only be able to produce frames of 352x288 resolution at 7 fps, and 640x480 resolution at 3 fps. The reason for this is that the video camera on the mobile device requires more computational resources (CPU and memory) for larger dimensions, and hence, more time to generate the larger video frames.

The issue is that the data sizes of video frames grow geometrically with increasing resolution. To double the apparent resolution of the video frame will result in a quadrupling of the data size. As the data size of the video frames increases geometrically, there is a similar geometric growth in the computational cost of compressing the frames. A video frame with 176x144 resolution contains 25,344 pixels, while a video frame of 352x288 resolution contains 101,376 pixels and will require four times as much computation for the video compression algorithm. As such, it becomes considerably more difficult for the video compression algorithm to completely compress these larger video frames in the time available between frames. Figure 4 shows how higher resolution frames can leave less time between frames for video compression in real time.

What happens if the video compression algorithm is not able to completely compress a video frame between the generation of two consecutive video frames? The result is that the camera will not generate the next video frame because the processor and memory necessary to do so are not available. There is simply not enough available computational resource to support in parallel the camera generating its next frame and the compression algorithm compressing the previous frame. The impact is that the computational resources of the mobile device coupled with the video compression algorithm become a bottleneck and effectively reduce the fps rate of the mobile device. Since existing video compression algorithms originally designed for desktop computers are computationally intensive, careful algorithm design must be employed to scale down video compression algorithms to be able to operate in real time within the computational constraints of mobile devices.

Figure 4. The time for generating video frames in 1 second. The computational cost of compressing video frames increase geometrically while the time in between video frames do not increase geometrically. a) 352x288 resolution video frames generated at 7 fps produces 101,376 pixels and 1,216,512 bits per video frame in the YUV 4:2:0 format. b) 176x144 resolution video frames generated at 15 fps produces 25,344 pixels and 304,128 bits per video frame in the YUV 4:2:0 format

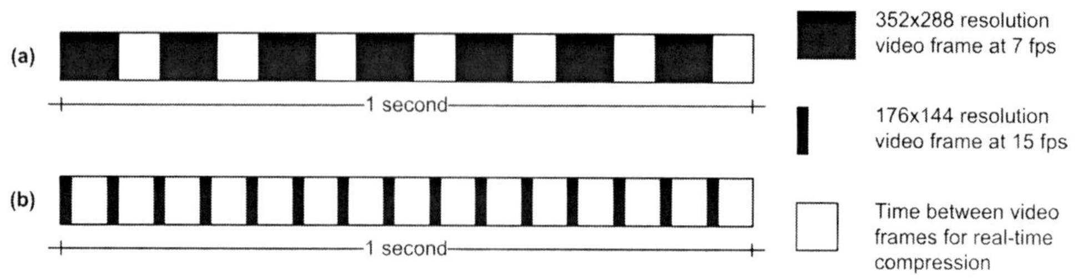

Although the YUV 4:2:0 video format can significantly reduce the data sizes of raw video frames, the uncompressed videos remain very large. Regardless of the color space of the raw video, the amount of data in each raw video frame is such that they must be compressed before the video can be streamed. Table 2 shows data sizes of video frames using different mobile video resolutions and formats. Higher resolution video can easily overwhelm the resources available for real time video compression and streaming from mobile devices.

To better illustrate the need for video compression, note that a 5 minute 176x144 video in YUV 4:2:0 format at 15 fps will require 163 MB of bandwidth if uncompressed.

$$5\ \mathrm{min} \times \frac{60\ \mathrm{sec}}{\mathrm{min}} \times \frac{15\ frames}{\mathrm{sec}} \times \frac{(176x144)\ pixels}{frame} \times \frac{12\ bits}{pixel} \times \frac{1\ byte}{8\ bits} \approx 163.15\ MB$$

Clearly, given the large data sizes of uncompressed video at even a low resolution (176x144), a very efficient video compression algorithm is necessary for compressing the video to a data size that will be suitable for real time streaming over cell phone networks.

Table 2. Data sizes of video frames in various formats and resolutions. The data size of the video frames can be obtained by multiplying the number of pixels in a video frame by the number of bits used to store each pixel

	Pixels in video frame	RGB 24 / YUV 4:4:4 (data size in bits)	YUV 4:2:0 (data size in bits)
120x120	14,400	345,600	172,800
176x144	25,344	608,256	304,128
176x176	30,976	743,424	371,712
240x320	76,800	1,843,200	921,600
320x240	76,800	1,843,200	921,600
400x240	96,000	2,304,000	1,152,000
352x288	101,376	2,433,024	1,216,512
640x480	307,200	7,372,800	3,686,400

COMPRESSION ALGORITHMS

The major types of video compression techniques are based on either *Discrete Cosine Transform (DCT)* or *Discrete Wavelet Transform (DWT)*. Both transforms have been adopted for use in standard image and video compression techniques. The DCT-based MPEG-4, and specifically part 10 which is also known as H.264/*Advanced Video Coding (AVC)*, is the latest and most widely used standard video compression technique (Marpe & Wiegand, 2006). DWT is the image compression technique used in JPEG2000 (JPEG2000, 2004). Building on the JPEG2000 image compression, Motion JPEG2000 is a commonly used video compression technique where each frame in the video is individually compressed using the JPEG2000 image compression. Each of these video compression techniques has their merits in terms of computational cost, adaptability, video quality, data sizes, and bandwidth usage. These video compression techniques have been extensively researched in the literature at different levels of specificity. The discussion in this section will be limited to those aspects of the video compression techniques that relate to the real time video streaming challenges described throughout this chapter.

There are several profiles in the MPEG-4 standard. The *constrained baseline profile* and the *baseline profile* are the least complex MPEG-4 profiles that can be used for videoconferencing and mobile applications. The MPEG-4 standard, however, only defines the video decoder (recipient of a streamed video signal) and not the encoder (which performs the compression prior to streaming). A solution to stream real time video from a mobile device using any of the MPEG-4 profiles *must* develop an encoder that will run on the mobile device to match the decoder defined in the standard. On the one hand, the integer forward

DCT of the MPEG-4 encoder can be implemented with additions and shifts and without the use of a multiplier (Wang, Kondi, Luthra, & Ci, 2009). The integer operations are easy to implement without any loss in precision. This is a very important consideration given the low computational capacity of mobile processors. On the other hand the motion estimation component of MPEG-4 is the most computationally intensive step of the encoder (Wang, Kondi, Luthra, & Ci, 2009), and can place excessive computational demands on mobile devices. The high computational complexity of MPEG-4 encoders is a significant reason for why real time MPEG-4 encoders are not commonly available for mobile devices. However, the baseline profile of MPEG-4 includes three error resilient tools (flexible macroblock ordering, arbitrary slice order, and redundant slices) to deal with the high bit error rate that can occur in cell phone networks. The development of low-complexity MPEG-4 encoders that can operate in real time on mobile devices and stream video over cell phone networks is a very challenging topic of research.

The MPEG-4 standard utilizes inter-frame compression techniques to reduce temporal redundancy across multiple video frames. There are four types of MPEG-4 video frames. The I frame is encoded independent of other frames. The P frame is encoded as the difference between the preceding or following I or P frame and the current frame. The B frame is encoded to rely on both the previous and following I and P frames. The *Bi-predicted Reference* (B_R) frame is a B frame that is used as a reference frame for preceding or subsequent frames. To reduce the complexity of the encoders, the B and B_R frames are not allowed in the constrained baseline and baseline profiles. A GoP is a sequence of frames starting with an I frame and its dependent P, and B frames. In a GoP, the I frame, which is encoded with no prediction information, is larger than P and B frames. Overall, the temporal redundancy reduced across multiple video frames results in increased video compression in a GoP.

There are two consequences of GoP with respect to the real time streaming of video over wireless links. First, the very important I frame, which is usually larger and contains many packets, can be lost or corrupted in the cell phone network. A lost or corrupt I frame will result in error propagation to *all* dependent frames in the GoP. This will intensify the effect of the already low bandwidth as even successfully received dependent frames will become erroneous. The second implication of GoP is that of adaptability. In streaming real time video from mobile devices, it is desirable that the video be adaptable to better utilize the variable bandwidth of cell phone networks. MPEG-4 encoders typically allow adaptability on the GoP level but due to the inter-frame dependencies, individual video frames in a GoP are not adjustable (Hassan & Krunz, 2007). The loss of frame-level adaptability makes it difficult to finely adapt an MPEG-4 video to match the available bandwidth. Also, the adjustment of the data size of a GoP to match a target size is usually done by the iterative adjustment of the quantization level of each video frame in the GoP (Hassan & Krunz, 2007). The additional computational cost that can be incurred in the repeated quantization of video frames does not cohere with the demands of real time video streaming from mobile devices.

The adaptability of the MPEG-4 standard continues to be improved. The *Scalable Video Coding* (*SVC*) is an extension of the MPEG-4 standard to provide adaptive video coding that allows the encoder to adapt various aspects of the video. For mobile applications, the scalable baseline profile of the SVC provides temporal, spatial, quality, and combined scalable coding for the video. Each scalable coding option provides adaptability by utilizing subsets of packets from the larger video stream. The adaptation of MPEG-4 video is an active area of research. See (Schwarz, Marpe, & Wiegand, 2007; Wien, Cazoulat, Graffunder, Hutter, & Amon, 2007). Researchers have proposed schemes to improve the quality of MPEG-4 video and particularly to give priority to I frames that are being streamed in the wireless networks (Ke & Chilamkurti, 2008). More information on H.264 and MPEG-4 can be found in (Richardson, 2003).

The lightweight daViKo (Cycon, Schmidt, Hege, Wählisch, & Palkow, 2008) is an example of an H.264/AVC commercial software implementation that can operate on mobile devices. The software is a scaled-down version of the highly optimized commercial daViKo desktop video conferencing tool. The lightweight daViKo software, however, only supports mobile peer-to-peer video conferencing over 802.11 WLAN/WiFi (Schmidt, Hege, Wählisch, Cycon, Palkow, & Marpe, 2010, online). Schmidt, et al., do not report streaming video over cell phone networks. As a commercial product, the software is not open source.

The JPEG2000 image compression technique provides progressive transmission in four dimensions (quality, resolution, spatial location, and component) to support image improvement as more bits are received (Bilgin & Marcellin, 2003). This technique allows the image quality and features to be continuously improved as more data is received. In the quality progressivity, the quality of the image improves with more data being decoded. For the resolution progressivity, the resolution of the decoded image grows by a factor of two from a thumbnail to a full size image. The spatial location progressivity allows an image to be transmitted using a *stripe* or *scan-based* approach, and is targeted for use by low memory printers and scanners. Component progressivity allows an image to be transmitted such that various components (grayscale, color, text) of the image can be decoded as more bits are received. Because it offers a variety of choices of progressive transmissions, the JPEG2000 offers a large degree of flexibility in terms of image adaptations.

As with MPEG-4, the JPEG2000 standard only specifies the decoder. The encoder must be implemented to comply with the decoder in the standard. JPEG2000 encoders can be computationally intensive for mobile devices depending on the particular wavelet filter selected for video compression. Wavelet filters have various lengths and can be implemented using floating point or integer operations. The 5-3 and 9-7 wavelet filters are the two choices available in the JPEG2000 standard. The 5-3 wavelet filter is short and can be implemented using integer operations while the 9-7 is longer and requires floating point operations (Bilgin & Marcellin, 2003). Although longer wavelet filters produce better compression, shorter integer wavelet filters are less computational intensive and are more suited to the computing resources on mobile devices. The Motion JPEG2000 video does not utilize inter-frame compression or motion estimation techniques. More information about JPEG2000 and Motion JPEG2000 can be found in (Rosenbaum, 2009; Taubman & Marcellin, 2002).

At a fundamental level, a major advantage of the DWT-based compression techniques over the DCT-based compression techniques is that image quality is much improved because the blocking artifacts that can result from DCT are absent in DWT images (Gonzalez & Woods, 2008). Generally, the image quality of DWT degrades more gracefully than that of DCT. DWT produces better quality images at very low bit rates. Instead of blocking artifacts, increased blurring is a side-effect of highly compressed DWT images. An exception can be found in the case of JPEG2000 where the DWT can be applied to small *tiles* of the image and can result in blocking artifacts in highly compressed images. The blocking artifacts result because DWT was not applied to the entire image. For wavelet adaptations, researchers have developed *spatial orientation tree* adaptation techniques that allow wavelet compressed images to be adapted to match a target data size (Shapiro, 2002; Said & Pearlman, 2002).

Wavelet video can also employ 3D wavelet and motion estimation techniques for inter-frame compression to reduce temporal redundancies across multiple frames. The additional computational cost and memory requirements of these techniques can be high for mobile devices. More information on 3D wavelet and motion compensation can be found in (Secker & Taubman, 2001; Secker & Taubman, 2003).

As can be inferred from this section, the selection of a compression technique for streaming video from mobile phones over cell phone networks must address numerous issues. These include the computational cost, degree of adaptability, video quality, resulting video frame size, bandwidth requirements, and the impact of lost frames.

THE CHALLENGES OF STREAMING REAL TIME VIDEO FROM MOBILE DEVICES

In this chapter, the definition for real time was selected as that of signal processing. That is, to stream video in real time, each video frame produced by the camera should be compressed and streamed prior to the production of the next video frame. With a camera capturing video frames at 15 fps, this introduces a time limit of 0.067 s to generate, compress and stream the frame. In Table 2 it was noted that a raw video frame can be as large as 7,372,800 bits (900 KB), 640x480 resolution RGB 24/YUV 4:4:4 format. As discussed in the section on bandwidth, it was noted that cell phone network bandwidth can be widely variable. In the example experiment, at 3 KB every 0.067 s, most video frames tend to make it through a 3G cell phone network with a minimal drop rate (about 95% of all frames can make it through). This of course will depend on a number of external factors however such as proximity from the base station, the amount of network traffic, which itself is partially based on the time of day and day of week, and environmental conditions. Thus, the mobile device must perform some form of video compression to reduce a file as large as 900 KB to something more like 3 KB during this 0.067 s window.

As discussed in the section on the computing resources of mobile devices, mobile devices are much more limited in computing power than desktop or laptop computers. Because of this, it is a mistake to think that video compression techniques used to stream real time video from computers over the Internet can be simply ported to mobile devices. The mobile devices have processors that are not only weaker than their desktop/laptop counterparts, but the amount of available processing power varies depending on background applications and battery management strategies. The battery management strategies, run by the mobile device operating systems, reduce processing speed as the battery begins to the drain. Turning off the battery management options can lead to better processor performance if one is aware enough to do this, however it does drain the mobile device battery more rapidly.

Mobile device memories are much smaller than those of desktop and laptop computers, leading to another obstacle for any mobile video compression algorithm. Although memory is typically expandable by using flash memory, this will slow down any process. Therefore, the larger flash memories may not be of much use during real time video compression.

Another factor in streaming real time video is that the data size of a compressed video frame itself can vary widely according to the picture content of the particular video being streamed. For instance, a video frame with a solid background can be more compressed than a video frame with a lot of texture. Thus, the bandwidth needed to stream the compressed video frame will vary significantly from frame to frame if the video has fast changing scenes.

Figure 5 shows the entire process by which the raw video frames from the camera of the mobile device undergo several stages including data reduction via RGB→YUV conversion, compression and adaptations before being streamed from the device through the cell phone data channel (3G, EDGE...) across the Internet to the receiving video decoder. The video frames are then decoded and displayed. The conversion from RGB to YUV and YUV to RGB can be optional, depending on the mobile device and the video application. Each video frame works its way through the process from being generated by the camera to being streamed into the cell phone network stream before the next video frame is generated. If the receiving device is another mobile device then the following factors should also be taken into consideration: screen size, memory, processor speed, and disk space if the video is to be saved.

And so the constraints of the problem become clear. The raw video must be compressed quickly but the reduction in size must be enough to permit the video frame to be streamed without a large chance of it being dropped by the cell phone network. Even a 176x144 resolution video frame in YUV 4:2:0 format will be 38 KB in size and therefore must be compressed significantly to be streamed with confidence of not being dropped. How can these factors be reconciled?

Any application that attempts to stream video in real time from mobile phones across cell phone networks should be able to do each of the following.

- Compress each video frame to a size that is manageable by the cell phone network.
    - First, convert the color space to an appropriately reduced format (e.g. from RGB 24 to YUV 4:2:0)
    - Apply a compression algorithm that requires minimal computation, preferably using *integer-only* operations to compress video frames substantially in size.
- Adapt the video transmission rate to match the available bandwidth at the time of the video streaming.
- Determine, based on the video being streamed, how much the video should

Figure 5. Real time streaming of video originating from mobile devices. The raw video frames from the camera of the mobile device are compressed and adapted before being streamed via the cell phone data channel across the Internet to the video decoder where the video frames are decoded and displayed

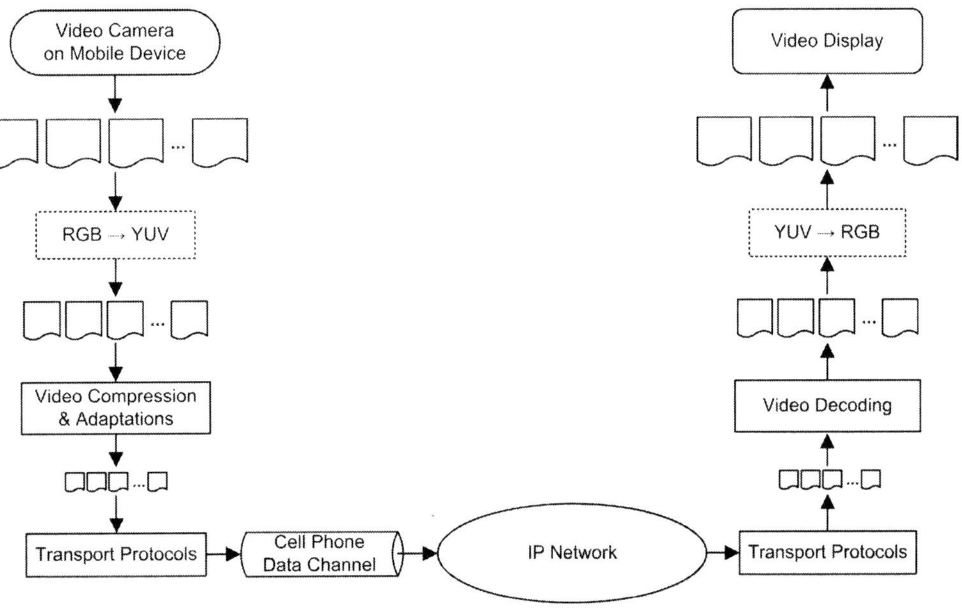

be compressed to reach a desired target size. That is, as the video itself changes, the compression level may also require changing.

The strategy described above goes beyond the capabilities of a video compression algorithm. The entire streaming process, which includes video compression, should also adapt the video transmission rate and/or control the video compression levels.

Applying this strategy, in (Uti N. V., 2009) a real time mobile video compression algorithm was developed using an integer implementation of the 5-3 DWT and the *Sub-band Aligned Integer Run-Length Encoding* (*SAIRLE*) scheme for video compression. The real time video compression algorithm was implemented on a real mobile phone to compress the raw video frames from the camera of the phone. The mobile phone had a 400 Mhz processor with 24.17 MB of memory available for video compression. The camera on the mobile phone generated raw video frames with a resolution of 240x320 pixels and at a rate of 15 fps. The 5-3 DWT and SAIRLE algorithm was efficient enough to compress the raw video frames from the camera to attain the maximum 15 fps rate. That is, the compression algorithm could perform its task within the 0.067 s window between the production of video frames in real time. The real time mobile video compression algorithm did not utilize 3D wavelet inter-frame compression or motion estimation techniques.

Following on from the above work, further research in (Uti & Fox, 2009) implemented a real time video streaming framework to adapt the fps transmission rate of the video frames for streaming. Experiments were conducted to stream real time video originating from the camera of a 400 Mhz mobile phone over the EDGE and 3G cell phone networks. In these experiments, the video frames had resolutions of 176x144 pixels. Compressed video frames can be dropped depending on the quality of bandwidth at the time of the video streaming. In some of the experiments, as much as 58% of the video frames were dropped in the cell phone network. Adaptation of the fps transmission rate was used in an attempt to match the video being streamed to the bandwidth of the cell phone networks at the time the video was being streamed. At 10 second intervals, the video transmission rate was adapted by increasing, maintaining, or decreasing the fps rate depending upon the percentage of dropped video frames from previous intervals and whether the data sizes of the compressed video frames were increasing, steady or decreasing. In the 25 experiments conducted, the fps adaptations resulted in improvements of between 18% to 43% cumulative video frame delivery rate with an average of 32% improvement. Only the fps video transmission rate was adapted, the video compression level was not adapted.

FUTURE RESEARCH DIRECTIONS

The streaming of real time video from the cameras of mobile devices is expected to have a significant impact on economic development, law enforcement, emergency medical response, and education. The problem of streaming real time video from mobile devices is a new and untapped area of research. This topic has only minimally been explored as mobile devices only are now attaining a level of maturity that can support very complex tasks such as the compression and streaming of real time video from the devices. However, a lot of the problems discussed in this chapter have already been addressed in the context of desktop computers and video streaming over stable and rich bandwidth environments. Video compression algorithms contain many complex components that have themselves been developed by experts specializing in very specific aspects of the algorithm. For example, experts on 3D wavelets tend to only work on their specialization.

One challenge is to select and scale the components of the video compression algorithms that are most suitable to the computational resources available on mobile devices. Because of the huge body of research spanning multiple fields which must be engaged, new researchers who are interested in the challenges of streaming video in real time from mobile devices should pick a specific area of interest and investigate how to scale it to operate on mobile devices. For example, the motion estimation component of MPEG-4 Encoders can be computationally intensive, even for desktops. Research resulting in the ability of this component to function efficiently on mobile devices in real time would represent a significant breakthrough with diverse applications.

Another direction of research is in the area of energy consumption. It has already been established that more computation results in greater energy consumption. However, it is not clear how the various components of compression algorithms influence the battery life of mobile devices. Investigations identifying the energy efficiency of various components of these video compression algorithms would be very useful and could identify which components should be preserved and which eliminated due to excessive energy consumption. More specifically, research could be undertaken to examine the tradeoff between the contributions of each component compared to its energy demands.

Future research should seek to exert greater control of the real time video being streamed over cell phone networks from the cameras of mobile devices. In addition to simulations in the labs, research should be conducted on real mobiles devices so that both the constraints of the mobile devices and the cell phone networks can be taken into account while scaling video compression algorithms to work within the real time constraints of mobile devices and wireless networks. Additional adaptation can be implemented to dynamically adjust the quality of video frames so as to increase or decrease the data of sizes of the compressed video frames being streamed. In addition to adapting the fps video transmission rate, the adaptation of the compression levels of video frames could further improve the video delivery rate, as the sizes of video frames in conjunction with the adapted fps rate will better match the available bandwidth.

Further, if the situation permits, the user preferences of higher fps rate or higher quality video frames could be taken into account in order to adapt the real time video to better suit the particular desires of the user. In certain situations, adapting the video may require taking into consideration privacy-related issues such as the blanking of background or foreground images to obscure unnecessary content. Many other possibilities exist for video adaptation.

The availability and usage of memory on mobile devices is also a fruitful area of research. Mobile device memory is limited and varies greatly between devices, and can vary from moment to moment depending on the demands of the operating system and other essential software. Research could be conducted to examine how much memory is available to the video compression algorithm for real time video compression and streaming and what type of adaptations are feasible on a given device. For example, if inter-frame compression is being employed, how many frames can be in a GoP at the time of video streaming?

Another area of potential innovation is the development of realistic simulations of cell phone network bandwidth. Because the bandwidth of wireless networks can vary significantly, researchers must make additional efforts to simulate correctly the conditions found in real world networks. The accurate modeling of wireless bandwidth behaviors would greatly facilitate the ability of researchers to test a variety of software and hardware issues in the lab.

Mobile phone processors themselves are targets for research. With limited computational capabilities, further benchmark tests should be run to determine their capabilities under a number of constraints. Power management systems, while critical to the function of these devices, can nonetheless alter processor power when the battery begins to drain. This topic should be examined to better understand how the reduction in processor speed can impact the performance of real time video compression algorithms and the streaming of video. Such information could lead to methods to mitigate the degradation of computational capacity from necessary power management functions.

CONCLUSION

This chapter has described the challenges and some of the benefits of streaming real time video from mobile devices across existing cell phone networks. While much of the research to date has concentrated on streaming video from computer to mobile device or from computer to computer over the Internet, there are those who are attempting this significantly harder problem. Most research use simulated networks and/or desktop environments instead of streaming real time video originating from the cameras of mobile devices, and so many of the challenges are being ignored. Instead, this chapter has described the various problems that are specific to the compression and streaming of real time video from mobile devices over cell phone networks.

The solution to compressing and streaming real time video from mobile phones over cell phone networks must integrate elements from a number of normally discrete fields of computer science and engineering, particularly wireless networking, mobile computing, data compression, signal processing, and real time systems. The challenge of undertaking such atypical research likely accounts for why there has been little academic research done on the specific topic of compressing and streaming real time video originating from mobile phones.

There are numerous open questions that continue to present challenges. Some of these questions include: Will a 4G network have more stable and higher bandwidth with features to support real time mobile video? Will the multimedia demand on the cell phone network increase with improved bandwidth? When will mobile phone processors reach a capability equivalent to modern desktop computers, for instance by employing superscalar pipelines? Will there be additional breakthroughs in compression technologies to better support streaming real time video under such tight conditions?

ACKNOWLEDGMENT

This work was supported in part by the National Science Foundation Graduate Research Fellowship and the Ford Foundation Pre-Doctoral Fellowship. The authors also thank Dr. Yizong Cheng for discussions and valuable feedback.

REFERENCES

Al-Turkistany, M., Helal, A., & Schmalz, M. (2009). Adaptive wireless thin-client model for mobile computing. *Wireless Communications and Mobile Computing*, *9*(1), 47–59. doi:10.1002/wcm.603

Albright, P. (2010, June 14). *IEEE Computer Society*. Retrieved on July 24, 2010, from Mobile More Important than Ever. Mobile phones poised to overtake PCs by 2013: http:// www.computer.org /portal/ web/ buildyourcareer/ HS16

Allen, I. E., & Seaman, J. (2010). *Learning on demand: Online education in the United States, 2009*. Babson Park, MA: Babson Survey Research Group.

Anokwa, Y., Hartung, C., Brunette, W., & Lerer, A. (2009, Oct.). Open source data collection in the developing world. *Computer*, 97–99. doi:10.1109/MC.2009.328

Bilgin, A., & Marcellin, M. W. (2003). JPEG2000: Highly scalable image compression. In Sayood, K. (Ed.), *Lossless compression handbook* (pp. 351–369). San Diego, CA: Academic Press. doi:10.1016/B978-012620861-0/50019-X

Boccignone, G., Esposito, L., & Marcelli, A. (1993). An experimental vision tool for real time quality control. In *Computer Analysis of Images and Patterns, 719*(1993), 706-710. Berlin / Heidelberg, Germany: Springer.

Bovik, A. (2005). Introduction to digital image and video processing. In Bovik, A. C. (Ed.), *Handbook of image & video processing* (pp. 3–18). Amsterdam, The Netherlands: Elsevier Academic Press. doi:10.1016/B978-012119792-6/50065-6

Cycon, H. L., Schmidt, T. C., Hege, G., Wählisch, M., & Palkow, M. (2008). An optimized H.264-based video conferencing software. *The 12th IEEE International Symposium on Consumer Electronics* (pp. 1-4). Piscataway, NJ: IEEE.

Fischer, M., Lim, Y., Lawrence, E., Ganguli, L., & Kargl, F. (2008). ReMoteCare: Health monitoring with streaming video. *7th International Conference on mBusiness* (pp. 280-286). Los Alamitos, CA: IEEE Computer.

Gass, R., & Diot, C. (2010). An experimental performance comparison of 3G and Wi-Fi. In *Passive and Active Measurement, 6032*(2010), 71-80. Berlin-Heidelberg, Germany: Springer.

Gonzalez, R., & Woods, R. (2008). *Digital image processing* (3rd ed., pp. 525–623). Upper Saddle River, NJ: Pearson Prentice Hall.

Hassan, M., & Krunz, M. (2007). Video streaming over wireless packet networks: An occupancy-based rate adaptation perspective. *IEEE Transactions on Circuits and Systems for Video Technology, 17*(8), 1017–1027. doi:10.1109/TCSVT.2007.903128

JPEG2000 (2004). *JPEG 2000 image coding system.* ISO/IEC 15444-1:2004.

Ke, C.-H., & Chilamkurti, N. (2008). A new framework for MPEG video delivery over heterogeneous networks. *Computer Communications, 31*(11), 2656–2668. doi:10.1016/j.comcom.2008.02.029

Kehtarnavaz, N. (2005). *Real time digital signal processing: Based on the TMS320C6000.* Amsterdam; Boston: Elsevier:Newnes.

Kehtarnavaz, N., & Gamadia, M. (2006). *Real time image and video processing: From research to reality*. San Rafael, CA: Morgan & Claypool Publishers.

Kurose, J., & Ross, K. (2009). *Computer networking: A top-down approach* (5th ed.). Boston, MA: Addison Wesley.

Leeson, P. (2007). Better off stateless: Somalia before and after government collapse. *Journal of Comparative Economics, 35*(4), 689–710. doi:10.1016/j.jce.2007.10.001

Lenders, V., & Martonosi, M. (2009). Repeatable and realistic experimentation in mobile wireless networks. *IEEE Transactions on Mobile Computing, 8*(12), 1718–1728. doi:10.1109/TMC.2009.87

Marpe, D., & Wiegand, T. S. (2006). The H.264 / MPEG4 advanced video coding standard and its applications. *IEEE Communications Magazine, 44*(8), 134–144. doi:10.1109/MCOM.2006.1678121

Molony, T. (2009). Trading places in Tanzania: Mobility and marginalization at a time of travel-saving technologies. In M. de Bruijn, F. Nyamnjoh, & I. Brinkman (Eds.), *Mobile phones: The new talking drums of everyday Africa* (pp. 92-109). Bamenda & Leiden, North West Region, Cameroon & The Netherlands: Langaa & African Studies Centre. MPEG-4 (2005). *H.264/AVC - Part 10: Advanced video coding.* ISO/IEC 14496-10:2005.

Richardson, I. E. (2003). *H.264 and MPEG-4 video compression: Video coding for next-generation multimedia*. Chichester, West Sussex: Wiley. doi:10.1002/0470869615

Rosenbaum, R. (2009). Mobile image communication: New concepts using JPEG2000. In Ibrahim, I. (Ed.), *Handbook of research on mobile multimedia* (2nd ed., pp. 201–216). Hershey, PA: IGI Global.

Said, A., & Pearlman, W. A. (2002). A new fast/efficient image codec based on set partitioning in hierarchical trees. In P. N. Topiwala (Ed.), *Wavelet Image and Video Compression, 450*, 157-170). New York, NY: Kluwer Academic Publishers.

Schmidt, T. C., Hege, G., Wählisch, M., Cycon, H. L., Palkow, M., & Marpe, D. (2010, online, March 16). Distributed SIP conference management with autonomously authenticated sources and its application to an H.264 videoconferencing software for mobiles. *Multimedia Tools and Applications*.

Schwarz, H., Marpe, D., & Wiegand, T. (2007). Overview of the scalable video coding extension of the H.264/AVC standard. *IEEE Transactions on Circuits and Systems for Video Technology*, *17*(9), 1103–1120. doi:10.1109/TCSVT.2007.905532

Secker, A., & Taubman, D. (2001). Motion-compensated highly scalable video compression using an adaptive 3D wavelet transform based on lifting. *IEEE International Conference on Image Processing* (pp. 1029-1032). Thessaloniki, Greece: IEEE.

Secker, A., & Taubman, D. (2003). Lifting-based invertible motion adaptive transform (LIMAT) framework for highly scalable video compression. *IEEE Transactions on Image Processing*, *12*(12), 1530–1542. doi:10.1109/TIP.2003.819433

Seeling, P., Fitzek, F. H., & Reisslein, M. (2007). *Video traces for network performance evaluation. Dordrecht*. Dutch, Netherlands: Springer.

Shapiro, J. M. (2002). Embedded image coding using zerotrees of wavelet coefficients. In P. N. Topiwala (Ed.), *Wavelet Image and Video Compression, 450*, 123-155. New York, NY: Kluwer Academic Publishers.

Srinivasan, S., & Velev, M. (2003). Formal Verification of an Intel XScale Processor Model with Scoreboarding, Specialized Execution Pipelines, and Impress Data-Memory Exceptions. *Proceedings of The First ACM and IEEE International Conference on Formal Methods and Models for Co-Design*, (pp. 65-74).

Stankovic, J. A., & Ramamritham, K. (1990). What is predictability for real time systems? *Real-Time Systems*, *2*(4), 247–254. doi:10.1007/BF01995673

Sutter, J. D. (2010, June 7). *CNN Tech*. Retrieved on July 24, 2010, from Will iPhone 4 usher in era of video calls?: http:// www.cnn.com/ 2010/ TECH/ mobile/ 06/07/ apple.iphone.video /index.html

Taubman, D. S., & Marcellin, M. W. (2002). *JPEG2000: Image compression fundamentals, standards and practice*. Boston, MA: Kluwer Academic Publishers.

Uti, N. V. (2009). Real time mobile video streaming using wavelet transformation and run-length coding. *Proceedings of The 2009 International Conference on Wireless Networks, 2 (ICWN '09)* (pp. 368-373). Las Vegas.

Uti, N. V., & Fox, R. (2009). Improving delivery rate for real time mobile streaming video. *Proceedings of The 21st IEEE International Conference on Tools with Artificial Intelligence (ICTAI)* (pp. 49-53). Newark: IEEE.

Uti, N. V., & Fox, R. (2010). Testing the computational capabilities of mobile device processors: Some interesting benchmark results. *9th IEEE/ACIS International Conference on Computer and Information Science* (pp. 477-481). Yamagata: IEEE.

Wang, H., Kondi, L., Luthra, A., & Ci, S. (2009). *4G wireless video communications*. Chichester, West Sussex: Wiley. doi:10.1002/9780470745168

Wien, M., Cazoulat, R., Graffunder, A., Hutter, A., & Amon, P. (2007). Real time system for adaptive video streaming based on SVC. *IEEE Transactions on Circuits and Systems for Video Technology*, *17*(9), 1227–1237. doi:10.1109/TCSVT.2007.905519

Wisely, D. (2009). *IP for 4G*. Chichester, West Sussex: Wiley. doi:10.1002/9780470986363

YUV (2010, July 22). Retrieved on July 24, 2010, from Wikipedia: http:// en.wikipedia.org /wiki/YUV

ADDITIONAL READING

Bovik, A. (Ed.). (2009). *The Essential Guide to Image Processing*. Burlington, MA: Academic Press.

Chow, C.-O., & Ishii, H. (2007). Enhancing real time video streaming over mobile ad-hoc networks using multipoint-to-point communication. *Computer Communications*, *30*, 1754–1764. doi:10.1016/j.comcom.2007.02.004

Flierl, M., & Girod, B. (2004). Video coding with motion-compensated lifted wavelet transforms. *Signal Processing Image Communication*, *19*(7), 561–575. doi:10.1016/j.image.2004.05.002

Fowler, J. E., & Pesquet-Popescu, B. (Eds.). (2007). *Wavelets in source coding, communications and networks*. Nasr City / New York: Hindawi Publishing Corporation.

Gibson, J. (Ed.). (2000). *Multimedia communications: Directions and innovations*. San Diego, CA: Academic Press.

Hellwagner, H., Kuschnig, R., Stütz, T., & Uhl, A. (2009). Efficient in-network adaptation of encrypted H.264/SVC content. *Signal Processing Image Communication*, *24*(9), 740–758. doi:10.1016/j.image.2009.07.002

Jack, K. (2007). *Video demystified: A handbook for the digital engineer* (5th, ed.). Burlington, MA: Newnes.

Jamalipour, A., & Lorenz, P. (2006). End-to-end QoS support for IP and multimedia traffic in heterogeneous mobile networks. *Computer Communications*, *29*(6), 671–682. doi:10.1016/j.comcom.2005.07.021

Ko, H., Yoo, K., & Sohn, K. (2009). Fast mode-decision for H.264/AVC based on inter-frame correlations. *Signal Processing Image Communication*, *24*(10), 803–813. doi:10.1016/j.image.2009.08.001

Kumar, S., Xu, L., Mandal, M. K., & Panchanathan, S. (2006). Error resiliency schemes in H.264/AVC standard. *Journal of Visual Communication and Image Representation*, *17*(2), 425–450. doi:10.1016/j.jvcir.2005.04.006

Lee, D.-S., & Kim, Y.-M. (2009). Efficient coding of motion vector and mode information for H.264/AVC. *Signal Processing Image Communication*, *24*(10), 834–839. doi:10.1016/j.image.2009.07.004

Lee, J. (2005). *Scalable continuous media streaming systems: Architecture, design, analysis and implementation*. Chichester, West Sussex: Wiley. doi:10.1002/047001539X

Liu, Y., & Ngan, N. K. (2007). Fast multiresolution motion estimation algorithms for wavelet-based scalable video coding. *Signal Processing Image Communication*, *22*(5), 448–465. doi:10.1016/j.image.2007.03.001

Moinuddin, A. A., Khan, E., & Ghanbari, M. (2010). The impact of tree structures on the performance of zerotree-based wavelet video codecs. *Signal Processing Image Communication*, *25*(3), 179–195. doi:10.1016/j.image.2009.12.004

Rupp, M. (Ed.). (2009). *Video and multimedia transmissions over cellular networks: Analysis, modelling and optimization in live 3G mobile networks*. Chichester, West Sussex: Wiley.

Sayood, K. (2006). *Introduction to data compression* (3rd ed.). San Francisco, CA: Morgan Kaufmann.

Seran, V., & Kondi, L. P. (2007). Drift controlled scalable wavelet based video coding in the overcomplete discrete wavelet transform domain. *Signal Processing Image Communication*, *22*(4), 389–402. doi:10.1016/j.image.2007.01.003

Setton, E., & Girod, B. (2009). *Peer-to-peer video streaming*. New York, NY: Springer.

Simpson, W. (2008). *Video over IP, 2nd Edition: IPTV, Internet video, H.264, P2P, Web TV, and streaming: A complete guide to understanding the technology*. Burlington, MA: Focal Press.

Skodras, A. N., Christopoulos, C. A., & Ebrahimi, T. (2001). JPEG2000: The upcoming still image compression standard. *Pattern Recognition Letters*, *22*(12), 1337–1345. doi:10.1016/S0167-8655(01)00079-4

Topiwala, P. (2002). Wavelet still image coding: A baseline MSE and HVS Approach. In Topiwala, P. (Ed.), *Wavelet image and video compression* (*Vol. 450*, pp. 95–109). New York, NY: Kluwer Academic Publishers. doi:10.1007/0-306-47043-8_6

Wei, Z., Li, H., & Ngan, K. N. (2008). An efficient intra-mode selection algorithm for H.264 based on edge classification and rate-distortion estimation. *Signal Processing Image Communication*, *23*(9), 699–710. doi:10.1016/j.image.2008.08.002

Wu, Y., Gao, Y., & Chen, Y. (2008). Bit-depth scalability compatible to H.264/AVC-scalable extension. *Journal of Visual Communication and Image Representation*, *19*(6), 372–381. doi:10.1016/j.jvcir.2008.06.003

Xiao, Y., Chen, C. L., & Wang, B. (2002). Bandwidth degradation QoS provisioning for adaptive multimedia in wireless/mobile networks. *Computer Communications*, 1153–1161. doi:10.1016/S0140-3664(02)00006-3

Chapter 2
Multimedia Services for Mobile WiMAX:
Data-Partitioned Single Connection and Multi-Connection Streaming

Martin Fleury
University of Essex, UK

Salah S. Al-Majeed
University of Essex, UK

Laith Al-Jobouri
University of Essex, UK

Mohammed Ghanbari
University of Essex, UK

ABSTRACT

Mobile WiMAX is suitable for video-on-demand as well as broadcast TV services. Delivery of video streams over an IP packet-switched network, which includes broadband wireless access, requires rate or congestion control along with methods to protect the compressed stream from the impact of channel conditions. The Chapter considers possible solutions for both these requirements: multi-connection streaming overcomes the risk of poor channel utilization when restraining rates due to detected packet loss, while data-partitioning is a form of error resilience which includes the possible of graceful quality degradation. In the Chapter, ratelesschannel coding is adaptively employed along with error control. The Chapter also surveys the WiMAX option, acknowledging demerits as well as merits of this technology, and considering what other investigations have taken place to deliver multimedia services over broadband wireless.

DOI: 10.4018/978-1-61350-144-3.ch002

INTRODUCTION

Multimedia services are anticipated to be a key application of broadband wireless access networks (Fazel & Kaiser, 2008), which are supported by both mobile *Worldwide Interoperability for Microwave Access* (*WiMAX*) (*Institute of Electrical and Electronic Engineers* (*IEEE*) 802.16e (IEEE, 2005)) and *Long Term Evolution* (*LTE*) (Dahlman et al., 2008). The focus of this Chapter is mobile WiMAX video streaming. Mobile WiMAX (Jain et al., 2008) provides broadband wireless access independently of a pre-existing cellular system. It is not reliant on hardware authentication (Rather than a *Subscriber Identity Module* (*SIM*) card, WiMAX employs digital certificates in a Public Key Infrastructure). Furthermore, it can also deliver data in a cost-effective way (So-In et al., 2010) at approximately 3–4 times the rate of current 3G cellular systems. One additional short-term advantage is that mobile WiMAX is currently deployed and *Four Generation* (*4G*) LTE was ratified in December, 2010. Thus, mobile WiMAX is a promising alternative way to deliver multimedia services, supported as it is by leading companies such as Intel and Freescale. As an example, Yota Telecom (Yota TV, 2010) already supports 25 TV channels over its 10 Mbps mobile WiMAX network, which extends to 20 million people in major Russian cities, though Yota are now also considering LTE deployment. The WiMAX Forum has catalogued operation in 140 countries by 475 operators. This Chapter examines video streaming techniques that are applied within a mobile WiMAX setting, though, in fact, the techniques are generally applicable within other wireless environments.

Two ways developed by the authors to achieve effective streaming over WiMAX are discussed. The first approaches streaming as a source coding problem, using data-partitioning of the compressed bitstream (Mys et al., 2006) to separate out more important data. An interesting observation was made of data-partitioned video by Stockhammer and Bystrom (2004) that while it does not necessarily supply better overall video quality than unicast video without partitioning, because of its scalable nature few instances of poor quality video occur. This is because it is possible to partially reconstruct data-partitioned video without the presence of the lower-priority partition-C. In a novel approach proposed in this Chapter, the more important partition-A and partition-B data output by an H.264/*Advanced Video Coding* (*AVC*) encoder (Wenger, 2003) are protected by their smaller packet sizes. Smaller packet sizes arise provided an appropriate *Quantization Parameter* (*QP*) is selected. *Forward error correction* (*FEC*) is applied at the application layer of the protocol stack in the form of adaptive rateless channel coding (MacKay, 2005). By exploiting the rateless nature of Raptor coding (Shokorallahi, 2006), a scheme is developed in this Chapter that, when required, piggybacks extra redundant data to better reconstruct the compressed bitstream.

The second of the approaches developed in this Chapter from the authors' experience is employed when mobile users are connected by WiMAX to a wired network infrastructure. For example, this could allow one mobile device user to stream over WiMAX through the Internet to another mobile user also connected to a WiMAX *Base Station* (*BS*). We have found that multi-connection streaming (Chen & Zakhor, 2006) removes the need for cross-layer interaction, which is a common way to cope with congestion control over the wireless portion of the streaming path. In multi-connection video streaming, a single video source is multiplexed onto several congestion-controlled connections across the wireless link in order to improve the wireless channel utilization, resulting in an increase in throughput. The scheme uses *Datagram Congestion Control Protocol* (*DCCP*) (Kohler et al., 2006), which as previously mentioned is an industry standard form of congestion control. DCCP's main role when congestion occurs is to reduce the video streaming data rate across the wired portion of the concatenated network but unless multi-connection DCCP is used unacceptable streaming delays can arise.

For both these forms of streaming, the IEEE 802.16e configuration and the presence of other traffic needs to be taken into account. Even though other traffic may occupy a different class of service queue, the time to service that queue can contribute to buffer overflows in another queue. In general, careful modeling of the wireless technology through simulation is necessary, because the variety of parameters makes an analytical solution unfeasible. For example, to select the best number of connections to achieve good wireless channel utilization, the effect on packet drops must be considered, as well as considering the start-up delay, packet end-to-end delay, and packet arrival jitter. As the data-partitioning scheme additionally employs intra-refresh M*acro Blocks* (*MBs*) to avoid the sudden data transmission spurts that arise with periodic intra-coded I-frames, another factor comes into play, as intra-refresh MBs increase the size of partition-B packets.

The remainder of this Chapter is organized as follows. Section 2 reviews WiMAX technology and its application to *Internet Protocol TV* (*IPTV*), multimedia streaming including multicast and broadcast and the need to control the rate of the streams, because of the presence of other traffic and because of the risk of self-congestion, i.e. exceeding the resources of the network by too rapid streaming rate. Section 3 is an analysis of error-resilience methods to protect video, which, together with application-layer FEC, are the main ways open to the developer to protect fragile compressed video against the impact of a wireless channel. In particular, the Section considers data-partitioning, which has some of the features of a scalable form of streaming and some features that potentially provide error resilience. Section 4 details the simulation configuration for two cases studies based upon on the authors' own work: multi-connection streaming and data-partitioned streaming respectively. All results in these Sections are those of this Chapter's authors. Section 5 reviews future developments in the field of broadband video streaming. Finally, Section 6 concludes the paper by summarizing the Chapter's contribution.

WIMAX TECHNOLOGY AND STREAMING

Mobile WiMAX (Li et al., 2007) was introduced in 2007, as part *e* of the IEEE 802.16 standard, to strengthen the fixed WiMAX part d standard of 2004. Like many recent wireless systems, part *d* utilized *Orthogonal Frequency Division Multiplexing* (*OFDM*) as a way of increasing symbol length to guard against multi-path interference. The sub-carriers inherent in OFDM were adapted for multi-user usage by means of *Orthogonal Frequency Division Multiple Access* (*OFDMA*), allowing subsets of the lower data-rate sub-carriers to be grouped for individual users. Sub-channel spectral allocation can range from 1.25 MHz to 20 MHz. At the *physical* (*PHY*) level, it includes multiple antennas and hybrid *Automatic Repeat request* (*ARQ*) to strengthen *non Line of Sight* (*NLOS*) transmission. Adaptive antenna systems and *Multiple Input Multiple Output* (*MIMO*) antennas can further improve coverage. However, IEEE 802.16 part e is not directly backwards compatible with part *d* and some doubts has been expressed that part m will be completely backwards-compatible with part *e*. Basic *Multicast and Broadcast Services* (*MBS*) are supported by mobile WiMAX, with further schemes under development. IEEE 802.16m (Ahmadi, 2010) is expected to increase data rates to 100 Mbps mobile and 1 Gbps fixed delivery. However, 802.16m is not backwards compatible with 802.16e, though it does support joint operation with it.

Compared to LTE (Ekstrom et al., 2006), a technical weakness of WiMAX in mobile systems is the use of OFDMA for the uplink, whereas LTE employs *Single Carrier Frequency Division Multiple Access* (*SC-FDMA*). SC-FDMA precedes OFDMA by an additional *Discrete Fourier Transforms* (*DFT*) stage for the uplink, causing de-correlation of the input data symbols. Both technologies use OFDMA for the downlink. As one technology introduces new features such as antenna diversity (Hanzo et al., 2010) in the course of its evolution so does the other.

For correlated data sources, OFDMA experiences a higher *Peak-To-Average Power Ratio* (*PAPR*) as it directly modulates the data symbols. SC-FDMA can mitigate the PAPR on the uplink and at the same time rely on the increased processing at the base station, which is required to equalize SC-FDMA's single carrier transmission. Though it might appear that LTE is more adapted to mobile transmission, it is important to notice that compressed video is already an uncorrelated data source and, therefore, it does not present high PAPRs. Thus, a WiMAX interactive video service is feasible, whereas LTE does not differ from WiMAX in the form of its modulation over the downlink, which represents the route taken by the majority of IPTV streaming services. WiMAX-based video services already exist, for example in Brazil (Meloni, 2008) and Korea (Kim et al., 2008) - *WiBro* in Korea is now harmonized with WiMAX-. Mobile WiMAX is attractive in areas in which there is no existing cellular infrastructure such as parts of Africa, and Central and South America. WiMAX is also cost effective in rural (Hillested, 2007) and suburban areas in some developed countries (Cicconnetti et al., 2006).

Application of WIMAX to IPTV

The success of IPTV services such as the BBC's iPlayer in the UK suggests that delivery of video streams to the user will be important. The iPlayer allows TV programs to be streamed on demand, either live programs or time-shifted TV. Though currently based on Adobe Flash Player technology, it could be that because of the limitations of *Transmission Control Protocol* (*TCP*) transport (unbounded delays and fluctuating bitrates) this technology will be superseded by streaming, especially if mobile TV will be supported on WiMAX (Lee et al., 2009). Video streaming rather than progressive download or file-cast, allows fast channel swapping or zapping (Gómez-Brown, 2010) to be implemented but requires error control when delivered from a remote server in the network core. *Networked Personal Video Recorder* (*nPVR*) is also under development, allowing TV programs to be remotely stored for later streaming. *Live Pause* is another streaming facility which allows digital video recorder facilities including skipping parts such as commercial breaks, while the paused frame continues to be displayed. Mobile TV services proliferate: Apart from LTE or WiMAX there are also terrestrial broadcast systems such as DVB-H, mediaFlo, and DMB (for an overview see (Kumar, 2007)). For mobile WiMAX, capacity studies (So-In et al., 2010) suggest up to 16 mobile TV users per cell in a 'lossy' channel depending on factors such as the form of scheduling and whether MIMO antennas are activated. However, given the predicted increase in data rates arising from IEEE 802.16m, the number of unicast video users (Oyman et al, 2010) with 4×2 *Multi User* (*MU*)-MIMO antennas, will be 44 at 384 kbps and 22 at 768 kbps in an urban environment. For a similar configuration but using 20 MHz rather than 80 MHz channels, Oyman et al. (2010) reported the number of unicast video users to be 11 and 6 depending on data rates.

An overview of how an IPTV system with WiMAX fixed or mobile delivery is presented in Uilecan et al. (2007). The system takes advantage of WiMAX's *Point To Multipoint* (*PMP*) mode for the broadcast of channels. *Moving Picture Experts Group* (*MPEG*)-*2-Transport System* (*TS*) packets are multiplexed onto *Real Time Protocol* (*RTP*)/*User Datagram Protocol* (*UDP*)/IP headers. Header suppression and compression techniques reduce the overhead. In Issa et al. (2010), IPTV streaming was evaluated on a WiMAX testbed for downlink delivery of TV channels and uplink delivery of either TV news reports or video surveillance; refer to Figure 1.a. However, that research did not consider the impact of the intervening core wired network connecting the WiMAX BSs. In (Degrande et al., 2008), ways to improve IPTV quality were discussed with the assumption that intelligent content management would bring popular video content nearer to the end viewer. The typical IPTV architecture considered, Figure 1.b, assumes a *Super Head-End* (*SHE*) distributor of content across a core network to regional *Video Hub Offices* (*VHOs*). VHOs are connected to *Video Serving Offices* (*VSOs*) over a regional metro network. It is a VSO that interacts with users over an access network.

WIMAX Streaming

Chatterjee et al. (2007) did an analysis of WiMAX packetization issues in respect to streaming. WiMAX provides flexible *Medium Access Control* (*MAC*) *Protocol Data Unit* (*MPDU*) sizes which can be varied according to channel conditions. Within each MPDU, one or more *MAC Service Data Units* (*MSDUs*) can be aggregated, with segmentation of individual MSDUs also being possible. Large MPDUs reduce header overhead but are more prone to channel error. An application layer hybrid ARQ scheme allowed retransmissions to be requested if doing so would not cause receiver buffer overflow and if the data contained in the MPDU are sufficiently important for reconstruction of the stream. The case study in Section 4.2, contributed by this Chapter's authors, can be viewed as an application of this general scheme to a data-partitioned video stream. Block channel

Figure 1. (a) Downlink and uplink streaming scenarios, (b) schematic IPTV distribution network

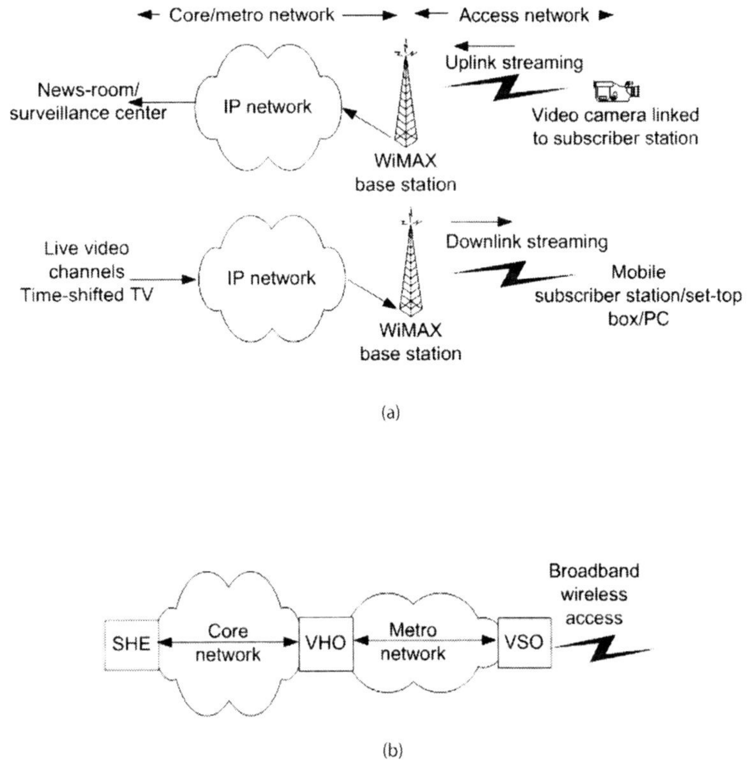

codes were employed in Chatterjee et al., 2007, whereas a more sophisticated rate less channel coding scheme is employed in Section 4.2.

MBS has received attention in respect to ways of minimizing energy consumption by waiting nodes. Hosein (2008) presented ways of allocating broadcast channels within WiMAX frames in order to minimize energy consumption at the mobile station. In WiMAX, an MBS zone is reserved for MBS bursts within a particular sub-channel. Therefore, if fixed-sized MBS MPDUs are used it will be possible for a mobile station to enter sleep mode when its TV channel is not being broadcast. This scheme is sufficient for *Constant Bit Rate* (*CBR*) streaming but is insufficient for constant quality *Variable Bit Rate* (*VBR*) streaming. To allow for streaming, one user is scheduled per WiMAX frame in (Hosein, 2008). In (Wang et al., 2007), super-frames are formed which contain the data for a group of mobile stations. Optimized allocation to the super-frame depends on spectral efficiency and video quality, with the latter regulated by temporal scalability. In Cohen et al., 2008, super-frames were also employed. Allocation to the super-frame is decided upon with an optimization algorithm according to the physical channel conditions of each user. This aims to manage power allocation.

The possibility of using the H.264/*SVC* extension for WiMAX streaming has interested a number of researchers. In (Reguant et al., 2008), *Digital Video Broadcasting-Terrestrial* (*DVB-T*) is combined with WiMAX so that different layers of a video stream can be transmitted simultaneously over the two delivery technologies. This allows a trade-off between the two systems depending on available capacity. Unfortunately, it also requires one provider to maintain two distribution networks, though in many cases DVB-T may already exist. In (Sharangi, 2010) scalable video is combined with MBS, when the allocation is determined based on the buffer size availability of the set of receivers. Another earlier use of SVC appeared in (Juan et al., 2007). In this research multi-connections were utilized, each connection carrying a different layer, with enhancement layers occupying less reliable connections.

Congestion Control for Wireless Video Streaming

Congestion control is the management of traffic including video streaming, ideally so that all traffic gains a fair share of the bandwidth capacity. On the wired Internet, TCP appears to dominate, due to its role as the underlying transport protocol for *Hyper Text Transfer Protocol* (*HTTP*). Historically, the stability of the Internet may have been maintained (Floyd & Fall, 1999) by TCP's co-operative form of congestion. However, this consensus has more recently been breached by peer-to-peer file download and even peer-to-peer streaming, which is likely to open multiple TCP connections.

These considerations are still relevant to wireless streaming, as the video server is likely to be remote and situated on a wired network. Volatile traffic load conditions on the Internet (Turletti & Huitema, 1996) create the need for video-specific measures and, on mobile wireless access networks, the presence of multiple mobile devices also leads to congestion at a BS's buffers. It is also possible for multiple applications such as automatic downloads, streaming, and web access, to share the link between the mobile station and the BS, again causing congestion. It is not necessary for packet queues to occur in the same service class queue, as delays in servicing one queue can lead to the build up of packets in another, eventually leading to buffer overflow. Therefore, a method of control that can negotiate congestion within a wired and wireless network is appropriate.

Unmodified TCP is normally considered unsuitable for video streams, because retransmissions introduce delay, which, apart from leading to display and/or decode deadlines to be missed, may delay feedback control messages. An additional cause of concern is that abrupt rate variations occur, due to TCP's congestion avoidance mechanism, which involves multiplicative reductions in rate. Both these problems can only be ameliorated (Krasic et al., 2001) by large receiver-side buffers, which are unsuitable for mobile devices and can lead to long delays. UDP transport is preferred for video transport and some type of application-layer congestion controller is provided to make up for UDP's "bare-bones" service.

Another sometimes understated reason for selecting a particular congestion controller is a desire by the manufacturers to conform to a standard. DCCP (Kohler et al., 2006) is along with *Stream Control Transmission Protocol* (*SCTP*) (Stewart, 2007) a standards-based way of providing congestion control for video streaming, without the disadvantages of TCP. That is, DCCP supports delay-sensitive streaming over UDP without TCP's delay-inducing reliability. It has two varieties: TCP-like congestion control and *TCP-Friendly Rate Control* (*TFRC*) (Handley et al., 2003), one variety of which is selected during initial handshaking. TFRC is employed, as it reduces the "saw-tooth" like rate changes associated with TCP congestion control mechanisms. Though SCTP mitigates other TCP shortcomings, such as lack of message structuring and exposure to *synchronize* (*SYN*) message flooding, it still essentially provides a TCP-like reliable service. However, an interesting feature of SCTP is support for multi-streaming with optional out-of-order delivery to avoid TCP's potential head-of-line blockages. This suggests that DCCP might be employed with multi-streaming, as explored later in this Chapter.

The main point of TFRC (Handley et al., 2003), i.e. effectively single-connection DCCP, is to maintain compatibility with TCP within the conventional wired Internet. TCP's congestion control is a co-operative process that seeks to avoid the acquisition of too much bandwidth by any one transport flow. The concept of TCP-friendliness (Floyd & Fall, 1999) on the conventional Internet was introduced to avoid the risk of congestion collapse, arising from the use of uncooperative protocols. As described in (Handley et al., 2003), TFRC is a receiver-based system in which the packet loss rate is found at the receiver and fed-back to the sender by acknowledgment messages. The sender calculates the *Round Trip Time* (*RTT*) from the acknowledgment messages and updates the packet sending rate. To alter the sending rate, TFRC changes the inter-packet sending time gap according to the TFRC equation. The throughput equation models TCP New Reno to find the sending rate:

$$TFRC(t_{rtt}, t_{rto}, s, p) = \frac{s}{t_{rtt}\sqrt{\frac{2bp}{3}} + t_{rto}\min\left(1, 3\sqrt{\frac{3bp}{8}}\right)p(1+32p^2)} \tag{1}$$

Where t_{rtt} is the RTT, t_{rto} is TCP's retransmission timeout, s is the segment size (TCP's unit of output) (herein set to the packet size), p is the normalized packet loss rate, and b is the number of packets acknowledged by each ACK. b is normally set to one and $t_{rto} = 4t_{rtt}$. Clearly packet loss and RTT cause the throughput to decrease in eq. 1, whereas other terms are dependent on these two variables in the denominator.

There have been many other attempts to improve the behavior of TCP by intervening at the interface between the wired and wireless Internet. For example, in the Snoop approach (Balakrishnan et al., 1997) a module resides on a base station. The Snoop module checks all traffic for TCP flows and carries out local retransmissions when a packet is lost on the wireless link. It is necessary for the Snoop module to suppress ACKs while retransmission is going on. Though the Snoop approach can be extended to TFRC it implies the introduction of additional delay to a real-time service by the introduction of retransmissions. It also increases the complexity of the implementation, especially if uplink rather than downlink streaming will take place. Another approach, for example (Cen et al., 2003), is to gather end-to-end statistics in order to distinguish congestion loss from wireless channel loss. Unfortunately, these techniques do not appear to be sufficiently accurate. A promising alternative is to modify TFRC to respond to explicit loss notification (Balakrishnan & Katz, 1998) of congestion losses and to ignore wireless channel losses.

In the research by Tappayuthpijam et al. (2010), single-connection TFRC was applied over the LTE wireless technology. The rationale was that, in this type of mobile network, retransmissions at the data-link layer effectively remove packet loss at the expense of end-to-end delay and throughput. TFRC was combined with H.264/SVC in such a way that layering was adapted to the bitrate. The results are promising compared to not using TFRC in terms of reduced packet losses, reduction of streaming interruptions, delay, and buffering. However, the approach by Tappayuthpijam et al. (2010) potentially re-introduces the risk of unbounded delay and head-of-line blocking, which make TCP unsuitable for real-time services. There is also the overhead of maintaining retransmission state at the transmitter and the delay arising if retransmissions are still occurring when a handoff occurs.

Other examples of single connection TFRC for video streaming over wireless include (Fu et al., 2006; Görkemli et al., 2008). In (Görkemli et al., 2008), SVC is again used in conjunction with DCCP in a scenario in which there are wireless access links to the core network at the sender and receiver sites. Therefore, this work investigates uplink as well as downlink streaming. Similarly to (Fu et al. 2006), lower-layer information is employed to modify TFRC's estimate of packet loss rates, in this case, by means of PHY-layer ARQ information. A similar cross-layer approach appeared in Yang et al., 2007, though TFRC was not used in that work. These methods are cross-layer modifications and, hence, suffer from the need to accord TFRC special treatment compared to other traffic.

DATA PARTITIONING AND OTHER FORMS OF ERROR RESILIENCE

Error resilience techniques are based around source coding. Their range has been expanded in an H.264/AVC codec (Wenger, 2003) compared to prior codecs in the MPEG series. Error resilience results in lower-delay than the inclusion of FEC, which increases the packet size. As such it is suitable for real-time, interactive video streaming, especially video-telephony and video conferencing. Application-layer FEC can be designed to act as an outer code after inner coding at the physical layer, in the manner of concatenated channel coding. In erasure coding, only correct data, information or redundant data are passed up from the physical layer. Then application layer coding can again attempt to reconstruct the missing data. This Section now reviews error resilience before specializing to data partitioning.

Error Resilience Methods

The simplest form of error resilience is through slicing. Compressed picture data are often split into a number of slices, each consisting of a set of MBs. Slice resynchronization markers in the slice header ensure that if a slice is lost then the entropy decoder will still be able to continue without loss of synchronization. An advanced feature of later codecs (MPEG-4 and H.263) is *Reversible Variable Length Coding* (*RVLC*) (Jeong et al., 2004), which allows partial decoding of a slice, even if part of it is damaged. Asymmetric RVLC produces better results but requires two look-up tables rather than the one of symmetric RVLC.

Returning to slicing, in the MPEG-2 codec, slices could only be constructed from a single row of MBs. Later codecs such as H.261 relaxed this requirement (Ghanbari, 2003), allowing contiguous MBs to form a slice. In an H.264/AVC codec, *Arbitrary Slice Ordering* (*ASO*) allows the decoder to reconstruct a picture if the slices arrive in a different order to the encoding order. A slice is a unit of error resilience and it is normally assumed that one slice forms a packet, after packing into a *Network Abstraction Layer unit* (*NALU*) in H.264/AVC. Each NALU is normally encapsulated in an RTP packet, though other transport options such as MPEG-2-TS are possible. For a given picture or frame, the more slices the picture is broken up into, the smaller the packet size and the less risk of packet loss through bit errors.

Additionally, in H.264/AVC, by varying the way in which the MBs are assigned to a slice (or rather group of slices), *Flexible Macroblock Ordering* (*FMO*) or slice groups gives a way of reconstructing a frame, even if one or more slices will be lost. Within a frame up to eight slice groups are possible. A simple FMO method is to continue a row of MBs to a second row but allow disjoint slice groups (Lambert et al., 2006). Regions of interest are supported by FMO. Furthermore, checkerboard or dispersed slice group selection allows one slice group to aid in the reconstruction of another slice group (if its packet will be lost) by temporal (accomplished by motion vector averaging) or spatial interpolation. Assignment of MBs to a slice group can be general (type 6) but the other six assignment types pre-define an assignment formula, thus removing the coding overhead from providing a full assignment map.

The insertion of *Intra-Refresh* (*IR*) MBs into pictures (Schreier & Rothermel, 2006) normally encoded through motion-compensated prediction allows temporal error propagation to be arrested if matching MBs in a previous picture will be lost. IR through periodic insertion of I-pictures (with all MBs encoded through spatial reference (intra-coded)) is the usual way of arresting error propagation. However, I-pictures cause periodic increases in the data rate when encoding at a variable bit rate, which increases are particularly unfavorable to wireless communication, because of the restricted bandwidth. They are also unnecessary if channel switching points or *Video Cassette Recorder* (*VCR*) functions will not be required. In the H.264/AVC JM implementation, various IR schemes exist such as random, which sets a maximum percentage of MBs, or cyclic, which replaces each line of the picture in turn in cyclic order. Notice that naturally encoded IR MBs are also inserted into predictively-coded P-pictures, when inter-coding brings limited or no advantage. For example, this may occur during rapid motion or when a new object that is not present in a previous picture is uncovered. It is also possible to vary the IR rate according to scene type or channel conditions (Liang et al, 2006).

Redundant slices (Baccichet et al., 2006) may also be inserted as a more severe form of error resilience. The redundant slice is normally encoded at a lower compression ratio (equivalently high QP) than regular slices and is discarded by the decoder if it is not required. Of course, there is a penalty in user perception from switching between different fidelities but this penalty is preferable to the freeze frame effect that would result if the primary redundant slice were to be lost. Additionally, as the encoder is unaware of a redundant slice substitution by the decoder, there will be error drift when a substitution occurs. Methods to refine the selection of redundant slices (Ferré et al., 2010) have also been designed.

Flexible reference frames in H.264/AVC allows (Wiegand et al., 1999) MBs in a single picture to reference multiple frames. Multiple referencing can reduce the need for sending intra-coded data. Another possibility made available by multiple referencing is for inter-prediction through different sub-sequences of prior frames. Another possibility is to use the switching frames, which allow a smooth transition between video streams encoded at different qualities (assuming VBR encoding), as a form of error resilience or method of adaptation (Stockhammer et al., 2006) to channel conditions.

Data Partitioning

A H.264/AVC codec conceptually separates the *Video Coding Layer* (*VCL*) from the *Network Abstraction Layer* (*NAL*). The VCL specifies the core compression features, while the NAL supports delivery over various types of network. In a communication channel, the *Quality of Service* (*QoS*) is affected by the two parameters of bandwidth and the probability of error. Therefore, as well as video compression efficiency, which is provided for through the VCL layer, adaptation to communication channels should be carefully considered. The concept of the NAL, together with the error-resilience features in H.264/AVC, allows communication over a variety of different channels. Table 1 is a summarized list of different NALU types. NALUs 1 to 5 contain different VCL data that will be described later. NALUs 6 to 12 are non-VCL units containing additional information such as parameter sets and supplemental information.

As previously mentioned, each frame can be divided into several slices; each of which contains a flexible number of MBs. *Variable Length Coding* (*VLC*) that is entropy coding of the compressed data takes place as the final stage of a hybrid codec. In H.264/AVC, arithmetic coding replaced other forms of entropy coding in earlier codecs. In each slice, the arithmetic coder is aligned and its predictions are reset. Hence, every slice in the frame is independently decodable. Each slice is placed within a separate NALU (see Table 1). The slices of an *Instantaneous Decoder Refresh* (*IDR*) - An IDR picture is confusedly equivalent to an

Table 1. NAL unit types

NALU type	Class	Content of NALU
0	--	Unspecified
1	VCL	Coded slice
2	VCL	Coded slice partition A
3	VCL	Coded slice partition B
4	VCL	Coded slice partition C
5	VCL	Coded slice of an IDR picture
6-12	VCL	Suppl. info., Parameter sets...
13-23	VCL	Reserved
24-31	Non-VCL	Unspecified

I-picture in previous standards. An I-picture in H.264/AVC allows predictive references beyond the boundary of a *Group of Pictures* (*GOP*) - or I-picture (i.e. a picture with all intra slices) are located in type 5 NALUs, while those belonging to a non-IDR or I-picture (P- or B-pictures) are placed in NALUs of type 1, and in types 2 to 4 when data partitioning mode is active, as now explained.

In type 1 and type 5 NALs, MB addresses, motion vectors and the transform coefficients of the blocks, are packed into the packet, in the order that they are generated by the encoder. In Type 5, all parts of the compressed bitstream are equally important, while in type 1, the MB addresses and motion vectors are much more important than the (integer) *Discrete Cosine Transform* (*DCT*) coefficients. In the event of errors in this type of packet, the fact that symbols appearing earlier in the bit-stream suffer less from errors than those which come later (because of the cumulative effect of VLC, symbols nearer the slice synchronization marker suffer less from errors than those that appear later in a bitstream) means that bringing the more important parts of the video data (such as headers and motion vectors) ahead of the less important data or separating the more important data altogether for better protection against errors can significantly reduce channel errors. In the standard video codecs, this is known as data partitioning.

However, in H.264/AVC when data partitioning is enabled, every slice is divided into three separate partitions and each partition is located in either of type 2 to type-4 NALUs. A NALU of type 2, also known as partition A, comprises the most important information of the compressed video bit stream of P- and B-pictures, including the MB addresses, motion vectors and essential headers. If any MBs in these pictures are intra-coded, their DCT coefficients are packed into the type-3 NALU, also known as partition B. Type 4 NAL, also known as partition C, carries the DCT coefficients of the motion-compensated inter-picture coded MBs.

Figure 2 is a comparison made by the authors of this Chapter between the relative sizes of the partitions according to QP. The test sequence was *Paris*, which is a studio scene with two three-quarter view images of presenters and moderate motion. The background is of high spatial coding complexity. This sequence is of a content type that *Quality of Experience* (*QoE*) subjective testing indicates is of interest to viewers on mobile devices (Agboma & Liotta, 2007). *Paris* was VBR encoded at *Common Intermediate Format* (*CIF*) (352×288 pixel/picture), with a GOP structure of IPPP..... at 30 fps, i.e. one initial I-picture followed by all predictive P-pictures. This arrangement removes the complexity of bi-predictive B-pictures at a cost in increased bit rate. From the Figure it is apparent that the partition-B's contribution increases as the percentage of IR MBs is increased, as a result making partition-B packets more vulnerable to congestion and channel error. The size of partition-C bearing packets declines with increasing QP, as a consequence of coarser quantization of DCT coefficients. The range of a QP in H.264/AVC is 0–51 with higher values corresponding to higher compression ratios and lower quality video.

In order to decode partition-B and -C, the decoder must know the location from which each MB was predicted, implying that partitions B and C cannot be reconstructed if partition-A will be lost. Though partition-A is independent of partitions B and C, *Constrained Intra Prediction* (*CIP*) should be set (Dhondt et al., 2007) to make partition-B independent of partition-C. By setting this option, partition-B MBs are no longer be predicted from neighboring inter-coded MBs, the prediction residuals of which reside in partition-C.

There is another dependency (Dhondt et al., 2007) arising from *Context-Adaptive VLC* (*CAVLC*) entropy coding, because the number of non-zero coefficients in one MB are predicted from the number in a neighboring MB. By design, when CAVLC is in operation, setting CIP also results in setting the number of non-zero coefficients in data-partitioned inter-coded MBs to zero to code intra-coded MBs. Thus, partition-B can be made independent of partition-C. It is not possible to employ the alternative *Context Adaptive Binary Arithmetic Coding* (*CABAC*), as this option is not supported in the extended profile of H.264/AVC, though this is the only profile in which data-partitioning is supported. As CAVLC still predicts from intra-coded MBs when coding partition-C's

Figure 2. Relative sizes of data partitions according to QP for the Paris video sequence, with (a) 5% IR MBs, and (b) 25% IR MBs

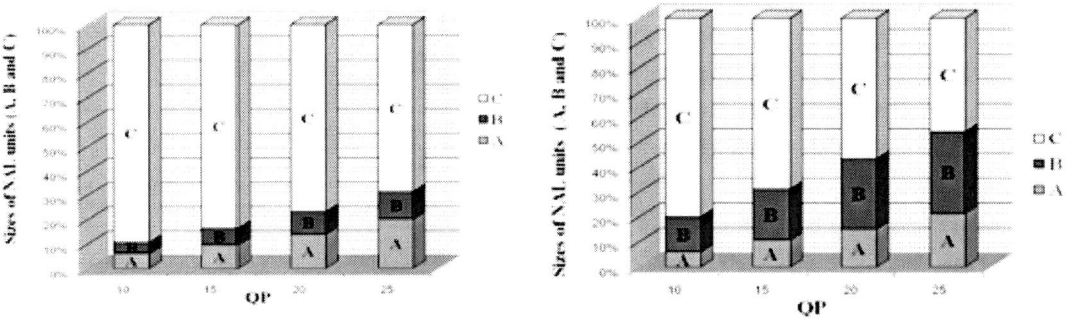

inter-coded MBs, partition-C cannot normally be made independent of partition-B.

SIMULATION RESULTS

In this Section we first present the common simulation environments configured for the two cases we simulate and then we present the particular characteristics of both cases their simulation results.

Common Simulation Environment Parameters

The common simulation environment configured had the following characteristics: The WiMAX system operating in PMP mode was simulated by well-known ns-2 simulator (v. 2.29 was employed), which was augmented by a WiMAX module (Tsai et al., 2006). That module supports mobile as well as the earlier fixed version of WiMAX. All simulation runs were repeated at least ten times with different random number seeds and the arithmetic mean was taken to form data points. The simulator was allowed to reach steady state before commencing runs.

The PHY settings selected for WiMAX simulation are given in Table 2. The range was set to 0.7 km to reduce the propagation delay component of latency; to better judge the queue waiting time. However, 0.7 km is a normal range in an urban environment. The *Downlink* (*DL*)/*Uplink* (*UL*) ratio is weighted in favor of the *Subscriber Stations* (*SSs*), whereas in practice the ratio would be reversed in order to allow the BS to allocate its capacity across a set of SSs. The antenna heights are typical ones taken from the Standard (IEEE, 2005). The antenna is modeled for comparison purposes as a half-wavelength dipole, whereas a sectored set of antenna on a mast might be used in practice to achieve directivity and, hence, better performance. The IEEE 802.16e *Time Division Duplex* (*TDD*) frame length was set to 5 ms in the second of the two studies (Section 4.3) but varied between 8 ms and 20 ms in the first case study (Section 4.3). While 5 ms is the only frame size value (Jain et al., 2008) supported in the WiMAX forum's simplification of the IEEE standard, the frame size is an important determinant of response latency. Increasing the frame size can allow the real-time service queue to be more completely serviced at any one polling instance, thus reducing latency. The data rate of 10.67 Mbps in Table 2 results from the use of one of the mandatory coding modes (IEEE, 2005) for a TDD DL/UL sub-frame ratio of 3:1.

As a test, the *Paris* clip was VBR-encoded with an H.264/AVC codec (Wiegand et al., 2003) at CIF resolution. 1065 frames were transmitted, corresponding to 35.5 s when at a frame rate of 30 Hz. *Paris* consists of two figures seated around a table in a TV studio setting. The background has high spatial coding complexity.

WiMAX networks support multiple service classes to accommodate heterogeneous traffic with varying requirements. Specifically, these service classes are *Unsolicited Grant Service* (*UGS*), *extended real-time Polling Service* (*ertPS*), *real-time Polling Service* (*rtPS*), *non-real-time Polling Service* (*nrtPS*), and *Best Effort* (*BE*). UGS is designed for synchronous services such

Table 2. WiMAX simulation parameters

Parameter	Value
PHY, Duplexing mode	OFDMA, TDD
Frequency, Modulation	5 GHz, 16-QAM 1/2
Frame length	5 ms, 8ms – 20 ms
Max. packet length	1024 B
Raw data-rate	10.67 Mbps
FFT size, Guard band ratio	1024,1/8
DL/UL ratio	1:3
Fragmentation	Yes
SS/BS transmit power	250 mW, 20 W
Range	0.7 km
SS/BS antenna heights	1.5/32 m
Antenna type, gain	Omni-directional, 0 dBD

as legacy voice without silence suppression. ertPS is intended for voice with silence suppression. *rtPS* is most suitable for real-time video services, particularly for VBR video, which is employed to maintain delivered video quality but may lead to 'bursty' arrival rates. Congesting traffic in the case studies is assumed to enter the *nrtPS* queue at a WiMAX SS. In the experiments for both queues, a drop-tail queuing discipline, which is the default in the Standard (IEEE, 2005), was simulated. Queue sizes were all set to a standard size of fifty packets. Particularly at mobile devices, larger buffer sizes, due to both active and passive energy consumption, can reduce battery recharge cycles and, hence, sizes are reduced.

In Figure 3, all links except a bottleneck link within the core network are set to 100 Mbps to easily accommodate the traffic flows entering and leaving the network. The link delays are minimal (2 ms) in order to avoid confusing propagation delay with re-ordering delay at the video sink, node C. A bottleneck link with capacity set to 5 Mbps was set up between the two routers, which are marked as R in Figure 2. The buffer size in each router was also set to 50 packets with drop-tail queue discipline. The network topology was not meant to physically correspond to a network layout but to represent the type of bottleneck that commonly lies at the core network edge before entry into a corporate or campus network. In the core network, node A sources to node B a CBR stream at 1.5 Mbps with packet size of 1 kB. Node A also sinks a continuous TCP FTP flow sourced at node B. Node B also sources an FTP flow to the WiMAX BS and a CBR stream at 1.5 Mbps with packet size 1 kB (see Table 2 downlink). As previously remarked, other SS sources apart from the video connections do not pass over the core network shown but are assumed to be routed elsewhere after passing the WiMAX BS.

A two-state Gilbert-Elliott channel model (Haßlinger & Hohlfeld, 2008) was introduced into the physical layer of the simulation to simulate the channel model for WiMAX. The Gilbert-Elliott is a type of hidden Markov model in which probabilities are set for going between the good and bad state and of remaining in each state. Once in a state, a statistical distribution models the risk of data loss.

The Gilbert-Elliott scheme though simple has been widely adopted, as it is thought to realistically model the burst errors that do occur and, more

Figure 3. IEEE 802.16 uplink service architecture

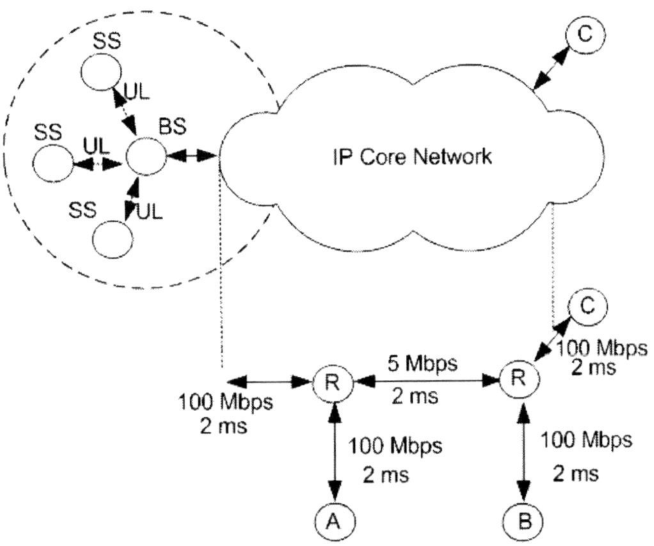

Table 3. Simulated traffic characteristics

SS-UL	Service type	Traffic type	Protocol	Packet size (B)
1	rtPS	VBR (video)	Multiple TFRC	Variable
		CBR	UDP	1000
	nrtPS	FTP	TCP	
2	rtPS	CBR	UDP	1000
	nrtPS	FTP	TCP	
3	rtPS	CBR	UDP	1000
	nrtPS	FTP	TCP	
SS-DL				
1,2, 3	rtPS	CBR	UDP	1000
	nrtPS	FTP	TCP	

significantly, can be particularly damaging to compressed video streams, because of the predictive nature of source coding. Therefore, the impact of 'bursty' errors (Liang et al., 2008) should be assessed in video-streaming applications.

Case Study 1: Streaming with Data Partitioning

We simulated an effective video streaming system for WiMAX that provides error resilience through data partitioning and which works *without* the need to apply privileged protection to the high-priority partitions. This is achieved in an H.264/AVC codec by setting the QP in such a way that lower-priority texture data, which can be replaced more easily at the decoder through error concealment, occupies a larger part of a frame's data. Thus, partition-A motion vectors are used in motion-copy error concealment when partition-C data are lost. Consequently, when packetized in a WiMAX MSDU within an MPDU (Fleury et al., 2009), partition-C is more likely to suffer error (assuming one MSDU per MPDU) than MSDUs bearing data from other partitions.

However, though no special protection is given, it is still necessary to protect the bitstream (without privileging the partition-A and -Bs) against the risk of packet loss, which was achieved through equal error protection using channel coding. Application-layer Raptor channel coding (Shokorallahi, 2006) was selected for its flexibility and its linear computational complexity at both the decoder and the encoder. In order to model Raptor 'rateless' coding, the following statistical model (Luby et al., 2007) was employed:

$$P_f(m,k) = \begin{cases} 1 & \text{if } m < k, \\ 0.85 \times 0.567^{m-k} & \text{if } m \geq k \end{cases} \quad (2)$$

where

$P_f(m,k)$

is the failure probability of the code with k source symbols if m symbols have been received. Luby et al., 2007 remark and show that for $k > 200$ the model almost perfectly models the performance of the code. When $m = k$ there is a failure rate of as much as 85%, but this failure rate reduces exponentially as the difference between m and k grows.

The particular environment simulation parameters used was the following:

- QP value was varied.
- Frame rate was, however, set at 15 Hz. This frame rate is typical of display rates on mobile devices. A GOP structure of IPPP.... was employed without periodic I-frames. It was, therefore, necessary to protect against temporal error propagation in the event of inter-coded P-frame slices being lost. H.264/AVC allows insertion of randomly-placed IR MBs to act as anchor points in the event of slice loss. The amount was specified as a maximum of 25%, which includes any existing naturally-encoded intra MBs. The Extended profile was employed,

as this is the only H.264/AVC that currently supports data partitioning.

- Video was transmitted over the downlink to a WiMAX SS with UDP transport. The station was simulated to move at a speed of 1 mps (2.2 mph), i.e. at strolling speed. In order to introduce sources of traffic congestion, an always available FTP source with TCP transport was sent to the SS. Likewise, a CBR source with packet size of 1000 B and inter-packet gap of 0.03 s was downloaded to the SS. While the CBR and FTP occupy the non-rtPS queue, rather than the rtPS queue, they can still contribute to packet drops in the rtPS queue for the video. Thus, if too many video packets occupy the rtPS buffer, while the nrtPS queue is being serviced, then packets arriving will be dropped due to buffer overflow.
- The probability of remaining in a good state (PGG) was set to 0.95, the probability of being in a bad state (PBB) was set to 0.96, the probability of losses in a good state (PG) was set to 0.02 and the probability of losses in a bad state (PB) was set to 0.165. Both states were modeled by a Uniform distribution.

In experiments, the percentage redundancy for the Raptor code was set to 10% similarly to the usage in (Ahmad et al., 2007) for Internet video streaming. The symbol size was set to bytes within a packet. Clearly, if instead 200 packets are accumulated before a rateless decoder can be applied (or at least equation (2) is relevant) there is a penalty in start-up delay for the video streaming and a cost in providing sufficient buffering. A single ARQ was made if the available data were insufficient to reconstruct a corrupted packet. Further, requests are ill-advised because of the delay that would result.

Rateless code decoding in traditional form operates by a belief-propagation algorithm (MacKay, 2005) which is reliant upon the identification of clean symbols. This latter function is performed by PHY-layer FEC which passes up correctly received blocks of data (identified through a Cyclic Redundancy Check) but suppresses erroneous data. For example, in IEEE 802.16e (Andrews et al., 2007), a binary, non-recursive convolutional encoder with a constraint length of 7 and a native rate of 1/2 operates at the PHY-layer. Upon receipt of the correctly received data, decoding of the information symbols is attempted, which will fail with a probability given by (1) for $k > 200$.

In a simulation, an estimate of the number of symbols successfully received is made by noting the loss probability given by the channel model and applying this probability to the packet payload length. This results in a value for m in equation (2). If $m \geq k$ then the decode probability is evaluated by means of equation (2). In a simulation the decidability decision is taken by comparing a Uniformly-distributed random variable's value with that of the probability given by (1). If a packet cannot be decoded because $m < k$ then extra redundant data are added to the next packet. From (2) it is implied that if less than k symbols (bytes) in the payload are successfully received then a further k-m + e redundant bytes can be sent to reduce the risk of failure. In tests in this Section, e was set to four, which reduces the risk of failure to below 9%.

Two types of erroneous packets were considered:

- Packet drops at the BS sender buffer and;
- Corrupted packets that were received but affected by Gilbert-Elliott channel noise.

If the packet could not be reconstructed as there was insufficient redundant data, an ARQ request was made for a fixed extra amount of redundant data (as previously described). Conversely, if there was sufficient redundant data present then equation (1) will determine if this data allowed recovery of the packet, bearing in mind the probabilistic nature of rateless code decoding.

Tests first checked whether partitioning a frame into three partitions was responsible for the gain from data-partitioning or whether the gain actually arose from the prioritized segregation of the information when using data partitioning. The comparison was made across QP values and across TDD frame sizes. Therefore, in the non-data-partitioning (non-DP) sliced tests, an encoded frame was geometrically divided into three equal-sized horizontal slices (simple slicing).

Comparing Tables 4 and 5, it is apparent that geometric slicing has an advantage in terms of the number of dropped packets for high-quality video (QP = 10). Otherwise, data-partitioning is preferable, because the relatively fewer partition-A packets corrupted allows better quality decoder reconstructions. The small fluctuations in the number of geometrically-sliced corrupted packets makes little difference to the objective video quality for QPs higher than 10. One can conclude that more partition-C packets than partition-A- and -B- bearing packets are corrupted, leading to the superior quality of the data-partitioned solution, even though no special protection is given to partition-A and -B packets (other than rateless coding equally weighted between the partitions).

A further comparison was made with the effect of not employing data-partitioning or slicing. In other words, the encoded video was configured with one slice per frame. If the single slice was larger than the maximum packet size of 1024 B then it was segmented at the network level. This also could occur with data-partitioning or slicing but is less likely as both these options result in slices smaller than the maximum packet size.

From Table 6, it is apparent that effectively increasing the packet size leads at low QP values to large numbers of dropped packets. This is because a lower QP setting leads to larger slices emerging from the encoder. The effect of the high number of dropped packets is that the decoder was unable to decode the video when multiple successive packets were dropped. As the percentage of corrupted packets was also over 10% in many tests, the objective video quality was lower than when data-partitioning was employed. This was the case even though rateless channel coding was used to protect the single-slice video data.

Table 4. Lost packets and video quality, when using geometric slicing with Raptor channel coding

QP	*Dropped packets (%)*		
	8 ms TDD	**10 ms TDD**	**20 ms TDD**
10	0	0	0
15	0	0	0
20	0	0	0
25	0	0	0
QP	*Corrupted packets (%)*		
	8 ms TDD	**10 ms TDD**	**20 ms TDD**
10	8.7	8.8	8.6
15	8.8	8.8	9.0
20	9.4	8.9	8.1
25	8.4	8.1	8.6
QP	*PSNR (dB)*		
	8 ms TDD	**10 ms TDD**	**20 ms TDD**
10	51.19	51.19	51.19
15	42.24	42.24	42.24
20	42.53	42.53	42.53
25	39.02	39.02	39.02

Table 5. Lost packets and video quality, when using data-partitioning with Raptor channel coding

QP	*Dropped packets (%)*		
	8 ms TDD	**10 ms TDD**	**20 ms TDD**
10	4.9	1.4	0.1
15	0	0	0
20	0	0	0
25	0	0	0
QP	*Corrupted packets (%)*		
	8 ms TDD	**10 ms TDD**	**20 ms TDD**
10	9.6	9.7	9.3
15	9.5	9.2	9.8
20	8.8	9.6	9.3
25	8.8	9.3	9.9
QP	*PSNR (dB)*		
	8 ms TDD	**10 ms TDD**	**20 ms TDD**
10	26.48	34.43	42.26
15	46.95	46.95	46.95
20	42.98	42.98	42.98
25	39.30	39.30	39.30

Table 6. Lost packets and video quality when using single slices with Raptor channel coding

QP	Dropped packets (%)		
	8 ms TDD	10 ms TDD	20 ms TDD
10	99.8	99.8	99.8
15	78.3	66.4	28.2
20	0.10	0.00	0.00
25	0.00	0.00	0.00
QP	Corrupted packets (%)		
	8 ms TDD	10 ms TDD	20 ms TDD
10	-	-	-
15	11.2	10.3	10.4
20	11.1	10.0	8.0
25	10.3	10.3	9.7
QP	PSNR (dB)		
	8 ms TDD	10 ms TDD	20 ms TDD
10	-	-	-
15	-	-	27.94
20	42.10	42.53	39.01
25	39.01	39.01	39.01

In the proposed scheme for WiMAX downlink video streaming, the compressed video bitstream is partitioned according to the information's importance to the decoding process. It was found that, provided rateless coding is used, data-partitioning is an improvement over simple geometric slicing or no slicing within a frame. This is because more partition-C packets tend to be lost because of their larger size, while the preservation of partition-A and partition-B slices still allows reconstruction through motion-copy error concealment at the decoder.

Case Study 2: Multi-Connection Streaming

Packet loss through channel error may be mistaken for congestion, leading to underutilization of the wireless link. In fact, burst errors, arising through slow fading or shadowing, are common in wireless networks, exacerbating the problem. When streaming takes place across a tandem network, one consisting of a wireless access network connected to a core wired network, then one proposed solution is to open multiple connections, multiplexing a single video stream between those connections.

The research in Chen and Zakhor (2005) proposed multi-connection MULTTFRC, originally as a form of downlink congestion control. Any single TFRC connection responds to packet loss by reducing its output rate as a result of increasing the inter-packet gap. By multiplexing a video stream in this way, it is hoped that the impact of packet loss on one or more of the connections will be mitigated by the data rate across the remaining connections. MULTTFRC represents a lightweight way to retain TFRC for the Internet path but avoid complex means of suppressing channel loss feedback to TFRC over the wireless link. Chen and Zakhor (2006) included a practical demonstration of MULTTFRC for IEEE 802.11. However, even in this most definitive account of MULTTFRC so far, there is no account of how a single video stream is multiplexed onto the multiple connections. Only generic packet loss and delay statistics were reported, even though the type of error pattern is known to affect video quality by several dBs. Therefore, after the initial development of the concept, it is now necessary to investigate how the scheme works for a particular wireless technology and what the QoS concerns are.

To systematically test the effect of multiple TFRC connections, rather than apply MULTTFRC's connection dynamic-switching mechanism (Chen & Zakhor, 2006), the number of TFRC connections was incrementally stepped up. In MULTTFRC itself, the number of connections is changed over time according to the average RTT of all the connections, but this hides the interpretability of results. A single stream was segmented into GOPs. Each connection was statically allocated its GOPs, which were taken in interleaved manner from the video sequence. Of course, this assumes that a re-ordering buffer is available at the receiver.

The particular environment simulation parameters used was the following:

- The initial QP was set to 30.
- H.264/AVC's Baseline profile was selected, as this is more easily supported by mobile devices because of its reduced computational overhead. The IR rate was every 30 frames with an IPPP…I frame structure. In RTP mode, an NALU from a H.264/AVC codec was encapsulated with an RTP header. After the addition of an IP header, this in turn formed a single WiMAX MPDU, that is variable-sized WiMAX packet. A trace file of the video input was input to ns-2 and packet losses recorded in the output. The output served to calculate the *Peak Signal-to-Noise Ratio* (*PSNR*). Video quality comparisons were made under the EvalVid environment (Klaue et al., 2003).
- A tandem network scenario was set up for uplink streaming from an SS, as illustrated in Figure 3. Table 2 records the simulated traffic characteristics for three SSs communicating with a BS within the WiMAX network. There were two SSs communicating to the BS, while the third SS streamed the *Paris* video sequence, multiplexing it between the multiple TFRC connections. The video stream is relayed from the BS to node C in Figure 2, whereas the other traffic flows remain internal to the WiMAX access network. The other SSs are introduced as sources of competing traffic across the wireless link and do not indicate the likely size of a WiMAX network. Potential SS capacities of IEEE 802.16e are reviewed in So-In et al, 2008 and Oyman et al., 2010. Coexisting rtPS queue CBR sources were all sent at 1500 kbps, i.e. the same target rate as the video source. The inter-packet gap was 0.03s for the CBR traffic. The FTP applications were set up out of convenience, as a way of occupying the nrtPS queues; otherwise a BE QoS queue might be more appropriate. Likewise, the DL traffic was selected to fully occupy the DL link capacity.
- The packet loss probability in the bad state was varied in order to judge the effect of the wireless channel. The PGG was set to 0.95 and PBB was set to 0.94, with both states again modeled by a Uniform distribution. The packet loss probability in the good state, PG, was fixed at 0.01 but was varied in the bad state as {0.01, 0.02, …, 0.1} in order to assess the effect of changing channel conditions.

Figure 4 shows the average data-rate over time when transmitting *Paris* over multiple connections. Clearly, TFRC is able to multiplex more data onto a link as the number of connections increases, though observation of a time-wise plot of throughput shows that during transmission TFRC quenches its overall sending rate in response to packet loss. There is a gradual decline in throughput as channel error conditions worsen.

In terms of PSNR, Figure 5's mean results show that reconstructed video quality for a four connection session is markedly below that of a single connection. This results from the packet drop pattern for a four connection streamin session, as consecutive packet losses are difficult to recover from within a GOP. While users are likely to tolerate PSNR of between 25 and 30 dB in the knowledge of the difficulty of transmitting over a mobile network (Sadka, 2006) PSNR levels below 20 dB result in unacceptable video quality. However, for low error rates two and sometimes three connections are within the desired range. This is fortunate, because from Table 7 it is evident that the sending period of single connection streaming is at least double the time to actually view the *Paris* video sequence, which is 35.5 s at 30 Hz. Put simply, the viewer would be faced with long freeze-frame times if only using a single connection because of the delays intro-

Figure 4. Average throughput for an increasing number of connections, according to channel error rate, showing 95% confidence intervals

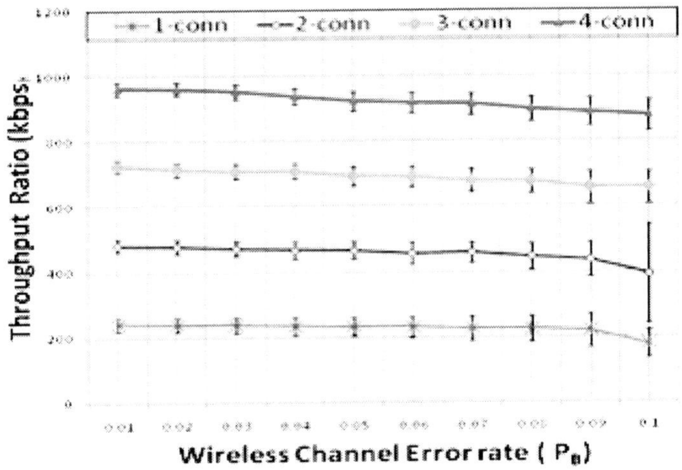

duced by DCCP/TFRC in response to wireless channel packet losses. Packet delay levels were very low, a few milliseconds in Figure 6. Latency declines with increasing channel error, presumably because TFRC reduces its sending rate and consequently reduces buffer occupancy and waiting time.

Multi-connection video streaming under TFRC congestion control is essentially a compromise between the number of connections and the resulting video quality. The technique is feasible, provided it is confined to a few connections and provided channel conditions are not too severe, as increasing the number of connections leads to unfavorable video quality.

Table 7. Multiple connection total sending period against throughput

	Sending period (s)	Average datarate (kbps)
1	71.4	216.69
2	35.8	437.45
3	23.3	662.77
4	17.4	889.49

FUTURE RESEARCH DIRECTIONS

Though mobile TV services have been available for cellular systems, 3G, or *Universal Telecommunications System* (*UMTS*), have largely been confined (Kumar, 2007) to Quarter CIF size at lower data rates (256 kbps) and limited numbers of users per cell (about six). Currently, *High Speed Data Packet Access* (*HSPDA*) is an evolution of 3G systems to rates of 550 kbps to 1 Mbps, which may soon be superseded by LTE. *Multimedia Broadcast and Multicast* (*MBMS*) may overcome the limitation of 3G networks to unicast delivery, and *TDtv* has been developed for UMTS TD-CDMA systems. One key advantage of networked mobile TV is the ability to supply both value-added unicast services as well as broadcast services. There is also time-switched TV and near Video on Demand. However, WiMAX systems with MBS are capable of supplying all these needs and WiMAX already exceeds the capability of HSPDA.

One area mobile WiMAX is particularly suited to video streaming for users on public transport systems, including buses, coaches and trains. It

Figure 5. Mean video quality for an increasing number of connections, according to channel error rate

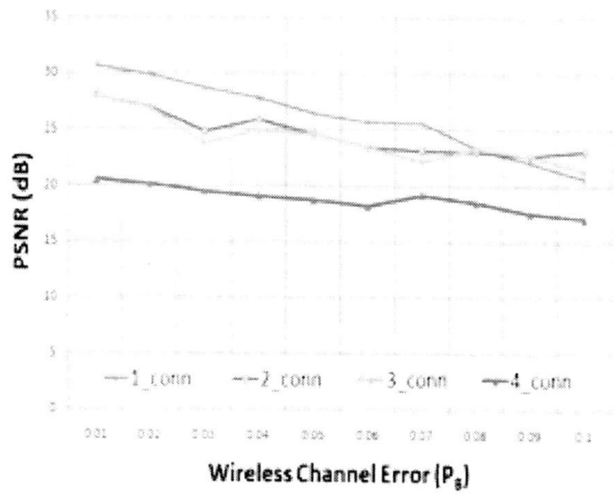

Figure 6. Average per packet delay for an increasing number of connections, according to channel error, with 95% confidence intervals

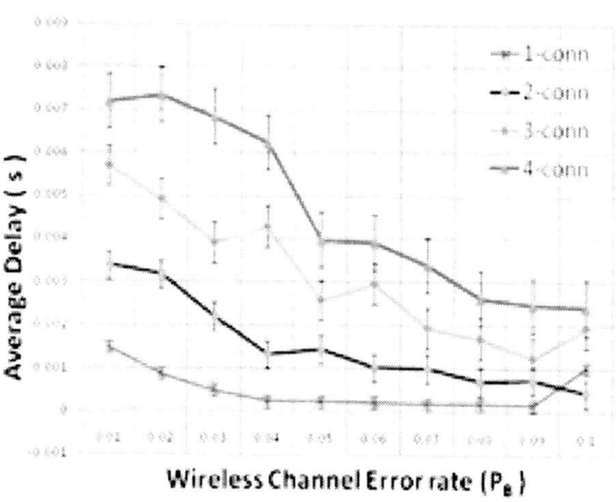

appears that issues of compatibility with IEEE 802.11 systems on board the vehicles have largely been solved. However, another issue in these networks is optimizing horizontal handoff, so as to cope with the movement of a user device from one WiMAX cell to another. There is also the need to build on previous work in optimizing the allocation of resources, such as data rates, channel selection, and energy consumption. WiMAX mesh mode is a comparatively unexplored research topic, as most attention has been given to PMP mode operation.

Increasingly also, video streaming and data services in general aim to support seamless mobility. In wireless access this implies an ability to go from outdoor WiMAX or LTE streaming to indoor IEEE 802.11 streaming. IEEE 802.21 aims at standardizing vertical handover between these networks. However, this standard does not solve the problem of the interruption to streaming, which is likely to be more severe than when a horizontal handoff occurs. Anticipation at the mobile device of handover allows pre-buffering of video but that must be at the cost of the existing stream. Alternatively, a stream can be sent in advance to the likely target network to which a handover is made.

CONCLUSION

The majority of investigations by the video coding community of video streaming over wireless channels have been conducted with IEEE 802.11 or 3G cellular systems in mind. Consequently, the focus has been on short-range, indoor networks with distributed medium access control (IEEE 802.11) or comparatively low bandwidth systems (cellular radio). The emergence of IPTV in the marketplace has switched the focus to broadband wireless access, along with the content distribution network that precedes the access network. There are comparatively few detailed studies of video streaming across such systems, though capacity studies and studies of the feasibility of streaming over WiMAX networks are available.

This Chapter has provided a review of broadband wireless access, with a concentration on mobile WiMAX and its application to video streaming in general and IPTV in particular. At the same time, the Chapter has also analyzed error resilience through data-partitioning. Error resilience represents the response of the source coding community to the problem of video transport over wireless networks. The network community has also tackled this problem. One of its solutions is to employ multi-connection streaming. To illustrate the discussion the error resilience route and the multi-connection route have both been illustrated by case studies and results arising from the authors' own work. Both solutions have their merits. In the case of error resilience, it was found that, provided rate less channel coding was used data-partitioning was an improvement over simple geometric slicing or no slicing within a frame. We are unaware of work by others in which rate less channel coding has been combined with data-partitioning, and, to the best of our knowledge adaptive rate less coding has not been used in this manner for broadband wireless transmission. In the case of multi-connections, it was found that the technique is feasible, provided it is confined to a few connections and provided channel conditions are not too severe. Though multi-connection streaming has been explored by others it has mainly been done so from a network perspective, whereas in this Chapter's contribution the impact on video quality and latency is assessed.

REFERENCES

Agboma, F., & Liotta, A. (2007). Addressing user expectations in mobile content delivery. *Mobile Information Systems*, *3*(3/4), 153–164.

Ahmad, S., Hamzaoui, R., & Al-Akaidi, M. (2007). Robust live unicast video streaming with rate less codes. In *International Packet Video Workshop*.

Ahmadi, S. (2010). *Mobile WiMAX: A systems approach to understanding IEEE 802.16m radio access technology*. London, UK: Academic Press.

Andrews, J. G., Gosh, A., & Muhamed, R. (2007). *Fundamentals of WiMAX: Understanding broadband wireless networking*. Upper Saddle River, NJ: Prentice Hall.

Baccichet, P., Rane, S., & Girod, B. (2006). Systematic lossy error protection based on H.264/AVC redundant slices and flexible macroblock ordering, In *IEEE International Packet Video Workshop*.

Balakrishnan, H., & Katz, R. (1998). Explicit loss notification and wireless Web performance. In *IEEE GLOBCOM Internet Mini-Conference*.

Balakrishnan, H., Padmanabhan, V., Sehan, S., & Katz, R. (1997). A comparison of mechanisms for improving TCP performance over wireless links. *IEEE/ACM Transactions on Networking*, *5*(6), 756–769. doi:10.1109/90.650137

Cen, S., Cosman, P., & Voelker, G. (2003). End-to-end differentiation of congestion and wireless losses. *IEEE/ACM Transactions on Networking*, *11*(5), 703–717. doi:10.1109/TNET.2003.818187

Chatterjee, M., Sengupta, S., & Ganguly, S. (2007). Feedback-based real-time streaming over WiMAX. *IEEE Wireless Communications*, *45*(2), 65–71.

Chen, M., & Zakhor, A. (2005). Rate control for streaming video over wireless. *IEEE Wireless Communications*, *12*(4), 32–14. doi:10.1109/MWC.2005.1497856

Chen, M., & Zakhor, A. (2006). Multiple TFRC connection based rate control for wireless networks. *IEEE Transactions on Multimedia*, *8*(5), 1045–1062. doi:10.1109/TMM.2006.879837

Cicconnetti, C., Lenzini, L., Mingozzi, E., & Eklund, C. (2006). Quality of service support in IEEE 802.16 networks. *IEEE Network*, *20*(2), 50–55. doi:10.1109/MNET.2006.1607896

Cohen, R., Katzir, L., & Rizzi, R. (2008). On the trade-off between energy and multicast efficiency in 802.16e networks. *IEEE Transactions on Mobile Computing*, *7*(3), 346–357. doi:10.1109/TMC.2007.70729

Dahlman, E., Parkvall, S., Sköld, J., & Beming, P. (2008). *3G evolution: HSPA and LTE for mobile broadband* (2nd ed.). Amsterdam, Netherlands: Academic Press.

Degrande, N., Laevens, K., & De Vleeschauwer, D. (2008). Increasing the user perceived quality for IPTV services. *IEEE Communications Magazine*, *46*(2), 94–100. doi:10.1109/MCOM.2008.4473090

Dhondt, Y., Mys, S., Vermeirsch, K., & Van de Walle, R. (2007). Constrained inter prediction: Removing dependencies between different data partitions, In *Advanced Concepts for Intelligent Visual Systems* (pp. 720-731).

Ekström, K., Furuskar, A., Karisson, J., Meyer, M., Parkvall, S., Tosner, J., & Wahlqvist, M. (2006). Technical solutions for the 3G Long-Term Evolution. *IEEE Communications Magazine*, *44*(3), 38-45. Digital Object Identifier: 10.1109/MCOM.2006.1607864

Fazel, K., & Kaiser, S. (2008). *Multi-carrier and spread spectrum systems: From OFDM and MC-CDMA to LTE and WiMAX* (2nd ed.). Chichester, UK: John Wiley & Sons.

Ferré, P., Agrafiotis, D., & Bull, D. (2010). A video error resilience redundant slices algorithm and its performance relative to other fixed redundancy schemes. *Image Communication*, *25*(3), 163–178.

Fleury, M., Razavi, R., Saleh Al-Majeed, S., Al-Jobouri, L., & Ghanbari, M. (2009). Enabling WiMAX video steaming. In *WiMAX* (pp. 213–238). Vukovar, Croatia: New Developments, In-Tech.

Floyd, S., & Fall, A. (1999). Promoting the use of end-to-end congestion control in the Internet. *IEEE/ACM Transactions on Networking*, *7*(4), 43–56. doi:10.1109/90.793002

Fu, Y., Hu, R., Tian, G., & Wang, T. (2006). TCP-friendly rate control for streaming service over 3G network. In *International Conference on Wireless Communications, Networking and Mobile Computing* (pp. 1–4).

Ghanbari, M. (2003). *Standard codecs: Image compression to advanced video coding*. London, UK: IET.

Gómez-Brown, D., Gozález, D., & Cardona, N. (2010). Application layer FEC for mobile TV delivery in IP datacast over DVB-H systems. *IEEE Transactions on Broadcasting*, *55*(2), 396–406. doi:10.1109/TBC.2008.2012024

Görkemli, B., Sunay, M. O., & Tekalp, A. M. (2008). Video streaming over wireless DCCP. In *IEEE International Conference on Image Processing* (pp. 2028–2031).

Handley, M., Pahdye, J., Floyd, & S., Widmer, J. (2003). TCP-friendly rate control (TFRC): Protocol specification. *Internet Engineering Task Force*, RFC 3448.

Hanzo, L., Akhtman, Y., Wang, L., & Jiang, M. (2010). *MIMO-OFDM for LTE, WIFI and WIMAX: Coherent versus non-coherent and cooperative turbo-transceivers. Blackwell*. Chichester, UK: Wiley. doi:10.1002/9780470711750

Haßlinger, G., & Hohlfeld, O. (2008). The Gilbert-Elliott model for packet loss in real time services on the Internet. In *14th GI/ITG Conference on Measurement, Modelling, and Evaluation of Computer and Communication Systems* (pp. 269-283).

Hillested, O., Perkis, A., Genc, V., Murphy, S., & Murphy, L. (2006). Delivery of on-demand video services in rural areas via IEEE 802.16 broadband wireless access networks. In *2nd International Workshop on Wireless Multimedia Networking and Performance Modeling* (pp. 43-52).

Hosein, P. (2008). Broadcasting VBR traffic in a WiMAX network. In *IEEE Vehicular Technology Conference* (pp. 1-5).

IEEE. 802.16e-2005 (2005). Air interface for fixed and mobile broadband wireless access systems. In *IEEE Standard for Local and Metropolitan Area Networks* (part 16).

Issa, O., Li, W., & Liu, H. (2010). Performance evaluation of TV over broadband wireless access networks. *IEEE Transactions on Broadcasting*, *56*(2), 201–210. doi:10.1109/TBC.2010.2046979

Jain, R., So-In, C., & Al-Tamimi, A.-K. (2008). System-level modeling of IEEE 802.16e mobile WiMAX networks. *IEEE Wireless Communications*, *15*(5), 73–79. doi:10.1109/MWC.2008.4653135

Jeong, W.-H., Yoon, Y.-S., & Ho, Y.-S. (2004). *Design of asymmetrical reversible variable-length codes and comparison of their robustness*. In European Signal Processing Conference.

Juan, H., Huang, H., Huang, C., & Chiang, T. (2007). Scalable video streaming over mobile WiMAX. In *International Symposium on Circuits and Systems* (pp. 3463-3466).

Kim, H., Lee, J., & Lee, B. G. (2008). WiBro – A 2.3 GHz MobileWiMAX: System design, network deployment, and services. In Chen, K.-C., & De Marca, J. R. B. (Eds.), *Mobile WiMAX* (pp. 257–290). Chichester, UK: Wiley & Sons. doi:10.1002/9780470723937.ch14

Klaue, J., Rathke, B., & Wolisz, A. (2003). EvalVid - A framework for video transmission and quality evaluation. In *International Conference on Modeling Techniques and Tools for Computer Performance* (pp. 255-272).

Kohler, E., Handley, M. & Floyd, S. (2006). Datagram congestion control protocol. *Internet Engineering Task Force*, RFC 4340.

Krasic, C., Walpole, J., Li, K., & Goel, A. (2001). The case for streaming multimedia with TCP. In *8th International Workshop on Interactive Distributed Multimedia Systems* (pp. 213-228).

Kumar, A. (2007). *Mobile TV: DVB-H, DMB, 3G systems and rich media applications*. Burlington, MA: Focal Press.

Lambert, P., de Neve, W., Dhondt, Y., & van de Walle, R. (2006). Flexible macroblock ordering in H.264/AVC. *Journal of Visual Communication, 17*, 358–375.

Lee, J. M., Park, H.-J., Choi, S. G., & Choi, J. K. (2009). Adaptive hybrid transmission mechanism for on-demand mobile IPTV over WiMAX. *IEEE Transactions on Broadcasting, 55*(2), 468–477. doi:10.1109/TBC.2009.2015605

Li, B., Qin, Y., Low, C. P., & Gwee, C. L. (2007). A survey on mobile WiMAX. *IEEE Communications Magazine, 45*(12), 70–75. doi:10.1109/MCOM.2007.4395368

Liang, Y. J., Apostolopoulos, J. G., & Girod, B. (2008). Analysis of packet loss for compressed video: Effect of burst losses and correlation between error frames. *IEEE Transactions on Circuits and Systems for Video Technology, 18*(7), 861–874. doi:10.1109/TCSVT.2008.923139

Liang, Y. J., El-Maleh, K., & Manjunath, S. (2006). *Upfront intra-refresh decision for low-complexity wireless video telephony*, In International Symposium on Circuits and Systems.

Luby, M., & Gasiba, T., Stockhammer, & Watson, M. (2007). Reliable multimedia download delivery in cellular broadcast networks. *IEEE Transactions on Broadcasting, 53*(1), 235–246. doi:10.1109/TBC.2007.891703

MacKay, D. J. C. (2005). Fountain codes. *IEE Proceedings. Communications, 152*(6), 1062–1068. doi:10.1049/ip-com:20050237

Meloni, L. G. P. (2008). A new WiMAX profile for DTV return channel and wireless access. In Chen, K.-C., & de Marca, J. R. B. (Eds.), *Mobile WiMAX* (pp. 291–392). Chichester, UK: Wiley & Sons. doi:10.1002/9780470723937.ch15

Mys, S., Lambert, P., & De Neve, W. (2006). SNR scalability in H.264/AVC using data partitioning. In *Pacific Rim Conference on Multimedia* (pp. 329-338).

Oyman, O., Foerster, J., Tcha, Y.-j., & Lee, S.-C. (2010). Towards enhanced mobile video services over WiMAX and LTE. *IEEE Communications Magazine, 48*(8), 68–76. doi:10.1109/MCOM.2010.5534589

Reguant, V., Prats, F., de Pozuelo, R., Margalef, F., & Ubiergo, G. (2008). *Delivery of H.264 SVC/MDC streams over WiMAX and DVB-T.* In IEEE International Conference on Consumer Electronics.

Sadka, A. H. (2006). *Compressed Video Communications*. Chichester, UK: Wiley & Sons.

Schreier, R. M., & Rothermel, A. (2006). Motion adaptive intra refresh for low-delay video coding. In *International Conference on Consumer Electronics* (pp. 453-454).

Sharangi, S., Krishnamurti, R., & Hefeeda, M. (2010). *Streaming scalable video over WiMAX networks*. In IEEE International Workshop on Quality of Service.

Shokorallahi, A. (2006). Raptor codes. *IEEE Transactions on Information Theory, 52*(6), 2551–2567. doi:10.1109/TIT.2006.874390

So-In, C., Jain, R., & Tamimi, A.-K. (2008). Capacity evaluation for IEEE 802.16e mobile WiMAX, *Journal of Computer Systems, Networks, and Communications, 2010* [online], 12 pages.

Stewart, R. (Ed.). (2007, September). Stream control transmission protocol. *Internet Engineering Task Force*, RFC 4960.

Stockhammer, T., & Bystrom, M. (2004). H.264/AVC data partitioning for mobile video communication. In *IEEE International Conference on Image Processing* (pp. 545-548).

Stockhammer, T., Liebl, G., & Walter, M. (2006). Optimized H.264/AVC-based bit stream switching for mobile video streaming [online]. *EURASIP Journal on Applied Signal Processing, 2006*. doi:10.1155/ASP/2006/91797

Tappayuthpijam, K., Liebl, G., Stockhammer, T., & Steinbach, T. (2009). Adaptive video streaming over a mobile network wth TCP-friendly rate control. In *International Conference on Wireless Communications and Mobile Computing* (pp. 1325–1329).

Tsai, F. C. D., Chen, J., Chang, C.-w., Lien, W.-j., Hung, C.-h., & Sum, J.-h. (2006). The design and implementation of WiMAX module for ns-2 simulator. In Workshop on Ns*2*: The IP Network Simulator (article no. 5).

Turletti, T., & Huitema, C. (1996). Videoconferencing on the Internet. *IEEE/ACM Transactions on Networking*, *4*(3), 340–351. doi:10.1109/90.502233

Uilecan, I. V., Zhou, C., & Atkin, G. E. (2007). Framework for delivering IPTV services over WiMAX wireless networks. In *IEEE International Conference on Electronic/Information Technology* (pp. 470-475).

Wang, J., Venkatachalam, M., & Fang, Y. (2007). System architecture and cross-layer optimization for video broadcast over WiMAX. *IEEE Journal on Selected Areas in Communications*, *24*(4), 712–722. doi:10.1109/JSAC.2007.070508

Wenger, S. (2003). H.264/AVC over IP. *IEEE Transactions on Circuits and Systems for Video Technology*, *13*(7), 645–655. doi:10.1109/TCSVT.2003.814966

Wiegand, T., Sullivan, G. J., Bjontegaard, G., & Luthra, A. (2003). Overview of the H.264/AVC video coding standard. *IEEE Transactions on Circuits and Systems for Video Technology*, *13*(7), 560–576. doi:10.1109/TCSVT.2003.815165

Wiegand, T., Zhang, X., & Girod, B. (1999). Long-term memory motion-compensated prediction. *IEEE Transactions on Circuits and Systems for Video Technology*, *9*(1), 70–84. doi:10.1109/76.744276

Yang, F., Zhang, Q., Zhu, W., & Zhang, Y.-Q. (2007). End-to-end TCP-friendly streaming protocol and bit allocation for scalable video over wireless Internet. *IEEE Journal on Selected Areas in Communications*, *22*(4), 777–790. doi:10.1109/JSAC.2004.826008

Yota, T. V. (2009). Mobile WiMAX. Retrieved on August 21st, 2010, on http://www.yota.ru/ en /info/ main/

ADDITIONAL READING

Ahson, A. S., & Ilyas, M. (Eds.). (2007). *WiMAX technology: Technologies, performance analysis, and QoS*. Boca Raton, FL: CRC Press.

Anderson, H. R. (2003). *Fixed broadband wireless: System design*. Chichester, UK: Wiley & Sons. doi:10.1002/0470861290

Bull, D., Canagarajah, N., & Nix, A. (Eds.). (1999). *Insights into mobile multimedia communications*. San Diego, CA: Academic Press.

Correia, L. M. (Ed.). (2006). *Mobile broadband multimedia networks*. Oxford, UK: Academic Press.

Etemard, K., & Lai, M.-Y. (2010). *WiMAX technology and network evolution*. Chichester, UK: Wiley & Sons. doi:10.1002/9780470633021

Furht, B., & Ilyas, M. (2003). *Wireless Internet handbook: Technologies, standards, and applications*. Boca Raton, FL: CRC Press.

Kumar, A. (2008). *Mobile broadcasting with WiMAX: Principles, technology, and applications*. Burlington, MA: Focal Press.

Lee, B. G., & Choi, S. (2010). *Broadband wireless access and local networks: Mobile WiMax and WiFi*. Norwood, MA: Artech House.

Nuaymi, L. (2007). *WiMAX: Technology for broadband wireless access*. New York, NY: Wiley & Sons. doi:10.1002/9780470319055

Patil, B., Iyer, J., Iyer, P., Premec, D., & Pazyannur, R. (2010). *WiMAX: End-to-end network architecture*. Chichester, UK: Wiley-Blackwell.

Schaar, Van der M., & Chou, P. (Eds.) (2007). *Multimedia over IP and wireless networks*. Amsterdam, Netherlands: Academic Press.

Chapter 3
High-Speed Cable Replacement for Multimedia Streaming in Wireless Personal and Local Area Networks

Chun-Ting Chou
National Taiwan University, Taiwan

ABSTRACT

The multimedia content is migrating promptly from standard quality to high-definition and even 3D. As a result, existing wireless technologies can no longer support multimedia streaming as their wired counterparts. To overcome this problem, new wireless technologies that support multi Gbps wireless transmission are desperately needed. In this chapter, we focus on the promising 60 GHz technology and investigate two important standards including ECMA-387 and IEEE 802.11ad standards. Key designs of the two standards are discussed and qualitatively evaluated. Based on our evaluation, one can select the solution that suits best for the targeted applications.

INTRODUCTION

The advance in wireless technologies has made multimedia streaming over wireless links a reality. Users at home now can wirelessly stream audio or video from their laptops or *Personal Computers* (*PCs*) to *Television* (*TV*) displays located in a different room via an *Institute of Electrical and Electronic Engineers* (*IEEE*) 802.11 wireless local area network (WLAN). Users on the move can also stream multimedia content from the Internet and replay it on real time via either *Third Generation* (*3G*) or the latest *Worldwide Interoperability for Microwave Access* (*WiMAX*)/*Long Term Evolution* (*LTE*) network. These wireless technologies,

DOI: 10.4018/978-1-61350-144-3.ch003

supporting data rates from several to tens of Mbps (or even a few hundred Mbps in the IEEE 802.11n network), provide users (to a certain degree) similar viewing/listening experiences as in the wired networks.

However, when the multimedia content migrates promptly from the standard quality to *High Definition* (*HD*) and even *Three Dimension* (*3D*), the existing wireless technologies are no longer be able to keep up with their wired counterparts. For example, *High Definition Multimedia Interface* (*HDMI*) 1.0 supports a nearly 5 Gbps transmission rate to carry uncompressed HD video (e.g. 1920x1080 with 60 frames per second) between a TV display and an HD/*Digital Video Disc* (*DVD*) player. Other high-speed cables such as *Digital Visual Interface* (*DVI*), *Universal Serial Bus* (*USB*) 3.0, *external Serial Advanced Technology Attachment* (*e-SATA*) and Gigabit Ethernet all support the same magnitude of transmission rates that are far beyond the reach of any existing wireless technology.

In our days, mobile telephones use IEEE 802.11 (*Wireless fidelity* (*WiFi*)) and *Bluetooth* wireless technologies for communication in a *Wireless Local Area Network* (*WLAN*) or *Wireless Personal Area Network* (*WPAN*). In the future this standards could be improved with new ones. For that reason it is interesting to review the state of actual technologies of communication that could be included in this kind of terminals. Moreover, in order to deliver HD multimedia content over wireless networks, new wireless technologies that match the transmission rate of the above high-speed cables are much needed. In this chapter, we investigate new developments in the 60 GHz band communication for this need. Major problems and challenges in using the 60 GHz technology for high-speed cable replacement are identified and studied. We then focus on two promising solutions, namely *European Computer Manufacturers Association* (*ECMA*) 387 and IEEE 802.11ad standards. Key designs of both standards are examined in different aspects of being a key technology enabler for HD multimedia distribution.

The rest of this chapter is organized as follows. In Section 2, we review the requirements of the existing high-speed cables that support distribution of various HD multimedia contents. In Section 3, major problems and challenges faced by the 60 GHz communication are discussed. Key designs of ECMA-387 and IEEE 802.11ad standards are investigated in Section 4 and Section 5, respectively, and compared in Section 6. In Section 7 we introduce the possible evolution in the next recent years of the standards reviewed in this chapter and, finally, conclusions are drawn in Section 8.

REPLACEMENT OF HIGH-SPEED CABLES USING WIRELESS LINKS

Among the cables to be replaced by wireless links, we focus on those that support HD streaming with transmission rates far beyond the reach of any existing wireless technology (e.g., IEEE 802.11n). These include:

- HDMI cables that connect TV displays to set-top boxes or HD/DVD players
- DVI cables that connect PC monitors to desktops or laptops
- The USB 2.0 (or 3.0) or eSATA cables that connect external disk drives, personal phones, handheld multimedia players, camcorders and digital cameras to desktops or laptops, and
- The Gigabit Ethernet cables that connect desktops in home or small office environments. These cables along with their targeted applications can be further divided into three categories including:
- Wireless display (for 1 and 2)
- Wireless sync-and-go (for 3), and
- WPAN or WLAN (for 3 and 4).

The requirements and typical usage scenarios of these three categories are discussed as follows.

The transmission rate in wireless display can easily exceed 1 Gbps. Take HD video over an HDMI cable as an example. The raw data rate of uncompressed HD video (with 1920x1080 pixels per frame, 24 bits per pixel, and 60 frames per second) is nearly 3 Gbps. Table 1 lists the data rates of several popular video formats. As the table shows, even the lightly compressed video cannot be transmitted via any mainstream wireless technologies.

The typical setting for wireless display would be a display/monitor placed within a few meters from a video source. The display and the source may be *Line Of Sight* (*LOS*) or non-LOS depending on user preference but the topology rarely changes once they are set up by the users. In any case, a transmission range of less than 10 m is usually sufficient for wireless display.

The transmission rates of mainstream high-speed cables (Sync-and-Go) are listed in Table 2. Given that external disk drives, handheld devices and mobile telephones now can store hundreds of GB worth of data, a wireless link that supports a multi Gbps transmission rate would be necessary to match the user experience of its wired counterparts.

Table 1. Transmission rates of compressed and uncompressed HD video (this table is reproduced and modified based on WiFi Alliance VHT usage model 2008)

Video Compression		Description	Rate (Mbps)
Uncompressed	720p	1280x720 pixels, 24 bits/pixel, 60 frames/s	1119
	1080i	1920x1080 pixels, 24 bits/pixel, 30 frames/s	1493
	1080p	1920x1080 pixels, 24 bits/pixel, 60 frames/s	2986
Lightly Compressed		Motion-JPEG2000	~150
Compressed		HD MPEG2	~20

The typical setting for wireless sync-and-go would be an external disk drive or handheld devices placed within less than 3 m from a data sink. With the handheld devices' portability, the topology formed by these devices may vary with time. Therefore, limited mobility support would be desirable for wireless sync-and-go. Nevertheless, the devices should be in LOS for most of the usage cases.

Wireless local area networking has been realized by the IEEE 802.11b/g technology. Take the IEEE 802.11g as an example. A maximum rate of 54 Mbps is provided in home or office environments. Although the IEEE 802.11g is very successful, the user experience of the existing WLAN still cannot match that in the high-speed Ethernet networks, especially in terms of quality of service provisioning. The latest IEEE 802.11n is developed to push the transmission rate to hundreds of Mbps with the help of *Multiple-Input-Multiple-Output* (*MIMO*) technologies and channel binding. However, squeezing more "juices" out of the crowded and relatively narrow *International Scientific and Medical* (*ISM*) bands is extremely difficult, if not impossible.

The typical WPAN/WLAN setting would be that desktops, mobile telephones, central data storages, and all sorts of networking devices in the same or different rooms, or even at different

Table 2. Transmission rates of mainstream high-speed cables

High-speed cable		Rate (Mbps)	Length (m)
Universal Serial Bus (USB)	2.0	480	3
	3.0	4800	3
FireWire 800		786	3
External SATA		3000	2

levels form a well-connected network. Therefore, a transmission range of tens of meters to a hundred meters would be needed. Unlike wireless display or wireless sync-and-go, the data flow is in general bi-directional. Moreover, the topology would be point-to-multipoint in most of the usage cases in contrast to point-to-point wireless display or wireless sync-and-go. (Note that a data sink such as a wireless USB host in wireless sync-and-go could also be point-to-multipoint.)

COMMUNICATION IN THE 60 GHZ BAND

The 60 GHz band has been considered as the promising land for the next-generation high-speed wireless communication. With its abundant bandwidth, the transmission rates required by wireless display, wireless sync-and-go and multi-Gbps WPAN/WLAN could be provided. In general, the 60 GHz band is referred to as the unlicensed millimeter-wave bands between 57 GHz and 66 GHz. Although the actual bands allocated by different regulatory bodies are different, a harmonized channelization has been defined for worldwide interoperability and coexistence as shown in Figure 1. Four channels have been specified, each with a bandwidth of 2160 MHz. With such a high bandwidth, a transmission rate of multiple Gbps over a single channel has been demonstrated by various technologies.

One major drawback for data communication in the 60 GHz band is severe signal attenuation due to strong oxygen absorption. The attenuation substantially reduces the transmission range even in the *LOS* transmission, and thus, limits the usability of the 60 GHz band. To overcome this problem, directional antennas are used at both transmitter and receiver sides so as to obtain sufficient link budget. Although the introduction of directional antennas extends the transmission range, many problems have been identified in the literature (Choudhury, 2002; Ko, 2000) and different algorithms were proposed (Ramanathan, 2001; Takai, 2002; Korakis, 2003; Vasudevan, 2005). In this chapter, we focus on the practical issues in the protocol and system design of a 60 GHz standard. These issues include:

- *Neighbor discovery*: In conventional wireless networks, transmission over a wireless link is omni-directional. That is, any device can "hear" the broadcast as long as it is in the transmission range of thc transmitter. The broadcasting nature is very useful for a wireless device to discover its neighbors and the intended recipient(s). Take IEEE 802.11 WLANs as an example. An IEEE 802.11 *Access Point* (*AP*) only needs to transmit periodic beacons and waits for stations in vicinity to receive its beacons and associate with it. However, in the 60 GHz band, transmissions including beacons are

Figure 1. Worldwide channelization in the 60 GHz band

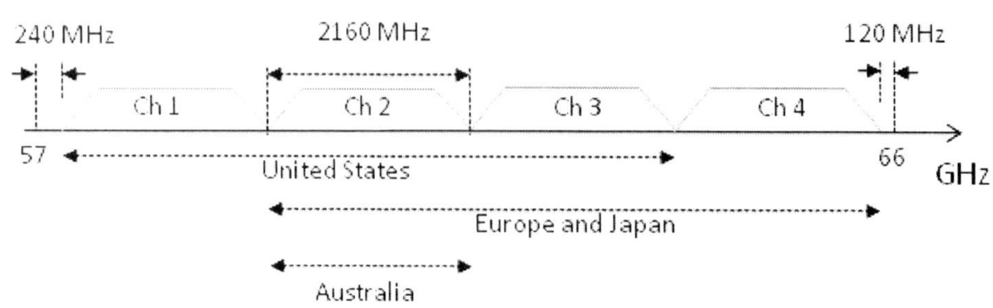

likely to be directional. Therefore, devices not knowing each other's antenna configuration may not be able to receive beacons to discover each other, let alone initiating antenna training for better link budget. In general, beacon sweeping (for omni-directional coverage) must be applied in the *Medium Access Control* (*MAC*) layer with the help of array or steerable antennas in the *Physical* (*PHY*) layer.

- *Collision avoidance*: In order for devices in vicinity not to transmit concurrently (thus, resulting in collisions at the common neighbor of the transmitting devices), *Carrier Sense* (*CS*) is usually applied, especially in distributed wireless ad hoc networks. The use of directional antennas may cause so-called directional deafness and defeat the purpose of CS as illustrated in Figure 2.a. In conventional CS operation, *Transmitter* (*TX*) 2 and *Receiver* (*RX*) 1 are in the coverage of TX 1. When TX 1 transmits data to RX 1, TX 2 detects the transmission via CS so that a collision is avoided. When using the directional antenna, TX 2 cannot detect the ongoing transmission from TX 1 to RX 1 via CS and may start a transmission, which results in a collision at RX 1. Although this problem can be solved by using omni-directional CS, doing so creates another problem due to asymmetric gain (K0, 2000). Using handshaking mechanisms such as exchange of *Ready To Send* (*RTS*) and *Clear To Send* (*CTS*) does not solve the problem as already being identified in (K0, 2000).
- *Spatial reuse*: Although directional deafness results in additional collisions and interference, it can improve spectrum efficiency if it will be used properly. As illustrated in Figure 2.b, TX 2 would not be able to communicate with RX 2 when TX 1 is communicating with RX 1, if all of them use omni-directional antennas. However, with directional deafness, these two pairs of devices can transmit concurrently without causing interference to each other. This is referred to as spatial reuse and should be exploited properly in the 60 GHz band to maximize the spectrum efficiency.
- *Link management*: Directional antennas transmit signals to a confined direction so as to achieve needed link budget. As a result, the signals traverse to the receiver via fewer paths (and less path diversity) and therefore, are more susceptible to being disrupted or blocked due to user mobility. Special designs such as link switching or cooperation with a third party relay may be needed in order to maintain certain *Quality of Service* (*QoS*) over the 60 GHz wireless link.

In views of the above challenges and opportunities, various new standards have been developed. For example, the WirelessHD Consortium has been developing WirelessHD specification for wireless display. The IEEE 802.15.3c TG developed a new PHY layer to support at least 1 Gbps wireless transmission. ECMA international has also developed ECMA-387 standard, targeting on HD streaming and multi Gbps WPANs. Finally, the IEEE 802.11ad TG is developing a new standard that supports Gbps in WLAN. In what follows, we focus on ECMA-387 and IEEE 802.11ad standards given that both standards cover a wider range of applications. We will explain the key designs in both standards and qualitatively analyze their pros and cons.

ECMA-387 STANDARD

In order to accommodate a wide range of applications while taking the cost/complexity into account, ECMA-387 standard defines two types of devices:

Figure 2. Directional deafness and spatial reuse in networks using directional antennas. a) Directional antennas, b) Spatial reuse

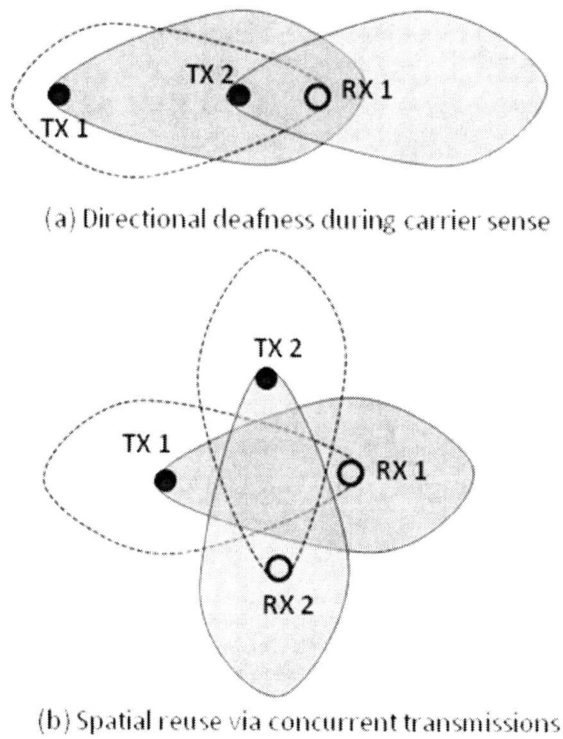

- The high-end device, referred to as Type-A device, supports both LOS and non-LOS, point-to-multipoint communication within 10m with trainable antennas.
- The economic device, referred to as Type-B device, supports only LOS, point-to-point communication within 3m with non-trainable antennas. Given its limited capabilities, a Type-B device supports only a subset of functionalities in both the MAC and PHY layers. Nevertheless, Type-B devices can operate independently without the presence of Type-A devices as we will explain later.
- Two types of channels are defined in ECMA-387 standard, including one discovery channel and a set of data channels:
- The discovery channel is mainly used by devices to discover each other, and by Type-A devices to perform antenna training. If no activity of neighbor discovery is detected in the discovery channel, devices will use the channel as a regular data channel, subject to giving it up if other devices initiate the neighbor discovery.
- The data channels are used by the devices that have found each other to exchange data or control frames. Given their different roles in the standard, the timing structures and channel access mechanisms in the discovery channel and data channels are very different.

Let us now review the principal elements of this standard:

Discovery Channel. In the discovery channel, time is divided into *Discovery Intervals* (*DI*). Within each DI, a device initiates one and only one

Figure 3. Timing structure of a Type-A DBS in ECMA-387 standard

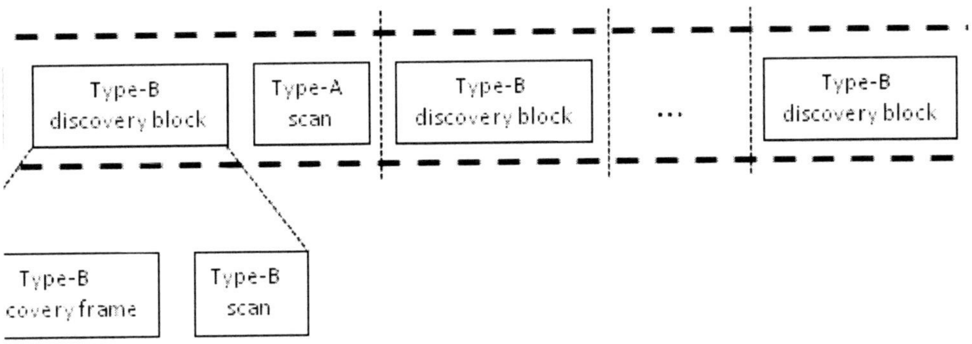

Discovery Block Set (*DBS*). Each DBS comprises a series of frame transmissions and channel scans. For a Type-A device, a DBS comprises a Type-A discovery frame, followed by at least one Type-B discovery block, and a Type-A scan. The scan is used by the Type-A device to receive responses from neighboring Type-A devices. Each Type-B discovery block comprises a Type-B discovery frame and a Type-B scan used by the Type-A device to receive responses from neighboring Type-B devices. The timing structure of Type-A DBS is illustrated in Figure 3.

In a Type-A DBS, the total number of Type-B discovery blocks is equal to the number of antenna sectors supported by the Type-A device. The first Type-B discovery block uses the same sector as the Type-A discovery frame and the Type-A scan, while the rest of the Type-B discovery blocks use up all other sectors, one after the other in a predetermined order.

Within a DI, a Type-A device waits for a random time before starting its DBS. To avoid colliding with the DBSs of other Type-A devices, the Type-A device performs CS using the first antenna sector chosen for the DBS. A Type-A device changes the first sector used in each DBS in a way such that all antenna sectors are used within a number of consecutive DIs equal to the number of antenna sectors supported by that device. This process is illustrated in Figure 4. In this figure, the Type-A device has three sectors: a, b and c. Within 3 consecutive DIs, the Type-A device has three DBSs, each using a different sector.

When a Type-A device is in the discovery channel but not within its DBS, the device scans the channel for Type-A discovery frames sent by other Type-A devices. The device scans the channel using the same sector for at least a number of consecutive DIs equal to the number of antenna sectors supported by the device, before changing to another sector. The scanning and corresponding sector change are also shown in Figure 4. Whenever a Type-A discovery frame is received, the Type-A device responds at the time specified in the received discovery frame. At that time, neighbor discovery is considered completed. The two

Figure 4. Neighbor Discovery using Type-A DBS in ECMA-387 standard

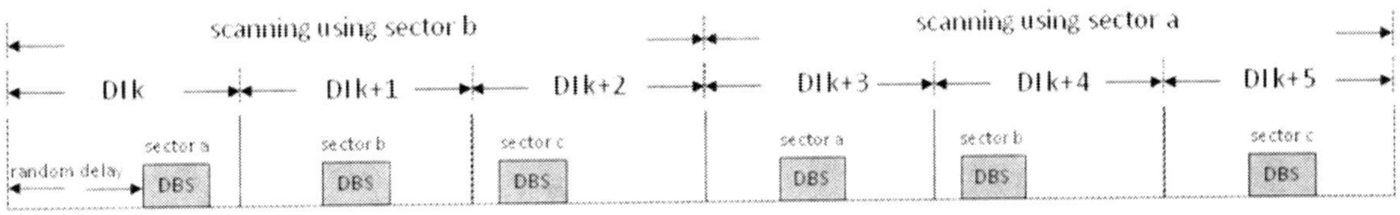

Figure 5. Superframe structure in a data channel of an ECMA-387 network

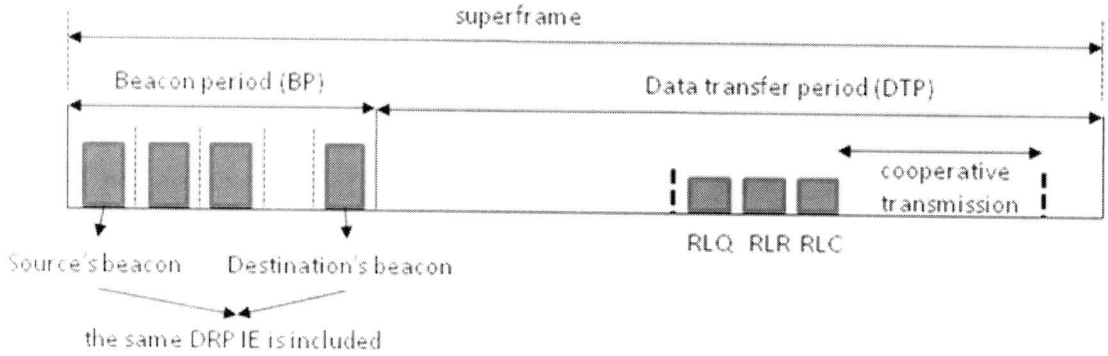

Type-A devices may stay in the discovery channel and perform antenna training. Otherwise, the two devices may select a data channel and initiate data transmission.

Type-B devices follows the same neighbor discovery procedure except that there is no Type-A frame transmission or Type-A scan in a Type-B DBS, as the Type-B device is unable to process the responses from any Type-A device. However, a Type-B device can still discover a Type-A device by responding to a Type-B discovery frame sent by that Type-A device.

Data Channel. Once the communicating devices discover each other in the discovery channel, they may switch to a data channel for data transmission. Note that the discovery channel can be regarded as a data channel, except that the use of discovery channel for data transmission may be preempted by devices performing neighbor discovery. Other than such a constraint, devices in both data and the discovery channels follow the same timing structure and channel access for data transmission.

The basic timing structure for data communication is superframe. A superframe is composed of a *Beacon Period* (*BP*) followed by a *Data Transfer Period* (*DTP*). In the BP, time is divided into beacon slots and devices choose their own slots for beacon transmission. In general, a device transmits a beacon in a fixed slot in every superframe. If a collision is detected, devices may choose a different slot. The superframe structure is shown in Figure 5.

To transmit a beacon, a device first scans the selected data channel for at least one superframe using the antenna sector determined during the neighbor discovery process. If no beacon is received, the device starts a BP by sending a beacon in the first beacon slot. If a beacon is received, the device joins the existing BP by transmitting a beacon in an available slot. The slot availability is indicated by the received beacons. Since all devices transmit beacons in the BP, no one "owns" the BP, not even the one that starts the BP. The timing reference is also not controlled by any specific device but rather jointly determined via mutual synchronization. Therefore, a network formed for data transmission is fully distributed.

Since a Type-B device cannot decode Type-A beacons (or any other Type-A frames), the Type-B is unable to locate available slots based on the received Type-A beacons and thus, may use the same slots as the existing Type-A devices. To avoid such collisions and improve coexistence between Type-A and Type-B devices, ECMA-387 standard requires a Type-B device to transmit a Type-A beacon frame, along with a regular Type-B beacon in two consecutive beacon slots. These two beacons carry identical information and are referred to as dual beacons. Given that the transceiver complexity lies in the receiver side (i.e. decoder) in the ECMA-387 PHY, transmitting a Type-A beacon does not incur significant complexity to a Type-B device. By transmitting the dual beacons, a Type-B device can inform neighboring Type-A devices of the beacon slot it is using. As a result, the Type-A devices can avoid using the same beacon slot.

A more intuitive way for coexistence between Type-A and Type-B devices would be that Type-A devices transmit the dual beacons to inform Type-B devices, since Type-A devices support full Type-B PHY. However, the transmission range achieved by Type-B PHY is much less than that by Type-A PHY (3 m vs. 10 m). By using the Type-B beacons as the "common" mode for coexistence, a Type-B device may not be able to receive the Type-B beacon in the dual beacon sent by a remote Type-A device but still suffer from interference caused by that Type-A device.

Relinquish Data Transmission for Neighbor Discovery/Antenna Training. When using the discovery channel for data transmission, devices are required to give up the channel access to the devices that perform neighbor discovery or antenna training. To request for access relinquishment, a device that needs to perform neighbor discovery joins the existing BP in the discovery channel by sending a beacon with status set to PREEMPTIVE. Once receiving a beacon with status set to PREEMPTIVE, every device in the same BP ceases its transmissions and may only restart until no neighbor-discovery or antenna-training activity is detected. At that time, the devices restart the operation as if they are powered up for the first time.

Based on the designs of the discovery channel and data channels, important operations in the 60 GHz band including antenna training, spatial reuse, interference mitigation, and link management can be implement in a fully distributed as explained in the following subsections.

Antenna Training and Tracking. After performing neighbor discovery in the discovery channel, a device pair may perform antenna training to acquire better antenna configuration for data transmission. Antenna training is initiated by exchange of RTS and CTS frames between the device pair in the discovery channel. Both the RTS and CTS senders should perform carrier sensing using the antenna sector determined in the neighbor discovery, before accessing the channel for RTS/CTS transmission. Once the RTS/CTS transmission is completed, the device pair performs antenna training via exchanging a series of tracking frames in order to obtain the optimal antenna configuration.

Once starting or joining a BP in a data channel, the device pair may perform antenna tracking to maintain the link quality. Antenna tracking is performed in the time period reserved by the devices using *Distributed Reservation Protocol* (*DRP*). During the reserved time period, the devices exchange a series of tracking frames as antenna training in the discovery channel.

Spatial Reuse and Interference Mitigation. When transmitting data in a data channel, a device must reserve certain time intervals in the DTP of a superframe using DRP. A DRP reservation is jointly made by the device and the intended recipient(s) based on their local channel utilization. The former is referred to as the owner of the reservation and the latter is referred to the target of the reservation. To make a DRP reservation, the owner first includes a DRP *Information Element* (*IE*) in its beacon. The DRP IE contains the information of time periods to be reserved by the owner along with the beam identifier that specifies the antenna to be used by the owner. Once receiving the DRP IE in the beacon(s) from the owner, the target should check whether or not the time periods proposed by the owner are available. If yes, the target includes the same DRP IE in the following beacons and the reservation is considered established.

To ensure interference-free communication, both the owner and target evaluate all DRP IEs received from the neighbor devices in the same

BP. If those devices are transmitting in the same time period using some antenna beams that may interfere with the owner or the target, that time period is considered as unavailable. Otherwise, the neighboring device pairs can use the same time period for concurrent transmission without causing interference to each other. With the help of DRP, spatial reuse is achieved in a fully distributed manner.

Link Management. When the direct path of a device pair is blocked and disrupted, an alternative path via a relay device can be used to promptly resume data transmission. In ECMA-387 standard, a relay device must be a Type-A device capable of performing amplify-and-forward functions. A relay device is associated with the source device (e.g., a DRP owner) and the destination device (e.g., a DRP target) by exchange of Relay IE via beacons. Once the association is completed, antenna training between the relay and the source/destination is performed before the relay operation is started.

IEEE 802.11AD STANDARD DRAFT

Being a standard in the IEEE 802.11 family, the 802.11ad supports various *Basic Service Sets* (*BSS*) for different applications. Given the unique natures in the 60 GHz bands, a *New Basic Service Set* (*BSS*), called *Personal BSS* (*PBSS*), is defined in the IEEE 802.11ad draft. PBSS is basically the same as *independent BSS* (*IBSS*) in the 802.11 standard families in the sense that they are both building blocks to form self-contained, ad hoc networks where stations can communicate directly with each other. However, in a PBSS, only one station can transmit beacons. This station, referred to as the *PBSS Central Point* (*PCP*), provides basic timing via transmission of beacons, and schedules most of the transmission between the member stations of the PBSS.

For each PBSS, time is divided into *Beacon Intervals* (*BI*). The BIs are maintained by the PCP via periodic transmission of beacons. Each BI can be further divided into access periods. Four access periods are defined in the IEEE 802.11ad draft, including *Beacon Time* (*BT*), *Association Beamforming Training* (*A-BFT*), *Announcement Time* (*AT*), and *Data Transfer Time* (*DTT*). Figure 6 illustrates an example of the BI structure with all four access periods. Within the DT, different numbers of *Service Periods* (*SP*) and *Contention-Based Periods* (*CBP*) can be scheduled by the PCP.

All access periods are scheduled by the PCP via periodic beacons or announcement frames. Transmission in each access period is generally controlled by the PCP and is subject to different access rules. In the BT, only beacon frames are transmitted and any member station other than the PCP cannot access the BT. In the A-BFT, beamforming is performed by the PCP. The training in the A-BFT allows the PCP to configure its antenna(s) and obtain better link budget when

Figure 6. BI and access periods in an IEEE 802.11ad PBSS

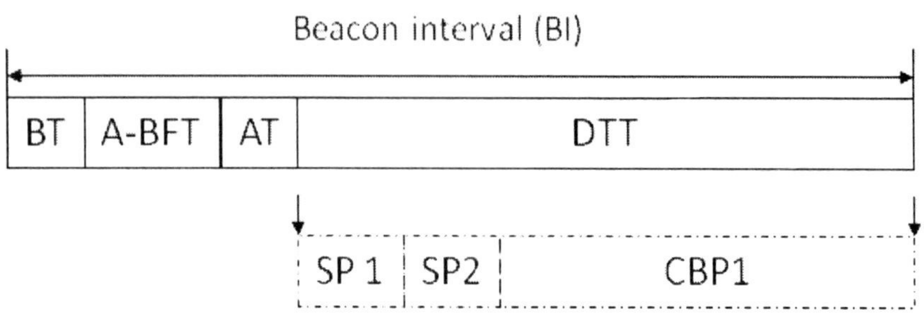

communicating with its member stations. In the AT, frames are exchanged between the PCP and its member stations on a request-and-response basis so that the member stations have opportunities to deliver non-data frames to the PCP. Finally, in the DTT, a member station exchanges data frames with the PCP or other member stations in a dedicated SP or in a CBP via contention. Note that only in the CBP, the stations use *Distributed Coordination Function* (*DCF*) or *Hybrid Coordination Function* (*HCF*) for channel access. In other words, the operation of IEEE 802.11ad PBSS in the 60 GHz band is essentially centralized and is very different from the existing IEEE 802.11 standard.

In order to deal with the issues resulting from the use of directional antennas in the 60 GHz band, the IEEE 802.11ad draft takes special consideration into account for its designs in beacon transmission, antenna training, interference mitigation, spatial sharing, and link management:

Beacon Transmission. Given that beacon frames provide crucial information for the operation of IEEE 802.11ad PBSS, timely reception of beacon frames is very important to all member stations in the PBSS. To ensure that, transmission of beacons frames are virtually "omni-directional" based on the draft. That is, at least one beacon is transmitted through each antenna configuration within a predefined time period by the PCP of the PBSS. Figure 7 illustrate an example where the PCP transmits four beacons, each to a different direction via a different antenna sector. By doing so, stations without prior information about the PCP's antenna configuration have opportunities to receive the PCP's beacons.

Antenna Training. In order to provide communicating stations with sufficient link budget, beamforming is supported by the IEEE 802.11ad draft via a two-stage process as shown in Figure 8. The first stage is referred to as sector level sweep (SLS), followed by an optional second stage referred to as *Beam Refinement Procedure* (*BRP*). The purpose of SLS is to enable two participating stations to communicate at the (lowest) control rate, while BRP enables participating stations to fine tune their antenna configurations to achieve a higher link budget.

The beamforming process is initiated by a station, referred to as the initiator, and followed by a second station, referred to as the responder. The BF process may take place in the BT, A-BFT or any SP. In the BT or A-BFT, the initiator is the PCP while in a SP, the initiator is the owner of the SP. The SLS is conducted via exchange of a series of beacon frames or sector sweep (SS) frames. Take the initiator's transmit sector sweep (TXSS) as an example. The initiator transmits some SS frames equal to the total number of sectors available to the initiator. If the initiator has more than one antenna, the transmission of SS frames must use up all sectors in all antennas in a single sweep.

During the sweep, the responder should have its receive antenna configured to a quasi-omnidirectional pattern. If the responder has more than

Figure 7. Beacon transmission in an IEEE 802.11 PBSS

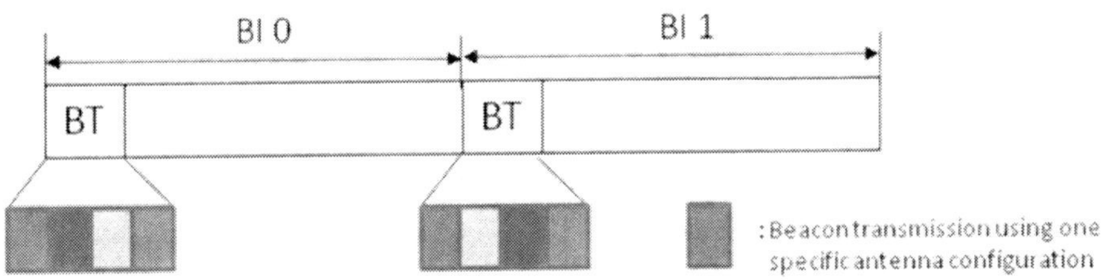

Figure 8. Two-stage beamforming using IEEE 802.11ad standard

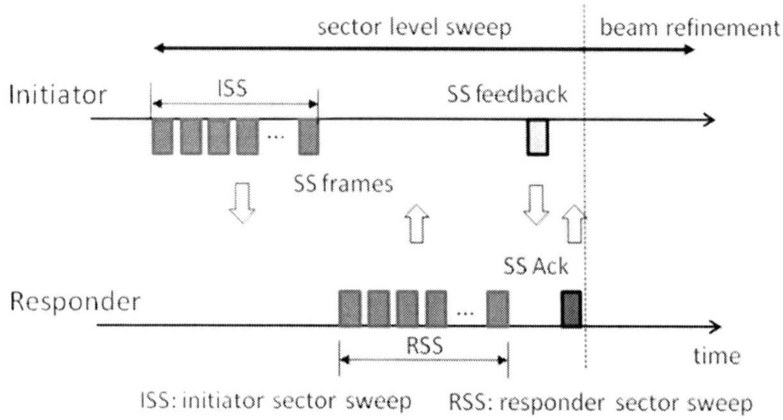

one receive antenna, the initiator should repeat the sector sweep equal to the number of the responder's receive antennas. By doing so, one can guarantee that the best combination of the initiator's and responder's sectors can be found. This is illustrated by a simple example in Figure 9. In this example, the initiator has 3 antennas, each with 2, 3, and 3 sectors respectively. The responder also has 3 antennas. Based on the setting, the initiator performs three sweeps, each with 8 SS frames transmitted using different sectors. At the end of the sweeps, the responder can find out the best receive antenna (i.e., RA1) and the best TX antenna of the initiator (i.e., TA1.2). The SLS is completed by the exchange of an SS feedback sent by the responder and an SS acknowledgement (ACK) sent by the initiator. The SS feedback and SS ACK inform the initiator and the responder, respectively, of the best antenna configuration learned from the sector sweep process.

Interference Mitigation. Given that SPs in a PBSS are scheduled by the PCP of the PBSS,

Figure 9. Transmit sweeps of the initiator in an IEEE 802.11ad PBSS

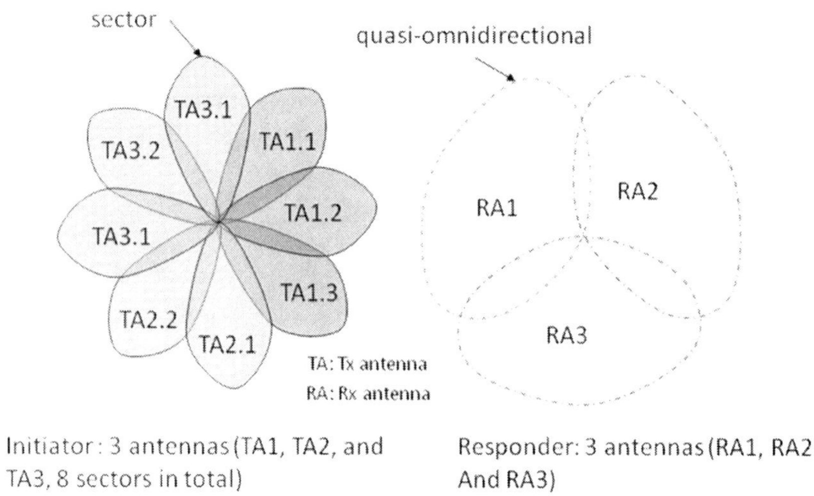

communicating station pairs in collocated PBSSs may be allocated with the SPs that overlap with each other. As a result, those stations may interfere with each other. To avoid such interference, a station may create so called protected periods in its SP. To create the protected periods, the source station sends an RTS and the recipient station responds with a CTS. The exchange of RTS and CTS allows the station pair to inform collocated stations that belong to a different PBSS of its scheduled SP. By do so, the collocated stations may restrain their transmission and may report to their PCPs about the potential interference during the overlapping SPs.

Another way to avoid interference is to prevent scheduling overlapping SPs by collocated PCPs beforehand. This is achieved by forming a PCP cluster. In a PCP cluster, one of the cluster members assumes the role of synchronization PCP (S-PCP), which provides synchronization among all member PCPs. Every member other than the S-PCP then selects a BI length that is an integer multiple of the BI length of the S-PCP. By doing so, the BTs of the member PCP do not overlap with every BT of the S-PCP in the same cluster. In addition, a member PCP must allocate periodic SPs for other cluster members to transmit their beacons. With beacons being fully exchanged in the cluster, the member PCPs can obtain the information of scheduled SPs or CBPs from each other and thus, may remove or reschedule conflicting SPs.

Spatial Sharing. To exploit spatial reusability in the 60 GHz channels, the IEEE 802.11ad supports so-called *Spatial Frequency Sharing* (*SFS*) within a PBSS. SFS allows stations in the same PBSS to use overlapping SPs in the same channel without causing harmful interference to each other. The SFS is initiated by the PCP via sending the Directional Channel Quality Requests to the stations involved in the SFS. Once receiving the requests, the stations measure the channel during the SPs of interest and report back to the PCP using the Directional Channel Quality Reports. Based on the measurement report, the PCP may schedule overlapping SPs to two communicating station pairs.

Link Management. Relay operation is also supported in the IEEE 802.11ad draft so that a source station can transmit frames to a destination station with the help of a relay station. Two types of relay operations, link switch and link cooperation, are specified. In the case that the link between the source and destinations is disrupted, the source station can send frames to the relay, which then forwards the received frames to the destination station. This is referred to as link switch, and is basically the conventional two-hop transmission. In the case of link cooperation, a frame transmitted by the source station is simultaneously transmitted by the relay station via *Amplify-And-Forward* (*AF*) or *Decode-And-Forward* (*DF*). By doing so, the signal quality at the destination station may be improved due to the additional spatial diversity gain.

Given the aforementioned designs in the IEEE 802.11ad draft, one can observe that the PCP of an IEEE 802.11ad PBSS plays a crucial role in nearly all operations of the PBSS (the only exception is data transmission in CBPs where DCF/HCF access is used). As a result, the resulting 802.11ad solution is intrinsically centralized and infrastructure-based, even though the IEEE 802.11ad standard aims to enable ad hoc communication in the 60 GHz band. As many other centralized solutions, the IEEE 802.11ad PBSS will suffer from the single-node failure problem and cause interference to collocated PBSSs (not in the same cluster). In ad hoc network environments, such problems will lead to more severe service disruption as we will explain next.

COMPARISON OF ECMA-387 AND IEEE 802.11AD STANDARDS

The key difference between ECMA-387 standard and IEEE 802.11ad draft is network architecture.

In ECMA-387 standard, no central control point is needed to manage any network operation. Type-A and Type-B devices jointly form a fully distributed network via exchange of periodic beacons. Even though Type-B devices are less capable in terms of transmission ranges and rates — mainly due to the use of simpler antennas, Type-B devices can function independently without the presence of Type-A devices. On the contrary, an IEEE 802.11ad PBSS is under the full control of the PCP. As a result, the IEEE 802.11ad PBSS is highly centralized by nature. This fundamental difference in network architecture differentiates ECMA-387 standard from IEEE 802.11ad draft in various ways, even though they both target on a similar set of applications.

Take the timing structure as an example. The timing structure of an IEEE 802.11ad PBSS is established and owned by the PCP. Based on the timing reference, the SP or CBP where data transmission or antenna training takes place are scheduled by the PCP. As any centralized network may suffer, IEEE 802.11ad PBSSs are susceptible to the single-node failure problem. That is, if the PCP stops function for whatever reasons, the entire network cannot operate normally until a new PCP is selected. The problem is less troublesome in the existing infrastructure 802.11 wireless LAN, given that the AP is mains powered and works as a gate way to the Internet (thus, rarely shut down). However, in wireless display or wireless sync-and-go, the network topology is ad hoc in nature. Take the scenario shown in Figure 10 as an example. A pair of handheld devices establishes a PBSS for data backup first, with Data Source being the PCP. In the meantime, PC and Monitor in vicinity are turned on and join the existing PBSS after receiving the beacons from Data Source. Both data backup and video streaming work as expected until the data backup is completed and Data Source is removed or turned off. At that time, the video stream is disrupted as the PBSS no longer exists (neither does the scheduled SP for PC and Monitor).

Another option for PC and Monitor is to establish a second PBSS of their own as shown in Figure 11. Since the two PBSSs are in the coverage of each other and may interfere with each other, the two PBSSs may form a cluster so that their BIs are synchronized with each other as we explained in Section 5. In the cluster, one of the PCPs assumes the role of synchronization PCP (S-PCP) to provide timing reference to all other member PCPs. The S-PCP can be considered as a "super PCP" in the newly-formed "super PBSS" (i.e., the cluster). If the S-PCP (say, Data Source) stops function, the second PBSS is disrupted until the so-called S-PCP handover is completed.

Figure 10. Single-node failure in a single IEEE 802.11ad PBSS

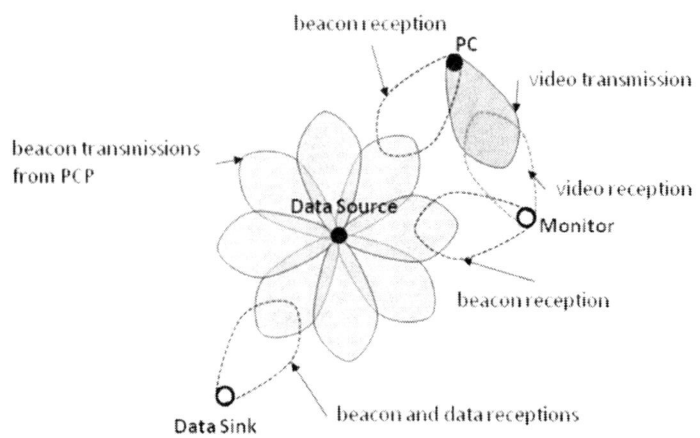

Figure 11. Single-node failure in an IEEE 802.11ad PBSS/PCP cluster

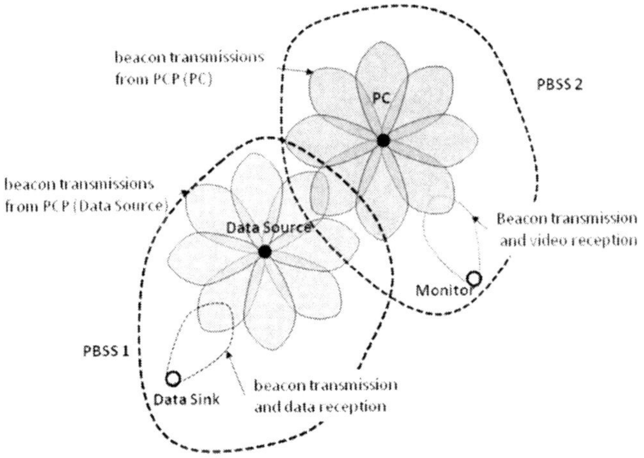

In other words, the impact of the single-node failure problem still remains even with the help of PCP clustering.

Compared to IEEE 802.11ad standard, ECMA-387 devices transmit beacons only using a single antenna sector. Take the scenario in Figure 10 as an example again. Only four beacons are transmitted in total, one by each device, as illustrated in Figure 12, when using ECMA-387 standard. No device has control over any other devices or provides network timing reference. In fact, PC and Monitor do not receive any beacon from Data Source or Data Sink as no beacon sweeping, unlike in the IEEE 802.11ad draft, is performed. As a result, these two pairs of devices work independently. The benefits of independent operation are reflected not only on the reduced number of transmitted beacons (8 vs. 4 in our example), but also on antenna tracking and link management. In Figure 10, PC or Monitor must perform antenna tracking training with each other as well as with Data Source. In Figure 12, antenna tracking is only needed for the video link between PC and Monitor. Considering the antenna tracking requires sector sweeping and may cause interference to other collocated station, fewer links to be tracked could improve the spatial reusability. Moreover, fewer links to be tracked implies simpler link management since any link failure in Figure 10 leads to service interruption and requires link switch or link cooperation.

ECMA-387 standard provides a more robust solution with the expense of one channel being dedicated as the discovery channel. Although the discovery channel can also be used as a data channel, transmission over the discovery channel has lower priority than the neighbor discovery process. Therefore, QoS may not be guaranteed in the discovery channel. Devices that seek QoS guarantee should then operate in the data channels; as a result, the data channels are prone to congestion.

The use of the dedicated discovery channel also imposes difficulties in forming point-to-multi-point networks. In point-to-multipoint networks, devices usually maintain multiple links with different devices. According to ECMA-387 standard, device pairs switch to a data channel after discovering each other and exchange beacons/data using only the antenna sectors determined during the neighbor discovery phase. When a new device powers up and seeks any device that has established connection in a data channel, it is difficult for the new device to receive beacons from the existing devices. ECMA-387 standard requires devices that have active data links to return to the discovery channel periodically to facilitate

Figure 12. Independent and distributed ECMA-387 networks

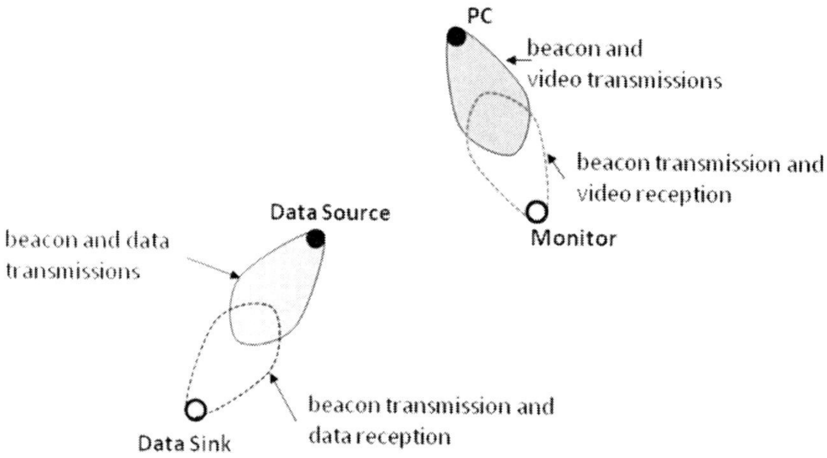

neighbor discovery in point-to-multiple networks. However, doing so may lower the achievable throughput in the data channels.

As a rule of thumb, one can conclude that ECMA-387 standard suits better for wireless display, wireless sync-and-go, or simple wireless PANs, given its distributed and robust design for ad hoc network settings. The IEEE 802.11ad suits better for larger wireless PANs or LANs composed of many devices in an infrastructural setting, given its centralized control and architecture. Nevertheless, one should always closely examine the application requirements and device complexity, and select the appropriate solutions accordingly.

FUTURE RESEARCH DIRECTIONS

There is a recent intense history in the development of WLAN and WPAN wireless technologies for short range wireless communications. The need to improve the communications in this land has been reviewed in this chapter. Related to the development of the reviewed standards we think they will substitute cables for different applications.

For applications that require data exchange between mobile devices in vicinity, Bluetooth or USB cables are usually the only two options. A killer application of the 60 GHz band technologies would be the wireless USB 2.0 that replaces its weird counterpart and the very slow Bluetooth. For such type of applications, the ECMA-387 Type-B device would be an ideal solution given its low power and infrastructure-free design. Mobile phones that support ECMA-387 Type-B operation could set up a direct link with minimal user intervene and obtain a transmission rate that matches the existing user experience or beyond. We believe that a multi-radio platform, with: ISM-band IEEE 802.11x for hotspot Internet access, 4G for mobile Internet and the 60 GHz for high speed data exchange in vicinity, would be the ideal hybrid solution for next-generation mobile multimedia applications.

CONCLUSION

In this chapter, we consider cable-replacement applications using the 60 GHz technologies. The problems and challenges of enabling cable replacements in the 60 GHz band — including wireless display, wireless sync-and-go, and Gbps WPAN/WLAN — are identified. We then investigate two promising standards, ECMA-387 and IEEE 802.11ad, and evaluate the advantages and drawbacks of the two standards based on their key designs.

In general, IEEE 802.11ad is developed based on the existing IEEE 802.11 WLAN technology and is a better solution for multi-Gbps, point-to-multiple points wireless networks. In this type of networks, power consumption is not a major issue since PCs, external hard disk drives and displays are usually mains powered. Advanced antenna training and tracking can then be performed to provide higher data rates or larger coverage while still maintaining QoS guarantees. However, handheld devices such as mobile phones will not be equipped with array or steerable antennas due to power saving and limitation in form factor. The mobile phones most likely will still use the existing ISM-band WLAN technology for access to Internet.

REFERENCES

Choudhury, R. R., Yang, X., Ramanathan, R., & Vaidya, N. (2002). Using directional antennas for medium access control in ad-hoc networks. In *ACM MOBICOM.*

IEEE. 802.11ad Draft 0.1. (2010). Retrieved from http://www.ieee802.org/ 11/TGad/ Draft%20 P802.11ad D0.1.pdf.

Ko, Y. B., Shankarkumar, V., & Vaidya, N. H. (2000). Medium access control protocols using directional antennas in ad-hoc networks. In *IEEE INFOCOM.*

Korakis, T., Jakllari, G., & Tassiulas, L. (2003). A Mac protocol for full exploitation of directional antennas in ad-hoc wireless networks. In *ACM MOBIHOC.*

Ramanathan, R. (2001). On the performance of ad-hoc networks with beamforming antennas. In *ACM MOBIHOC.*

Standard ECMA-387. (2008). *High rate 60 GHz PHY, MAC and HDMI PAL.* Retrieved from http:// www.ecma-international.org/ publications/ standards/ ECMA-387.htm.

Takai, M., Martin, J., Ren, A., & Bagrodia, R. (2002). Directional virtual carrier sensing for directional antennas in mobile ad-hoc networks. In *ACM MOBIHOC.*

Vasudevan, S., Kurose, J., & Towsley, D. (2005). On neighbor discovery in wireless networks with directional antennas, In *IEEE INFOCOM.*

WiFi Alliance (WFA). *(2008).* Very High Throughput (VHT) Study Group Usage Models. *Retrieved from* https://mentor.ieee.org/ 802.11/ dcn/07/ 11-07-2988-04-0000-liaison- from- wi-fi- alliance-to -802-11 -regarding-wfa-vht-study-group- consolidation-of- usage- models.ppt.

ADDITIONAL READING

Akkermans, J. (2010). *Planar beam-forming antenna array for 60 GHz broadband communication: Theory and application.* Germany: LAP Lambert Academic Publishing.

Bertsekas, A., & Gallager, R. (1992). *Data Networks.* Upper Saddle River, NJ: Prentice-Hall.

Standard ECMA-368. (n. d.). *High Rate Ultra Wideband PHY and MAC Standard.* Swiss. *ECMA International.*

Tanenbaum, A. (2003). *Computer Networks.* Upper Saddle River, NJ: Prentice-Hall.

Yong, S., Xia, P., & Carcia, A. (2011). *60GHz Technology for Gbps WLAN and WPAN: From Theory to Practice.* USA: Wiley.

Chapter 4
Codec Adaptation for Wireless Multimedia Streaming

Ping-Cheng Yeh
National Taiwan University, Taiwan

Yen-Chi Lee
National Taiwan University, Taiwan

Hung-Yun Hsieh
National Taiwan University, Taiwan

Chun-Cheng Chiang
National Taiwan University, Taiwan

Zhung-Han Wu
National Taiwan University, Taiwan

Antonio Ou
National Taiwan University, Taiwan

ABSTRACT

Due to the time-varying nature of wireless channels and the Internet backbone traffic, it is a challenging task to maintain the quality of wireless multimedia streaming throughout the transmission. An effective solution is to adapt the codec setting based on the wireless channel condition or the Internet backbone state. In this chapter, we present three cross-layer codec adaptation algorithms that adjust the codec setting in real-time based on media access control frame error rate, received signal strength indication, and path bandwidth respectively. Results show that the algorithms are effective in achieving good video quality for wireless multimedia streaming over wireless links.

INTRODUCTION

In the past decade, high-speed wireless data services have become increasingly popular following the explosive growth of the *Wireless Local Area Network* (*WLAN*) technology and the development of various mobile broadband technologies such as *3rd-Generation* (*3G*) and *Worldwide Interoperability for Microwave Access* (*WiMAX*) (Andrews, 2007). The ability to support high data rates in such wireless data services makes real-time wireless multimedia streaming a practical application nowadays. However, wireless channels are notorious for their time-varying nature due to the fading effect resulted from the multipath nature of wireless transmissions. Researchers of the physical layer technology have been working hard for decades to provide reliable data transmissions over time-varying wireless channels. Nevertheless, due to various compression techniques applied to multimedia data before transmission, wireless multimedia streaming is more sensitive to channel variation than typical data transmissions. As

DOI: 10.4018/978-1-61350-144-3.ch004

a result, it is a very challenging task to provide wireless multimedia streaming service with good video quality in wireless environments.

To provide good video quality in time-varying wireless environments, it is clear to see that the video codec setting has to be optimally adjusted in real-time according to channel variation. The optimality here is defined in the sense that the *Peak Signal-to-Noise Ratio* (*PSNR*) (Huynh-Thu, 2008) performance at the receiver side be maximized by adapting the codec setting. In wireless communications, the *Media Access Control* (*MAC*) *Frame Error Rate* (*FER*) is directly related to the channel quality. The FER statistic is also generally accessible which makes it desirable to design codec adaptation scheme according to the FER variation. In this chapter, we first demonstrate a novel FER based codec adaptation scheme which takes the dynamic of channel variation into consideration and chooses the setting that can achieve the best possible expected PSNR performance in the future, as opposed to existing schemes which generally choose the setting that works best only for the moment without considering the channel variation in the future. To find the optimal codec switching thresholds for FER based codec adaptation, we start by developing a new theoretical framework for analyzing the PSNR of multimedia streaming over wireless channels. Then the optimal codec switching thresholds can be found accordingly by maximizing the PSNR. The proposed FER based codec adaptation is shown to achieve great performance with PSNR gain of 3 to 5 dB.

In addition to the FER statistics at the MAC layer, the performance of multimedia streaming is also closely related to the channel *Received Signal Strength Indication* (*RSSI*) of the wireless link (physical layer information) and the bandwidth of the end-to-end communication path (network layer information). To further demonstrate the benefits of using cross-layer information for codec adaptation, we also present two heuristic schemes in this chapter for the server to adapt its codec setting based on channel RSSI and path bandwidth accordingly. Evaluation results show that codec adaptation using cross-layer information can effectively enhance the performance of multimedia streaming over time-varying wireless environments.

BACKGROUND

In the literature, there have been several works proposed for multimedia streaming over time-varying wireless channels. Different cross-layer approaches are designed to improve the PSNR performance for wireless multimedia streaming as summarized in the following.

Bucciol et al. (2004) propose that video packets that contain perceptually important frames should have higher priority during retransmission. The proposed scheme improves the resulting PSNR by 1.5 dB compared to the conventional *Automatic Repeat reQuest* (*ARQ*) schemes. However, the priority of every packet has to be calculated beforehand, which makes it impractical for real-time multimedia streaming. Also the perceptually important packets still suffer from the channel variation during transmission. Khan et al. (2006) consider the case of a multimedia server serving multiple clients. The distortions experienced by the different clients under different combinations of *Physical Layer* (*PHY*)/codec settings at clients' ends are simulated. By finding the combination that jointly optimizes the PSNR for everyone, the PHY/codec settings of all clients are determined. However, the scheme requires extensive simulations for all *PHY*/codec settings of all clients. This makes it difficult to be applied when the number of clients increases. Ruiz et al. (2002) adapt the codec settings according to the *Quality of Service* (*QoS*) report fed back from the client. The codec setting is upgraded when the QoS report is good, and downgraded when receiving a bad QoS report. Haratcherev et al. (2006) propose an algorithm where the codec setting is adjusted for the resultant bit-rate to match the

MAC transmission rate determined by the *Institute of Electrical and Electronics Engineers* (*IEEE*) 802.11 link-adaptation algorithm. However, for real-time multimedia streaming, it is impossible for one to know beforehand about the resultant bit-rates of specific video frames encoded using different codec settings. To know the resultant bit-rates, one has to encode the video frame in real-time using different codec settings, which introduces significant amount of extra complexity for real-time multimedia streaming.

In addition to the previous works, Schaar et al. (2003) propose a cross-layer optimization scheme which jointly considers the effects of MAC retransmission strategy, application-layer *Forward Error Correction* (*FEC*), packet size and codec settings. The optimal system setting is determined by using a linear PSNR approximation. Similarly, a joint source-channel coding scheme is considered (Argyrious, 2008) to optimize the video quality. The scheme optimizes *PSNR* by finding the optimal combination of the source rate, application-layer FER and PHY transmission rate for given average channel *Signal-to-Noise Ratio* (*SNR*). It shows that the PSNR performance can be improved if the PHY transmission is determined jointly with the source rate. Fallah et al. (2007) explore the effects of adaptive fragmentations in different layers for streaming over IEEE 802.11. Reducing the packet size in the bad channel state helps reduce the *Bit Error Rate* (*BER*). The paper shows that better PSNR can be achieved if fragmentation is done at the application layer (i.e. video encoding) than at the MAC layer. Sayantan et al. (2007) propose an adaptation scheme which adjusts the payload length and data rate according to the channel SNR. However, the optimization target is the throughput performance rather than the video quality (PSNR).

FER BASED CODEC ADAPTATION

For all of the approaches mentioned above, none of them considers the dynamic of channel variation in the near future. The codec setting is determined by choosing the setting that gives the best possible performance for the current moment. However, due to the oscillating nature of channel variation, it is possible to over-tune the codec by making decision solely on instantaneous or short-term averaged channel observations. In our opinion, the channel variation behavior in the near future should also be considered when adapting the codec setting. In our previous work (Chiang, 2009), an FER based codec adaptation scheme is proposed by switching the codec settings according to the periodic MAC FER fed back from the client. By modeling the wireless channel variation using the *Finite State Markov Channel* (*FSMC*) model (Guha, 2006) (Arauz, 2003) (Zhang, 1999), an induced Markov chain is proposed to jointly model the channel variation and the codec adaptation. The model enables us to take the dynamic of channel variation into consideration and thus to find the codec switching thresholds that optimize the expected PSNR for the future. Significant PSNR gain can be achieved by the scheme as we detailed in the following.

SYSTEM MODEL

In this section, we first analyze the protocol stack, and then introduce the channel model and FER model respectively.

In Figure 1, the system model of an H.264 (Richardson, 2003) video transmission system is depicted. At the media server, the H.264 codec compresses the raw data. The compressed data is then fragmented and combined with the *Real-time Transport Protocol* (*RTP*) header for delivery as RTP packets. Afterwards, the RTP packets are then sent to the client side through the *User Datagram Protocol/Internet Protocol* (*UDP/IP*) layer on top of the 802.11 MAC layer. When the data link layer detects *Cyclic Redundancy Check* (*CRC*) errors of the MAC frame, the ARQ mechanism of the 802.11 MAC retransmits the frame up to three times. Finally, the video frames are decoded at the client end using the inverse procedures. We assume availability of different codec settings for achieving codec adaptation in the encoder. For example, the *Quantization Parameter* (*QP*) is used prevalently in existing video codec standards such as *Moving Picture Experts Group-4* (*MPEG-4*) and H.264. It determines the quantization level of encoding parameters (e.g. *Discrete Cosine Transform* (*DCT*) coefficients), where a larger QP value results in fewer encoding bytes and hence coarse resolution of the content. Based on the feedback from the client, the server can perform codec adaptation through appropriate change of the QP values for subsequent video bitstreams. The mapping between the codec setting (e.g. QP value) and the target adaptation metric (e.g. bit rate) can be determined *a priori* based on, say, rate-distortion analysis.

Obviously the PSNR performance of the aforementioned H.264 video streaming is related to the FER at the MAC layer. To design the desired codec adaptation algorithm based on MAC FER, it is necessary to explore the relation between PSNR and MAC FER. In Figure 2, simulated PSNR versus MAC FER is plotted for different settings of H.264. Note that the MAC FER here counts all frame errors observed during regular frame transmissions and retransmissions (i.e. each MAC frame may be counted as erroneous frames multiple times depending on the number of re-transmissions). The definition of PSNR is:

$$PSNR = 10\log\frac{255^2}{\frac{1}{UV}\sum_{i=0}^{U-1}\sum_{j=0}^{V-1}(x_{ij}-\hat{x}_{ij})^2} \quad (1)$$

where U and V denote the frame size in pixels, and $x_{ij}, \hat{x}_{ij}$ represent the original pixel value and the reconstructed pixel value at location (*i, j*), respectively. It is clear to see from the figure that PSNR decreases as the MAC FER increases. The crossovers of these curves indicate that it is possible to achieve PSNR improvement by FER based codec adaptation.

In wireless communications, the fading effect is often characterized by the probability distribu-

Figure 1. The cross-layer optimized protocol stack for wireless H.264 video transmissions

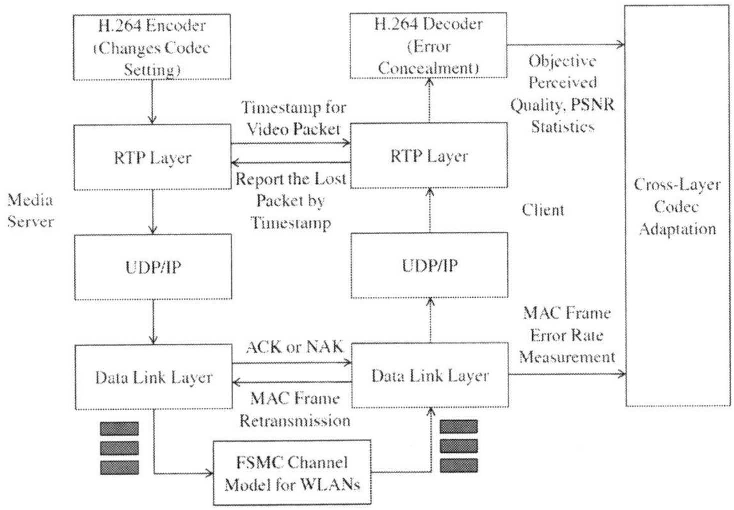

Figure 2. PSNR of the streamed Foreman clip (IPPP structure, 29.97 fps, GOP size = 15, packet size = 1462 bytes) versus MAC frame error rate

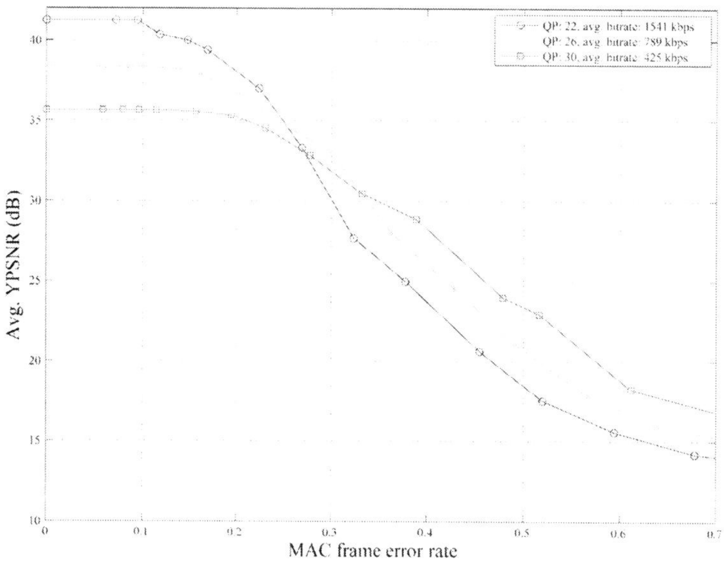

tion of the channel gain. If there exists Line-of-Sight (LOS) path between the transmitter and the receiver, the probability distribution is usually modeled by Ricean distribution; otherwise modeled by Rayleigh distribution. However, due to the changes of the wireless environment over time, such as the moving vehicles or pedestrians, the channel gain often varies with time. While the aforementioned Ricean or Rayleigh distributions suffice to describe the stochastic behavior of the channel gain at certain time instance, they are not capable of modeling the time-varying behavior of the channel gain. In the literature, the FSMC model is often used to model the time-varying channels. In this work, we consider the FSMC channel model proposed in (Zhang, 1999) that is commonly used to model the wireless channel of WLANs. Related work has shown that the FSMC model matches well with the actual channel measurement results.

As shown in Figure 3, the channel SNR γ is quantized to a number of channel states in the FSMC model. We say that the channel is in state w_k if $\gamma \in \left[\Gamma_k^w, \Gamma_{k+1}^w\right]$, where the dividing points $\{\Gamma_k^w\}$ can be determined based on different criterions. Following (Zhang, 1999), we set $\{\Gamma_k^w\}$ so that each channel state k has the same average sojourn time. In our numerical experiments, it is observed that the codec is often stuck to certain setting when the channel condition is poor and codec adaptation has little contribution to the expected PSNR. Thus in our design, the FER based codec adaptation is activated only when the channel SNR γ is above certain threshold value Γ_k^w. By replacing the *Probability Density Function (pdf)* $p(\gamma)$ of the channel SNR (Zhang, 1999) with the conditional pdf:

$$p_t(\gamma) = p(\gamma \mid \gamma > \Gamma_T^w) = \frac{p(\gamma)}{\int_{\Gamma_T^w}^{\infty} p(\gamma) d\gamma} \tag{2}$$

The channel state transition probabilities can be expressed as:

Figure 3. The FSMC Model of the wireless channel

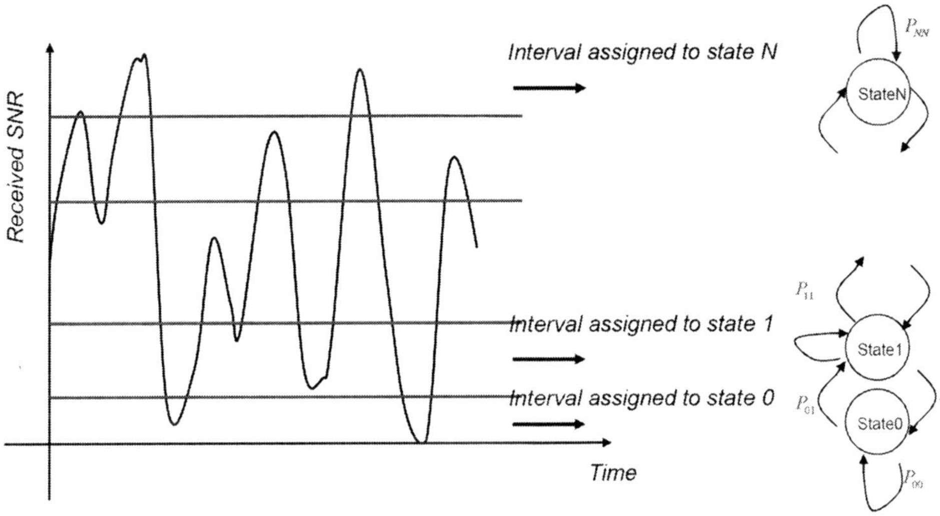

$$P^{w}_{k,k+1} = \frac{N(\Gamma^{w}_{k+1})T_p}{\pi^{w}_{k}}, k = 1,2,...,N-1$$
$$P^{w}_{k,k-1} = \frac{N(\Gamma^{w}_{k})T_p}{\pi^{w}_{k}}, k = 2,3,...,N \tag{3}$$

Where the function $N(\Gamma^{w}_{k+1})$ denotes the level-crossing rate at Γ^{w}_{k+1}, and π^{w}_{k} is the steady state probability of the form:

$$\pi^{w}_{k} = \int_{\Gamma^{w}_{k}}^{\Gamma^{w}_{k+1}} p(\gamma)d\gamma = \exp(-\frac{\Gamma^{w}_{k}}{\gamma_0}) - \exp(-\frac{\Gamma^{w}_{k+1}}{\gamma_0}) \tag{4}$$

The level-crossing rate $N(\Gamma)$ at SNR level Γ has the form:

$$N(\Gamma) = \sqrt{\frac{2\pi\Gamma}{\gamma_0}} f_m \exp(-\frac{\Gamma}{\gamma_0}) \tag{5}$$

Where γ_0 is the average SNR, f_m denotes the maximum Doppler frequency caused by the user motion, and T_p denotes the time between two consecutive MAC frame transmissions. It is assumed that T_p remains constant throughout the multimedia transmission.

The FER during frame transmissions is determined by the PHY transmission scheme, i.e. the coding and modulation setup, and the received channel SNR. The *Bit Interleaved Coded Modulation Scheme* (*BICM*) is often used for PHY transmissions in modern wireless communications. The BICM provides a simple way to combine the use of channel coding with various types of modulation schemes while achieving near optimal decoding performance. The flexibility of pairing channel coding with different modulation schemes provides a variety of choices for the transmission rates in PHY. This allows the system to have more adaptability when transmitting in time-varying channel fading. As a result, BICM is widely applied in modern wireless communications.

In this work, we use the BICM that combines the rate-1/2 [133,171] convolutional code with *16-Quadrature Amplitude Modulation* (*16-QAM*) for PHY transmissions. The instantaneous FER, when channel SNR equals γ, is denoted by $P_e(\gamma)$. The average FER E^{w}_{k}, when the channel is in state w_k, can then be derived as:

$$E_k^w = \frac{\int_{\Gamma_k^w}^{\Gamma_{k+1}^w} P_e(\gamma) p_t(\gamma) d\gamma}{\int_{\Gamma_k^w}^{\Gamma_{k+1}^w} p_t(\gamma) d\gamma} \quad (6)$$

Note that the expression in (6) can be computed numerically by curve-fitting the simulated $P_e(\gamma)$ using exponential function of the form $\exp\{a\gamma + b\gamma^2\}$. The simulated FER curve of the BICM system using rate-1/2[133,171] convolutional code with 16-QAM modulation is shown in Figure 4 which is curve fitted to compute (6) numerically.

CODEC ADAPTATION ALGORITHM

We introduce in this section the operations of the FER based cross-layer codec adaptation algorithm. Assume that the codec is free to choose any of the Q codec settings $\{o_1, o_2, ..., o_Q\}$ listed in descending order of the generated bit rate (and thus the video quality). For each observation period which consists of a fixed number of MAC frame transmissions, the observed FER is compared to some predetermined threshold values to decide which codec setting to use in the following observation period. If a low FER is observed, this indicates that the wireless channel is in good condition and it is possible to support codec setting of higher rates for better video quality. On the other hand, if we observe a high FER, it means that the channel is in bad condition. The codec should be switched to the low-rate setting so frame losses can be reduced in exchange of the video quality. In Figure 5, we show the operation of the FER based codec adaptation algorithm. The algorithm is described in details as follows:

- Arbitrarily choose a codec setting o_i to initiate the multimedia streaming.

  Repeat
  - Record the number of MAC frame errors during the following M MAC frame transmissions (one observation period). Compute FER ($=\frac{l}{M}$).
  - Switch the codec to setting o_{i+1} if $\frac{l}{M} > T_{i,i+1}$; switch to o_{i-1} if

Figure 4. The simulated FER of BICM usingrate-1/2 [133,171] convolutional code with 16-QAM modulation in Additive White Gaussian Noise (AWGN) channel, frame size 1462 bytes

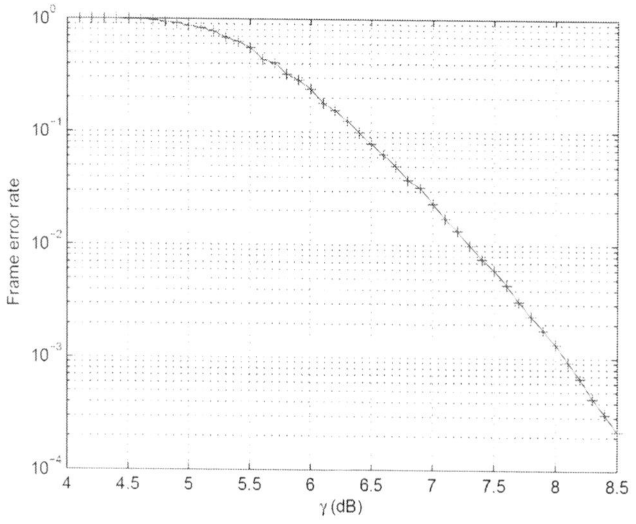

Figure 5. The FER based codec adaptation algorithm

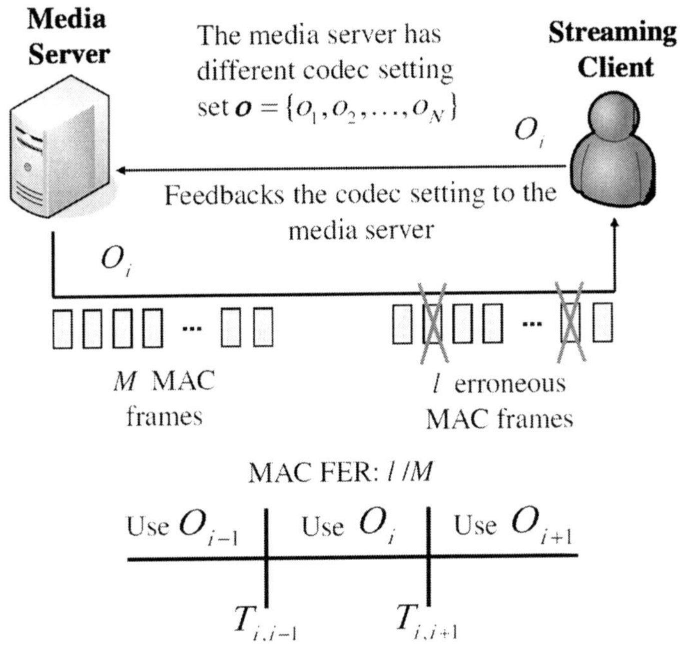

$\frac{l}{M} < T_{i,i-1}$; otherwise the codec setting remains unchanged.

Until received the last video frame.

PSNR ANALYSIS

In this section, we first introduce the induced Markov chain used in our analysis, and then the derivation of $F_{i,j}^{w}(l)$, which denotes the probability of observing l frame errors among M frame transmissions under certain channel states w_i and w_j.

One can easily see that the FER based codec adaptation algorithm is characterized by the codec switching thresholds $\{T_{i,i-1}, T_{i,i+1}\}$. Now the question is how to determine these threshold values to achieve the best possible video quality. To achieve the goal, we need to first derive the relation between the expected PSNR and the threshold values. From the relation, we can then set the values of $\{T_{i,i-1}, T_{i,i+1}\}$ to maximize the expected PSNR. The details are given below. Without loss of generality, we set the number of codec settings available $Q = 3$ in the following. Recall that the PSNR curves of codec settings o_1, o_2, o_3 are plotted against MAC FER in Figure 2. For the PSNR analysis, each PSNR curve is numerically fitted by a function $\Upsilon_i(x)$ to denote the PSNR of the codec setting o_i at a given FER x.

Before we derive the expected PSNR, we need to construct the induced Markov chain which can jointly capture the behaviors of channel state variation and the codec adaptation. The induced Markov chain is depicted in Figure 6. Since the codec setting is adjusted after each observation period, the state transition time is thus set to be one observation period, i.e., M MAC frame transmissions. The state $o_i w_j$ of the induced Markov chain indicates that the codec is operating using setting o_i in the current observation period while the wireless channel is in state w_j during the first MAC frame transmission in the observation period. It should be noted that the wireless channel might transit to other channel states during the remaining *M-1* frame transmissions of the observation period following the FSMC model presented earlier.

Under the operation of the FER based codec adaptation algorithm mentioned previously, we can derive the transition probability as:

$$P_{o_1w_i,o_1w_j}=\sum_{k=1}^{3}\sum_{l=0}^{\lfloor MT_{1,2}\rfloor}F_{i,k}^{w}(l)P_{k,j}^{w},\quad P_{o_1w_i,o_2w_j}=\sum_{k=1}^{3}\sum_{l=\lfloor MT_{1,2}\rfloor+1}^{M}F_{i,k}^{w}(l)P_{k,j}^{w},$$

$$P_{o_2w_i,o_1w_j}=\sum_{k=1}^{3}\sum_{l=0}^{\lfloor MT_{2,1}\rfloor}F_{i,k}^{w}(l)P_{k,j}^{w},\quad P_{o_2w_i,o_2w_j}=\sum_{k=1}^{3}\sum_{l=\lfloor MT_{2,1}\rfloor+1}^{\lfloor MT_{2,3}\rfloor}F_{i,k}^{w}(l)P_{k,j}^{w},$$

$$P_{o_2w_i,o_3w_j}=\sum_{k=1}^{3}\sum_{l=\lfloor MT_{2,3}\rfloor+1}^{M}F_{i,k}^{w}(l)P_{k,j}^{w},\quad P_{o_3w_i,o_2w_j}=\sum_{k=1}^{3}\sum_{l=0}^{\lfloor MT_{3,2}\rfloor}F_{i,k}^{w}(l)P_{k,j}^{w},$$

$$P_{o_3w_i,o_3w_j}=\sum_{k=1}^{3}\sum_{l=\lfloor MT_{3,2}\rfloor+1}^{M}F_{i,k}^{w}(l)P_{k,j}^{w} \tag{7}$$

Where $F_{i,j}^{w}(l)$ is defined to be the probability of observing *l* frame errors among the *M* frame transmissions, given that the first of the *M* frame transmissions encounters channel state w_i while the last frame transmission encounters channel state w_j. The detailed derivation of $F_{i,k}^{w}(l)$ is further illustrated later. To see why the transition probability expressions have the form in (7), let us take $P_{o_1w_i,o_1w_j}$ for example. If the codec is operating under setting o_1 in the current observation period and the codec setting remains to be o_1 in the next observation period, this implies that there must be less than $\lfloor MT_{1,2}\rfloor$ error frames observed during the *M* frame transmissions so that the resultant FER can be smaller than $T_{1,2}$ to keep the codec setting unchanged. This explains the summation over *l* from 0 to $\lfloor MT_{1,2}\rfloor$ in the expression. On the other hand, for the wireless channel state to transit from w_i in the first frame of the current observation period to w_j in the first frame of the next observation period, the channel state can transit to any state $\{w_k\}$ at the end of the current

Figure 6. The induced Markov chain for codec adaptation

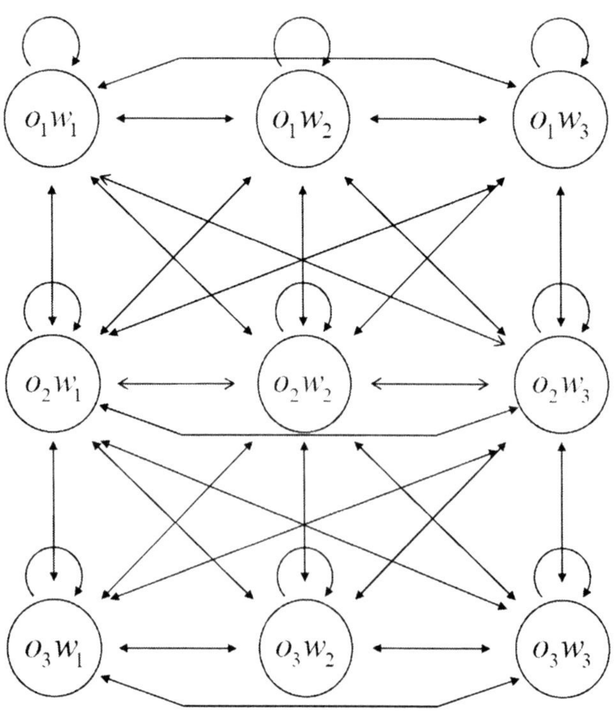

observation period (the M-th MAC frame) and then transits to w_j in one frame duration. This explains why we sum up $F^w_{i,k}(l)P^w_{k,j}$ over all possible k in the expression. The expressions of other transition probabilities can be easily interpreted following similar intuition.

Once we compute $F^w_{i,j}(l)$, $i,j \in \{1,2,3\}$ following the derivation given later, the transition probabilities in (7) can be directly computed. As a result, we can solve for the steady state probabilities of the induced Markov chain. Denote the vector of the steady state probabilities as:

$$\pi = [\pi_{o_1w_1}\ \pi_{o_1w_2}\ \pi_{o_1w_3}\ \pi_{o_2w_1}\ \pi_{o_2w_2}\ \pi_{o_2w_3}\ \pi_{o_3w_1}\ \pi_{o_3w_2}\ \pi_{o_3w_3}] \quad (8)$$

Using the PSNR function $\Upsilon_i(x)$ obtained earlier for each codec setting, we can compute the expected PSNR as:

$$\begin{aligned}\overline{PSNR} = {} & \pi_{o_1w_1}(\sum_{i=1}^{3}\sum_{j=0}^{M}F^w_{1,i}(j)\Upsilon_1(j/M) + \pi_{o_1w_2}(\sum_{i=1}^{3}\sum_{j=0}^{M}F^w_{2,i}(j)\Upsilon_1(j/M) \\ & +\pi_{o_1w_3}(\sum_{i=1}^{3}\sum_{j=0}^{M}F^w_{3,i}(j)\Upsilon_1(j/M) + \pi_{o_2w_1}(\sum_{i=1}^{3}\sum_{j=0}^{M}F^w_{1,i}(j)\Upsilon_2(j/M) \\ & +\pi_{o_2w_2}(\sum_{i=1}^{3}\sum_{j=0}^{M}F^w_{2,i}(j)\Upsilon_2(j/M) + \pi_{o_2w_3}(\sum_{i=1}^{3}\sum_{j=0}^{M}F^w_{3,i}(j)\Upsilon_2(j/M) \\ & +\pi_{o_3w_1}(\sum_{i=1}^{3}\sum_{j=0}^{M}F^w_{1,i}(j)\Upsilon_3(j/M) + \pi_{o_3w_2}(\sum_{i=1}^{3}\sum_{j=0}^{M}F^w_{2,i}(j)\Upsilon_3(j/M) \\ & +\pi_{o_3w_3}(\sum_{i=1}^{3}\sum_{j=0}^{M}F^w_{3,i}(j)\Upsilon_3(j/M)\end{aligned} \quad (9)$$

Note that the expected PSNR in (9) is related to the codec switching thresholds $\{T_{i,i-1}, T_{i,i+1}\}$ indirectly through the transition probabilities in (8). Hence, by applying numerical optimization methods to (8) and (9), the optimal switching thresholds $\{T_{i,i-1}, T_{i,i+1}\}$ can be obtained for the FER based cross-layer codec adaptation algorithm.

Now, we derive the probability distribution $F^w_{i,j}(\cdot)$. Let us define random variable $X_j = 0$ to denote the success of the j-th frame transmission, and $X_j = 1$ vice versa. Note that $F^w_{i,j}(l)$ is defined as the probability of observing l frame errors among the M frame transmissions, given that the first of the M frame transmissions encounters channel state w_i while the last frame transmission encounters channel state w_j. In other words, we need to find the conditional distribution of $X_1 + X_2 + \ldots + X_M$ given the initial channel state w_i and the ending channel state w_j of the observation period (M frames). For any discrete random variable Y, one can always characterize the distribution of Y by finding its *Factorial Moment Generating Function (FMGF)* $\phi_Y(\omega) = E[\omega^Y] = \sum_{y=-\infty}^{\infty}\omega^y$. Thus if we can derive the FMGF of $X_1 + X_2 + \ldots + X_M$ condition on the initial channel state w_i and the ending channel state w_j of the observation period, the probability distribution $F^w_{i,j}(\cdot)$ can then be obtained.

Recall that the MAC FER in the channel state w_k is denoted by E^w_k. If the wireless channel is in state w_k, we can derive the conditional FMGF of $X_j \mid w_k$ as:

$$\phi_{X_j|w_k}(\omega) = E[\omega^{X_j} \mid w_k] = \sum_{x=0,1}\omega^x P\{X_j = x \mid w_k\} = (1-E^w_k) + \omega\, E^w_k,\ k=1,2,3 \quad (10)$$

Define

$$\psi_{w_k}(\omega) = (1-E^w_k) + \omega\, E^w_k,\ k=1,2,3.$$

Also define random vector $\mathbf{w} = [w_{i1}, w_{i2}, \ldots, w_{iM}]$ to denote the channel states experienced by the M MAC frame transmissions in the observation period. We can derive the conditional FMGF for the total number of frame errors observed during the observation period as:

$$\begin{aligned}&\phi_{X_1+X_2+\ldots+X_M|\mathbf{w}}(\omega)\\&=\phi_{X_1|w_{i1}}(\omega)\,\phi_{X_2|w_{i2}}(\omega)\,\ldots\,\phi_{X_M|w_{iM}}(\omega)\\&=\psi_{w_{i1}}(\omega)\,\psi_{w_{i2}}(\omega)\,\ldots\,\psi_{w_{iM}}(\omega)\end{aligned} \tag{11}$$

One can see that the conditional FMGF only depends on the channel vector **w** experienced during the frame transmission. To compute the unconditional FMGF, we need to take the expectation of the conditional FMGF with respect to the probability of each possible realization of **w**, i.e., the M channel states visited in one observation period. We first define the matrix function $G(\omega)$ to denote following matrix:

$$G(\omega)\equiv P\cdot\psi(\omega)=$$
$$\begin{bmatrix}P^w_{1.1}\psi_{w_1}(\omega) & P^w_{1.2}\psi_{w_2}(\omega) & 0\\ P^w_{2.1}\psi_{w_1}(\omega) & P^w_{2.2}\psi_{w_2}(\omega) & P^w_{2.3}\psi_{w_3}(\omega)\\ 0 & P^w_{3.2}\psi_{w_2}(\omega) & P^w_{3.3}\psi_{w_3}(\omega)\end{bmatrix} \tag{12}$$

Where

$$\psi(\omega)\equiv$$
$$\begin{bmatrix}\psi_{w_1}(\omega) & 0 & 0\\ 0 & \psi_{w_2}(\omega) & 0\\ 0 & 0 & \psi_{w_3}(\omega)\end{bmatrix} \tag{13}$$

Now let us separate the M channel states visited in one observation period into the initial channel state and the following *M-1* states after *M-1* channel transitions. By taking advantage of the multiplication form in (11), we can find the unconditional FMGF of $X_1+X_2+\ldots+X_M$, i.e., the total number of frame errors in one observation period as:

$$H(\omega)\equiv\psi(\omega)G(\omega)^{M-1}=$$
$$\begin{bmatrix}H_{11}(\omega) & H_{12}(\omega) & H_{13}(\omega)\\ H_{21}(\omega) & H_{22}(\omega) & H_{23}(\omega)\\ H_{31}(\omega) & H_{32}(\omega) & H_{33}(\omega)\end{bmatrix} \tag{14}$$

Note that the resultant $H_{i,j}(\omega)$ is simply the FMGF of $F^w_{i,j}(l)$. Thus the values of $\{F^w_{i,j}(l)\}$ needed for the PSNR analysis can be easily obtained from $H_{i,j}(\omega)$ through the FMGF:

$$H_{i,j}(\omega)=\sum_{l=0}^{M}F^w_{i,j}(l)\omega^l,i,j\in\{1,2,3\} \tag{15}$$

NUMERICAL RESULTS

To implement the proposed FER based codec adaptation scheme on personal computers, one needs to have the access to the MAC FER information from the WLAN adaptors. Note that the FER here is the primitive FER that counts all frame transmission errors occurred during the transmissions, whether in the original frame transmissions or ARQ retransmissions. Unfortunately, such FER information is not accessible from the drivers of the WLAN adaptors in the current market. The MAC FER provided by the existing drivers only count the number of frames that fail to be successfully transmitted after reaching the limit of the maximum number of retransmissions. As a result, the performance of the proposed scheme is evaluated via simulations

In Table 1, the parameters used to simulate WLAN are given. The MAC frame duration T_p, which is used to compute the transition probabilities of FSMC in (3), can be computed as:

$$T_p=\frac{FrameSize}{Tx.Rate}=\frac{1462(bytes)\cdot 8}{24(Mbps)}=0.487(ms) \tag{16}$$

Table 1. Simulation parameters

Transmission Rate	**24 Mbps**
MAC Frame Size	**1462 bytes**
Maximal SW ARQ Retransmission	**3**

Throughout the numerical experiments, the codec adaptation activation SNR threshold Γ_T^w is set to be *-4* dB. In our numerical experiments, video streaming simulations are conducted for various combinations of Doppler frequencies (f_m = 3, 5, 8 Hz) and transmit SNR values (γ_0 = *7, 8* dB) to explore the performance of the FER based codec adaptation algorithm in different channel environments. Note that the Doppler frequency is proportional to the relative velocity between the transmitter and the receiver. When the velocity is high, that means the spatial nature of the wireless channel changes very fast, which results in the fast varying channel fading. Therefore, Doppler frequency is often used as an index of how fast the channel fading varies with time. On the other hand, the three H.264 codec settings used in the simulations are $\{o_1, o_2, o_3\}$ ={*QP22, QP26, QP30*}.

From (Zhang, 1999), we can compute the SNR threshold values needed for the FSMC channel model. The SNR threshold values used for our simulations are listed in Table 2. On the other hand, the transition probabilities of the WLAN channels under different f_m values are shown in Table 3. For transmitting SNR γ_0=*7* and *8* dB, the MAC FER's under different wireless channel states are listed in Table 4. Throughout the simulations, the observation period is set to be *48* MAC frames (*M=48*). Following the analysis in the previous section, the optimal codec switching threshold are shown in Table 5 and 6 for different combinations of f_m and γ_0.

First notice that the optimal thresholds $T_{i,i+1}$ are not equal to $T_{i+1,i}$, $i = 1,2$. This is because

Table 2. SNR thresholds of FSMC at $\Gamma_T^w = -4$ dB

Γ_{w1}	-4.0000
Γ_{w2}	1.1893
Γ_{w3}	4.7465
Γ_{w4}	8.3934

Table 3. Channel transition probability matrices under different f_m values

Transition prob. (P) $\begin{pmatrix} 0.9982 & 0.0018 & 0 \\ 0.0034 & 0.9956 & 0.0010 \\ 0 & 0.0043 & 0.9957 \end{pmatrix}$
(a) $f_m = 8Hz$
Transition prob. (P) $\begin{pmatrix} 0.9988 & 0.0012 & 0 \\ 0.0021 & 0.9973 & 0.006 \\ 0 & 0.0027 & 0.9973 \end{pmatrix}$
(b) $f_m = 5Hz$
Transition prob. (P) $\begin{pmatrix} 0.9993 & 0.0007 & 0 \\ 0.0012 & 0.9984 & 0.0004 \\ 0 & 0.0016 & 0.9984 \end{pmatrix}$
(c) $f_m = 3Hz$

Table 4. MAC FER of each channel state

Error rate	**Error rate**
E_1^w 0.7941 E_2^w 0.0048 E_3^w 0.0000	E_1^w 0.5948 E_2^w 0.0006 E_3^w 0.0000
γ_0=7 dB	γ_0 = 8 dB

Table 5. Optimal codec switching thresholds for γ_0 = 7dB and different f_m values

γ_0 = 7 dB					
Threshold		**Threshold**		**Threshold**	
$T_{1,2}$	0.0833	$T_{1,2}$	0.1042	$T_{1,2}$	0.1042
$T_{2,1}$	0.0833	$T_{2,1}$	0.0625	$T_{2,1}$	0.1458
$T_{2,3}$	0.6875	$T_{2,3}$	0.6458	$T_{2,3}$	0.5208
$T_{3,2}$	0.6458	$T_{3,2}$	0.6458	$T_{3,2}$	0.625
(a) $f_m = 8Hz$ (b) $f_m = 5Hz$ (c) $f_m = 3Hz$					

Table 6. Optimal codec switching thresholds for γ_0 = 8dB and different f_m values

γ_0 = 8 dB					
Threshold		**Threshold**		**Threshold**	
$T_{1,2}$	0.0625	$T_{1,2}$	0.0625	$T_{1,2}$	0.0625
$T_{2,1}$	0.0417	$T_{2,1}$	0.0417	$T_{2,1}$	0.0417
$T_{2,3}$	0.4375	$T_{2,3}$	0.3750	$T_{2,3}$	0.3958
$T_{3,2}$	0.4167	$T_{3,2}$	0.3958	$T_{3,2}$	0.3958
(a) $f_m = 8Hz$ (b) $f_m = 5Hz$ (c) $f_m = 3Hz$					

the numbers of frame errors in wireless states w_i and w_{i+1} have different distributions. If we mark the optimal codec switching thresholds on the PSNR-versus-FER plot of the three codec settings as shown in Figure 7, it is interesting to see none of the optimal switching thresholds takes place at the crossover points of these PSNR curves. From Table 5, we can observe that the codec switching thresholds get closer to the crossover points when the channel has less variation (f_m =3) and the thresholds get further away from the crossover points as the channel variation increases (f_m=8). The observation implies that the codec should not be switched just for the short-term channel variations. The channel transition behaviors in the long run should be taken into account, which is the major merit of our proposed algorithm. Both Tables 5 and 6 show that the FER range of codec o_2 becomes larger as f_m increases. This indicates that the codec has the tendency to stay with the codec setting of medium performance in a fast-varying channel. We also observe the increase of the switching thresholds as γ_0 decreases. The decrease in SNR results in higher FER, which causes the switching thresholds to increase accordingly.

For the simulations of H.264 video transmissions, we use the Foreman test sequence to trans-

Figure 7. Optimal codec switching thresholds with PSNR versus FER under codec setting adaptation for $\gamma_0 = 7$ dB and $f_m = 8$ Hz

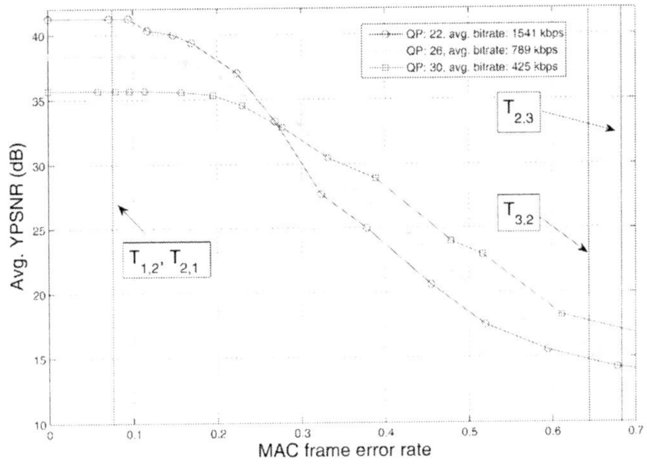

mit repeatedly in each simulation run. The frame-by-frame *Luminance Peak Signal-to-Noise Ratio* (*YPSNR*) of one particular video transmission under f_m=*8* and γ_0=*8* dB is depicted in Figure 8. It is clear that the FER based codec adaption outperforms non-adaptive algorithms using any of the three codec settings. Also note that the upper envelope of the three dashed curves is in fact the optimal PSNR performance of all codec adaptation schemes, since the envelope is equivalent to the performance of using the best codec setting for any given time instance. It is clear from Figure 8 that the PSNR curve of our FER based codec adaptation scheme is very close to the upper envelope, i.e., the optimal PSNR performance of all codec adaptation schemes. This is really remarkable considering we only adapt the codec setting once every *48* frames.

Figure 8. Frame-by-frame YPSNR under codec setting adaptation for $\gamma_0 = 8$ dB and $f_m = 8$ Hz

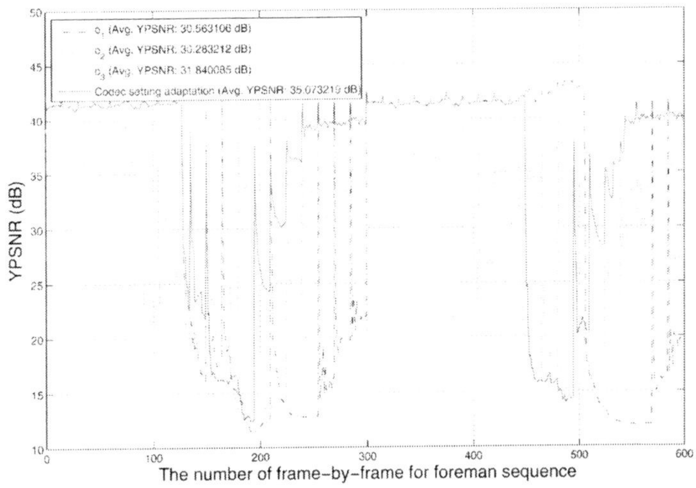

Table 7. Average YPSNR for video transmission over FSMC channel for γ_0=8 dB and f_m = 8Hz

Codec setting	Avg. YPSNR(dB)
o_1	29.5410
o_2	28.8931
o_3	29.1122
Codec setting adaptation	34.3292

In Table 7, we list the average YPSNR under each setup after transmitting the Foreman test sequence for *20* times. The average YPSNR of the proposed FER based codec adaptation has significant gains of *4.7882* dB, *5.4361* dB, and *5.2170* dB, over the three codec settings, respectively.

In Figure 9, we plot the simulated PSNR and the PSNR analysis results together for the cases of f_m=*3,5,8* Hz and γ_0=*7,8* dB. The PSNR analysis is observed to match well with the simulated PSNR. This implies that the proposed PSNR analysis can be useful to explore the PSNR performance of wireless multimedia streaming analytically.

RSSI AND BANDWIDTH BASED CODEC ADAPTATION

In the FER based codec adaptation algorithm presented earlier, the client observes and reports the FER back to the streaming server periodically for adaptation of codec parameters. Since FER is measured by the client at the MAC layer, FER based codec adaptation is a cross-layer scheme that relies on the interaction between the application and MAC layer. As we observed in previous sections, such a MAC layer metric can timely track the variation of the channel quality and trigger the adaptation process at the server for enhanced multimedia streaming.

In addition to the MAC layer, codec adaptation can also interact with other layers of the protocol stack to further enhance the performance of multimedia streaming over time-varying wireless environments. In particular, for wireless channels the RSSI measured at the physical layer is a key metric determining the BER (and hence FER)

Figure 9. PSNR analysis/simulation comparison at γ_0 = 7 and 8 dB for f_m = 3, 5, 8 Hz

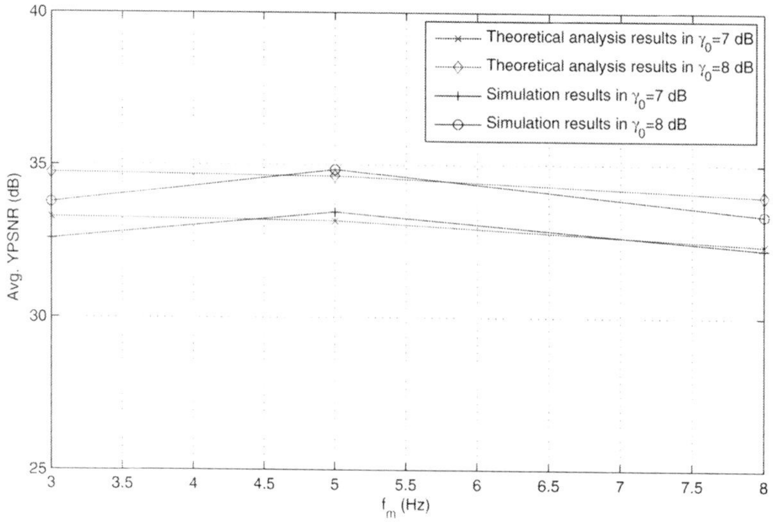

of the received data. It can therefore be used by the client to convey the first-hand channel quality information back to the server. On the other hand, since the streaming server may be located in the backbone Internet, the communication path between the client and server may extend beyond the wireless link. In this case, network layer information such as path bandwidth and delay can be used to allow for more intelligent codec adaptation against variation of the entire network environments.

In the following, we present two more cross-layer codec adaptation schemes based on physical layer and network layer information respectively. In particular, we focus on codec adaptation using *RSSI* and path bandwidth. As we show in Figure 10, while information may be measured by the client at different layers, such cross-layer information can be sent back to the server using *Real Time Control Protocol* (*RTCP*) in a unified protocol framework.

CODEC ADAPTATION BASED ON PHYSICAL LAYER INFORMATION

Other than the MAC FER, RSSI is another measure that is directly related to the PHY channel condition. Also since RSSI is commonly accessible on WLAN adaptors, it is thus desirable to design a codec adaptation scheme based on RSSI. The proposed approach here is to find the PSNR performance of different codec settings versus RSSI through extensive experiments. From the crossover points of the PSNR curves, we can then determine the codec switching thresholds for the RSSI based codec adaptation. The codec can thus make decision on the codec setting by comparing the measured RSSI to the switching thresholds in real-time.

The main challenge for the experimental approach is the measurement of RSSI experienced during the experiments so we can capture the PSNR performance of all RSSI values. The setup of our experiment is shown in Figure 11. With the help of the programmable attenuator, we are able to have the control over the RSSI experienced at the client end. Note that the use of the programmable attenuator is to emulate a wide range of channel conditions that might be experienced at the client in real life. In the experiment, we transmit a *19*-second TV video clip of 704×480 resolution using the MPEG-4 codec. Different combinations of QP values (*6, 8, 10, 15*) are tested for *5* runs under different levels of attenuation (*0dB, 3dB, ..., 27dB*). The *Group of Pictures* (*GOP*) size is set to be *15*. The experiment result is shown in Figure 12.

In Figure 12, we can clearly see the crossover points of the PSNR curves under different codec settings. From the crossover points, we can obtain the switching thresholds for the RSSI based codec adaptation as shown in Table 8. The operations

Figure 10. Feedback of cross-layer information

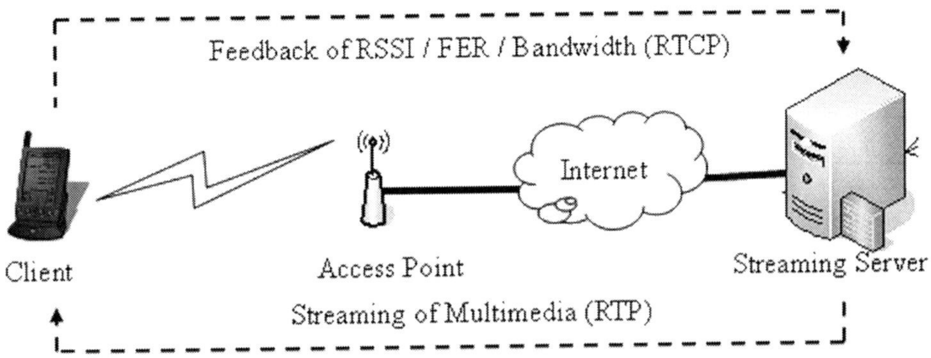

Figure 11. Experiment setup for PSNR versus RSSI measurement

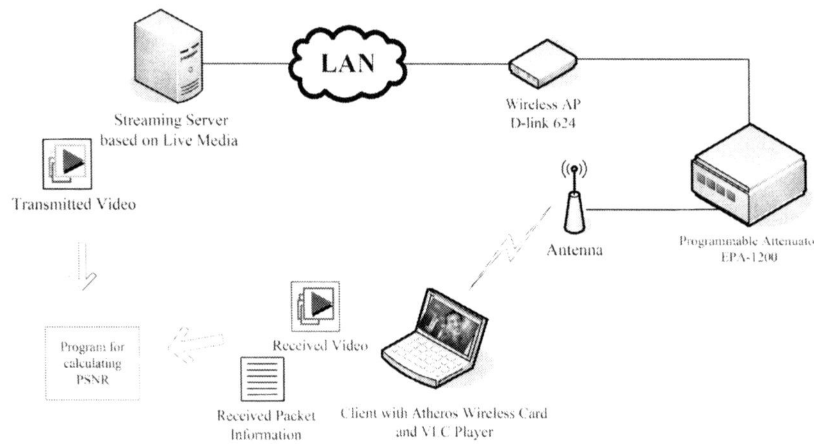

Figure 12. The PSNR performance of various codec settings versus the received SNR

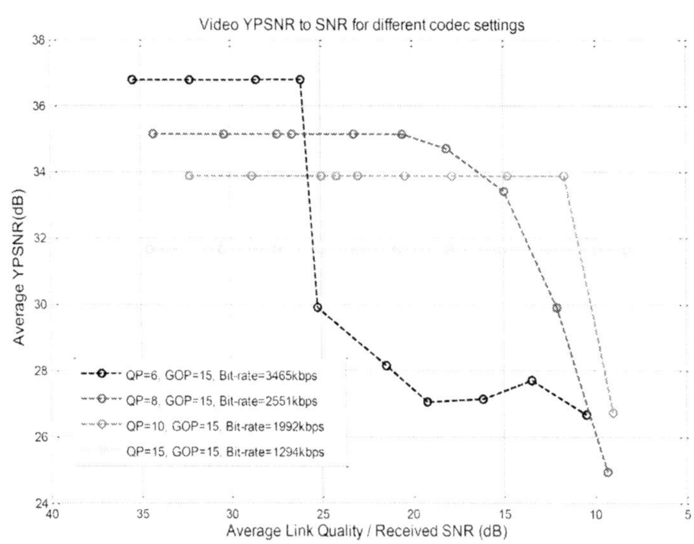

Table 8. Switching thresholds for RSSI based codec adaptation

Codec setting	The Range of RSSI
$QP = 6, GOP = 15$	$28 \leq SNR$
$QP = 8, GOP = 15$	$17 \leq SNR < 28$
$QP = 10, GOP = 15$	$13 \leq SNR < 17$
$QP = 15, GOP = 15$	$SNR < 13$

of the RSSI based codec adaptation algorithm can be summarized as follows:

- Arbitrarily choose a codec setting to initiate the multimedia streaming.
  Repeat
  - Measure the RSSI at the client end and send SNR feedback to the media server.

- From Table 8, the media server decides what codec setting to change to.

Until received the last video frame.

To examine the performance of the proposed RSSI based codec adaptation, we conduct video streaming of real-time TV for *120* seconds (*3600* frames) using the testbed in Figure 11. By using the programmable attenuator, the RSSI is controlled to have *4* different values (*30*dB, *12*dB, *21*dB, *30*dB) during different periods of the streaming. The frame-by-frame YPSNR is plotted in Figure 13 for three different codec settings without codec adaptation as well as the case of RSSI based codec adaptation. We can see that the case with codec adaptation (green curve) gives the most stable PSNR performance and the performance is generally better than the cases without codec adaptation. The experiment results show that the proposed RSSI based codec adaptation is indeed effective in combating channel variation.

CODEC ADAPTATION BASED ON NETWORK LAYER INFORMATION

For multimedia streaming over wireless channels, the client is connected to the server through the wireless interface. The server, however, may be located in the backbone Internet and hence the communication path between the client and server may involve both the wireless and wired links. Network layer information that can track the characteristics of the entire communication path can therefore be used to allow for more intelligent codec adaptation at the server.

The quality of multimedia streaming depends on several network layer characteristics, including bandwidth, delay, jitter, and packet loss rate. Related work (Chou and van de Schaar, 2007) has also investigated the length of the transmit queue as indication of network conditions. Among these metrics, path bandwidth plays a pivotal role since it poses an upper bound on the maximum streaming rate. While ideally a higher streaming rate can allow for better streaming quality,

Figure 13. Frame-by-frame YPSNR performance of various codec settings without codec adaptation and the case of optimal RSSI based codec adaptation

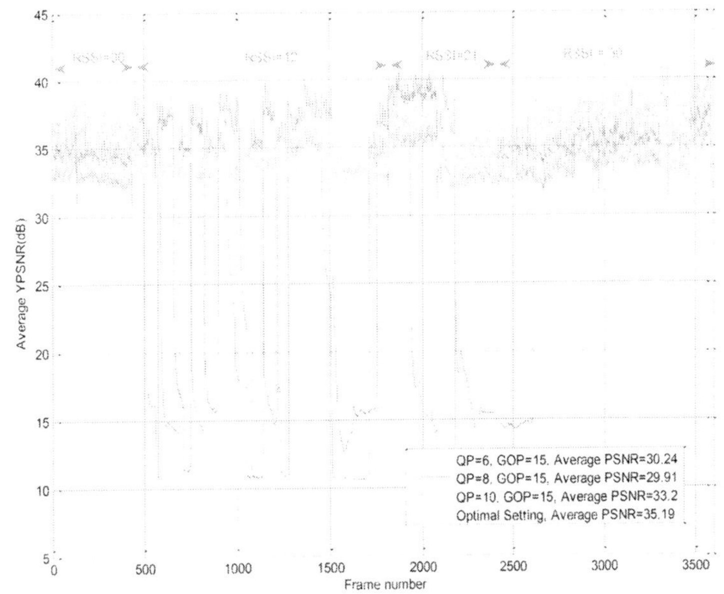

streaming multimedia at a rate more than the path can sustain would inevitably introduce excessive queuing delay and packet loss, thus degrading the streaming quality. In a time-varying network environment, if the clients can periodically feedback the available bandwidth to the server, the latter can perform codec adaptation appropriately for enhanced multimedia streaming.

To estimate the available path bandwidth, an obvious approach is to use RTCP reports as shown in Figure 14 for calculating the number of bytes received within a time interval. While such a "passive" approach can effectively track throughput change during streaming, the problem is its inability to identify extra bandwidth that has not been utilized by the streaming application. To address this problem, "active" bandwidth probing such as the packet train can be used.

In active bandwidth probing using the packet train, a sequence (train) of N packets is sent back to back at the sender along with the streamed multimedia. The receiver then measures the delay dispersion of the received packets and estimates the bottleneck bandwidth using:

$$\text{Bandwidth} = \frac{PacketSize * PacketNumber}{Latency}$$

where *PacketSize* is the size of each packet, *PacketNumber* is the number of received packets, and *Latency* is the delay dispersion between the first and last packets. The number of packets N in each train and the frequency of bursting the packet train can be appropriately determined based on the nature of bandwidth fluctuations and the tolerable overheads (e.g. tolerance of extra bandwidth for bandwidth probing). Figure 15, for example, shows the estimated bandwidth using a train of 5 packets. Note that since RTCP is used along with RTP for transport and control of multimedia stream, the server can use the *Sender Report* (*SR*) as the probing packet and the client can use the *Receiver Report* (*RR*) to feedback the measured bandwidth.

Using the network layer information sent back by the client, the server can perform codec adaptation as the path bandwidth changes. To adapt the streaming rate based on the currently available bandwidth, the server can change QP values in the encoding module.

Figure 14. RTCP reports

Sender Report

```
 0                   1                   2                   3
 0 1 2 3 4 5 6 7 8 9 0 1 2 3 4 5 6 7 8 9 0 1 2 3 4 5 6 7 8 9 0 1
+-+-+-+-+-+-+-+-+-+-+-+-+-+-+-+-+-+-+-+-+-+-+-+-+-+-+-+-+-+-+-+-+
|V=2|P|    RC   |   PT=SR=200   |             length            | header
+-+-+-+-+-+-+-+-+-+-+-+-+-+-+-+-+-+-+-+-+-+-+-+-+-+-+-+-+-+-+-+-+
|                         SSRC of sender                        |
+=+=+=+=+=+=+=+=+=+=+=+=+=+=+=+=+=+=+=+=+=+=+=+=+=+=+=+=+=+=+=+=+
|              NTP timestamp, most significant word             | sender
+-+-+-+-+-+-+-+-+-+-+-+-+-+-+-+-+-+-+-+-+-+-+-+-+-+-+-+-+-+-+-+-+ info
|             NTP timestamp, least significant word             |
+-+-+-+-+-+-+-+-+-+-+-+-+-+-+-+-+-+-+-+-+-+-+-+-+-+-+-+-+-+-+-+-+
|                         RTP timestamp                         |
+-+-+-+-+-+-+-+-+-+-+-+-+-+-+-+-+-+-+-+-+-+-+-+-+-+-+-+-+-+-+-+-+
|                     sender's packet count                     |
+-+-+-+-+-+-+-+-+-+-+-+-+-+-+-+-+-+-+-+-+-+-+-+-+-+-+-+-+-+-+-+-+
|                      sender's octet count                     |
+=+=+=+=+=+=+=+=+=+=+=+=+=+=+=+=+=+=+=+=+=+=+=+=+=+=+=+=+=+=+=+=+
|                 SSRC_1 (SSRC of first source)                 | report
+-+-+-+-+-+-+-+-+-+-+-+-+-+-+-+-+-+-+-+-+-+-+-+-+-+-+-+-+-+-+-+-+ block
| fraction lost |       cumulative number of packets lost       |   1
-+-+-+-+-+-+-+-+-+-+-+-+-+-+-+-+-+-+-+-+-+-+-+-+-+-+-+-+-+-+-+-+
|           extended highest sequence number received           |
+-+-+-+-+-+-+-+-+-+-+-+-+-+-+-+-+-+-+-+-+-+-+-+-+-+-+-+-+-+-+-+-+
|                      interarrival jitter                      |
+-+-+-+-+-+-+-+-+-+-+-+-+-+-+-+-+-+-+-+-+-+-+-+-+-+-+-+-+-+-+-+-+
|                         last SR (LSR)                         |
+-+-+-+-+-+-+-+-+-+-+-+-+-+-+-+-+-+-+-+-+-+-+-+-+-+-+-+-+-+-+-+-+
|                   delay since last SR (DLSR)                  |
+=+=+=+=+=+=+=+=+=+=+=+=+=+=+=+=+=+=+=+=+=+=+=+=+=+=+=+=+=+=+=+=+
```

Receiver Report

```
 0                   1                   2                   3
 0 1 2 3 4 5 6 7 8 9 0 1 2 3 4 5 6 7 8 9 0 1 2 3 4 5 6 7 8 9 0 1
+-+-+-+-+-+-+-+-+-+-+-+-+-+-+-+-+-+-+-+-+-+-+-+-+-+-+-+-+-+-+-+-+
|V=2|P|    RC   |   PT=RR=201   |             length            | header
+-+-+-+-+-+-+-+-+-+-+-+-+-+-+-+-+-+-+-+-+-+-+-+-+-+-+-+-+-+-+-+-+
|                     SSRC of packet sender                     |
+=+=+=+=+=+=+=+=+=+=+=+=+=+=+=+=+=+=+=+=+=+=+=+=+=+=+=+=+=+=+=+=+
|                 SSRC_1 (SSRC of first source)                 | report
+-+-+-+-+-+-+-+-+-+-+-+-+-+-+-+-+-+-+-+-+-+-+-+-+-+-+-+-+-+-+-+-+ block
| fraction lost |       cumulative number of packets lost       |   1
+-+-+-+-+-+-+-+-+-+-+-+-+-+-+-+-+-+-+-+-+-+-+-+-+-+-+-+-+-+-+-+-+
|           extended highest sequence number received           |
+-+-+-+-+-+-+-+-+-+-+-+-+-+-+-+-+-+-+-+-+-+-+-+-+-+-+-+-+-+-+-+-+
|                      interarrival jitter                      |
+-+-+-+-+-+-+-+-+-+-+-+-+-+-+-+-+-+-+-+-+-+-+-+-+-+-+-+-+-+-+-+-+
|                         last SR (LSR)                         |
+-+-+-+-+-+-+-+-+-+-+-+-+-+-+-+-+-+-+-+-+-+-+-+-+-+-+-+-+-+-+-+-+
|                   delay since last SR (DLSR)                  |
+=+=+=+=+=+=+=+=+=+=+=+=+=+=+=+=+=+=+=+=+=+=+=+=+=+=+=+=+=+=+=+=+
```

Figure 15. Bandwidth estimation using the packet train

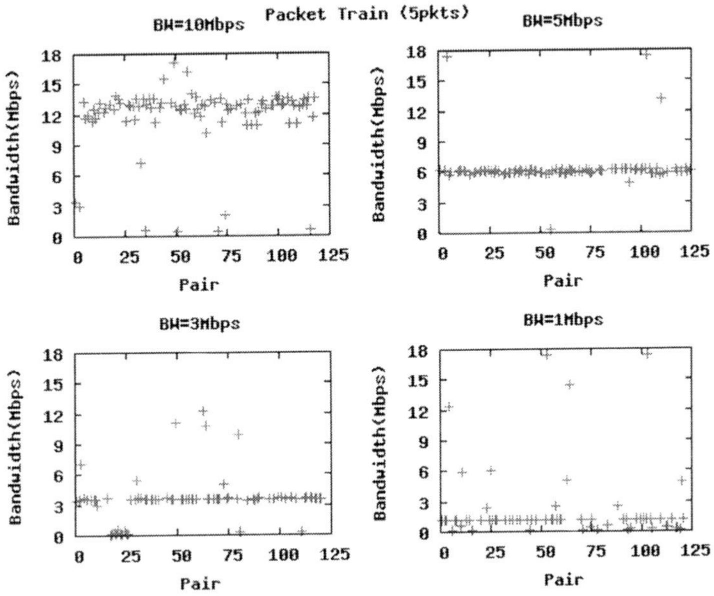

To understand the benefits of network-layer based codec adaptation, observe the result shown in Figure 16. During streaming of an MPEG-4 video stream, the path bandwidth is changed from *5Mbps* to *1Mbps*. As we observe from the figure, for the non-adaptive scheme when the path bandwidth decreases downward to *1Mbps*, a significant portion of packets sent by the server is not received at the client due to buffer overflow and/or late arrivals. The PSNR at the client thus suffers from significant degradation. The adaptive scheme, on the other hand, increases the value of QP when the path bandwidth decreases. While increasing QP value results in coarse resolution and hence decreased PSNR, the overall PSNR is much better than the non-adaptive scheme since the streaming rate is appropriately reduced to fit the available bandwidth. Therefore, network layer information indeed allows the server to perform intelligent codec adaptation for enhanced multimedia streaming (Table 9).

FUTURE RESEARCH DIRECTIONS

For FER based codec adaptation proposed in this chapter, the scheme chooses the codec setting that gives the optimal streaming performance for the induced Markov chain in the steady state. However the amount of time needed for the induced Markov chain to enter steady state depends on the channel. If it is longer than the observation period, this implies that the codec setting would be switched again before the system reaches the steady state. Our scheme could be less effective under such circumstances. Currently we are developing new codec adaptation scheme to optimize the transient performance instead of the steady state performance for the induced Markov chain. Preliminary results show that such scheme is promising for wireless multimedia streaming.

For codec adaptation based on network information, other methods for estimating the bandwidth of the end-to-end path based on various packet-dispersion techniques such as packet train can be used. Evaluation of the proposed codec adaptation schemes can be performed in

Table 9. Intelligent codec adaptation for enhanced multimedia streaming

Time(second)	0~40	40~80	80~120	Average
Bandwidth (Mbps)	5	3	1	3
PSNR (Adaptive)	42.73	36.83	32.46	37.30
PSNR (Non-adaptive)	42.73	36.08	12.33	30.27

a more sophisticated network testbed such as outdoor *Wireless Fidelity* (*WiFi*), WiMAX, or 3G systems to further profile the performance gain of the proposed schemes. Finally, in addition to simulation validation and empirical evaluation, mathematical analysis such as rate-distortion optimization of codec parameters can also be applied for a more complete evaluation framework of codec adaptation for wireless streaming.

CONCLUSION

In this chapter, different cross-layer designs for optimizing video streaming performance have been proposed. For codec adaptation based on the FER at the MAC layer, the major merit resides in the fact that the codec switching thresholds are determined by taking the possible future channel transitions into consideration. The induced Markov model that jointly characterizes the random process of the codec adaptation and the wireless channel variation has been constructed. The expected PSNR analysis has been derived which enables us to find the optimal codec switching thresholds for codec adaptation. Numerical results show that the FER based codec adaptation provides significant gain in PSNR performance.

In addition to the MAC FER, the video streaming performance is also sensitive to RSSI of the PHY channel and the bandwidth of the Internet backbone. Two different codec adaptations based on the physical layer and network layer information thus have also been developed. Experiment results show that the RSSI based codec adaptation is very effective in providing stable PSNR performance in varying wireless environments, whereas the bandwidth based adaptation works great in combating the bandwidth fluctuation of the Internet backbone.

Figure 16. Performance of bandwidth based codec adaptation

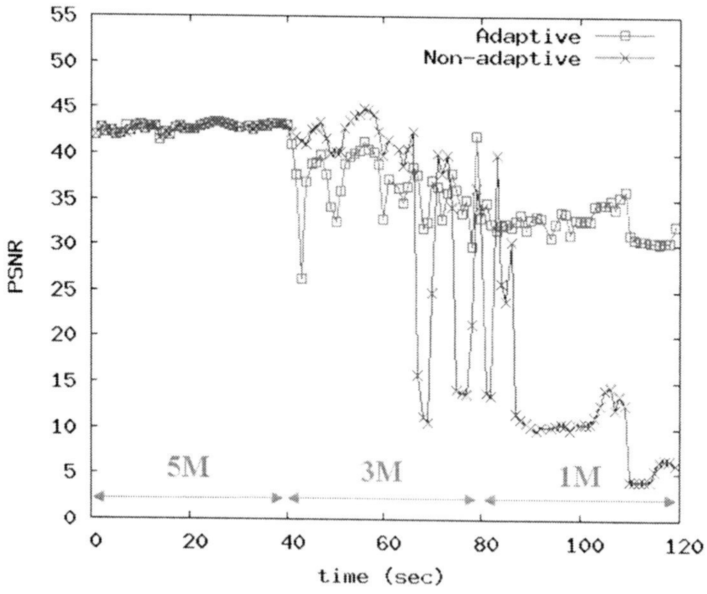

REFERENCES

Andrews, J. G., & Ghosh, A. (2007). *Fundamentals of WiMAX*. Upper Saddle River, NJ, USA: Prentice Hall.

Ar'auz, J., & Krishnamurthy, P. (2003). *Markov modelling of 802.11 channels*. In IEEE 58th Vehicular Technology Conference. (pp. 771-775). Amsterdam, The Netherlands: IEEE Press.

Argyrious, A. (2008). *Distortion-optimized video encoding and streaming in multi-rate wireless LANs*. In IEEE Intl. Conf. Acoustics, Speech and Signal Processing (pp. 2169-2172). Amsterdam, The Netherlands: IEEE Press.

Bucciol, P., Davini, G., Masala, E., Filippi, E., & De Martin, J. C. (2004). *Cross-layer perceptual ARQ for H.264 video streaming over 802.11 wireless networks*. In IEEE Global Telecommunications Conference (pp. 3027-3031). Amsterdam, The Netherlands: IEEE Press.

Chiang, C. C., Wu, Z. H., Shih, C. Y., Yeh, P. C., & Hsieh, H. Y. (2009). *MAC FER based codec adaptation for multimedia streaming over wireless networks*. In IEEE Intl. Symp. a World of Wireless Mobile and Multimedia Networks (pp. 1-9). Amsterdam, The Netherlands: IEEE Press.

Chou, P. A., & van der Schaar, M. (2007). *Multimedia over IP and wireless networks*. Burlington, MA, USA: Academic Press.

Fallah, Y., Koskinen, D., Shahabi, A., Karim, F., & Nasiopoulos, P. (2007). *A cross layer optimization mechanism to improve H.264 video transmission over WLANs*. In IEEE 4th Consumer Communications and Networking Conference (pp. 875-879). Amsterdam, The Netherlands: IEEE Press.

Guha, R. K., & Sarkar, S. (2006). *Characterizing temporal SNR variation in 802.11 networks*. In IEEE Wireless Communication and Networking Conference (pp. 1408-1413). Amsterdam, The Netherlands: IEEE Press.

Haratcherev, I., Taal, J., Langendoen, K., Lagendijk, R., & Sips, H. (2006). Optimized video streaming over 802.11 by cross-layer signaling. *IEEE Communications Magazine*, *44*, 115–121. doi:10.1109/MCOM.2006.1580941

Huynh-Thu, Q., & Ghanbari, M. (2008). Scope of validity of PSNR in image/video quality assessment. *Electronics Letters*, *44*(13), 800–801. doi:10.1049/el:20080522

Khan, S., Peng, Y., Steinbach, B., Sgroi, M., & Kellerer, W. (2006). Application-driven cross-layer optimization for video streaming over wireless networks. *IEEE Communications Magazine*, *44*, 122–130. doi:10.1109/MCOM.2006.1580942

Richardson, I. E. G. (2003). *H.264 and MPEG-4 video compression: Video coding for next-generation multimedia*. Chichester West Sussex, (UK): John Wiley & Sons. doi:10.1002/0470869615

Ruiz, P., & Garcia, E. (2002). *Improving user-perceived QoS in mobile and wireless IP networks using real-time adaptive multimedia applications*. In The 13th IEEE Int. Symp. Personal, Indoor and Mobile Radio Communications (pp. 1467-1471). Amsterdam, The Netherlands: IEEE Press.

Sayantan, C., & Gibson, J. (2007). Payload length and rate adaptation for multimedia communications in wireless LANs. *IEEE Journal on Selected Areas in Communications*, *25*, 796–807. doi:10.1109/JSAC.2007.070515

Van der Schaar, M., Krishnamachari, S., Choi, S., & Xu, X. (2003). Adaptive cross-layer protection strategies for robust scalable video transmission over 802.11 WLANs. *IEEE Journal on Selected Areas in Communications*, *21*(10), 1752–1763. doi:10.1109/JSAC.2003.815231

Zhang, Q., & Kassam, S. A. (1999). Finite-state Markov model for Rayleigh fading channels. *IEEE Transactions on Communications*, *47*, 1688–1692. doi:10.1109/26.803503

ADDTIONAL READING

Argyriou, A. (2008). Error-resilient video encoding and transmission in multi-rate wireless LANs. *IEEE Transactions on Multimedia, 10*(5), 691–700. doi:10.1109/TMM.2008.922776

Chang, Y. C., Salim, M., & Tang, T. F. (2003). Performance evaluation of MPEG-4 visual error resilient tools over a mobile channel. *IEEE Transactions on Consumer Electronics, 49*(1), 6–13. doi:10.1109/TCE.2003.1205449

Li, Z. G., Zhu, C., Ling, N., Yang, X. K., Feng, G. N., Wu, S., & Pan, F. (2003). A unified architecture for real-time video-coding systems. *IEEE Transactions on Circuits and Systems for Video Technology, 13*(6), 472–487. doi:10.1109/TCSVT.2003.813420

Liang, Y. J., Apostolopoulos, J. G., & Girod, B. (2008). Analysis of packet loss for compressed video: Effect of burst losses and correlation between error frames. *IEEE Transactions on Circuits and Systems for Video Technology, 18*(7), 861–874. doi:10.1109/TCSVT.2008.923139

Pyun, J. Y. (2008). Error concealment aware streaming video system over packet based mobile networks. *IEEE Transactions on Consumer Electronics, 54*(4), 1705–1713. doi:10.1109/TCE.2008.4711224

Stockhammer, T., & Hannuksela, M. M. (2005). H.264/AVC video for wireless transmission. *IEEE Wireless Communications, 12*(4), 6–13. doi:10.1109/MWC.2005.1497853

Talluri, R. (1998). Error-resilient video coding in the ISO MPEG-4 standard. *IEEE Communications Magazine, 36*(6), 112–119. doi:10.1109/35.685373

Vass, J., Zhuang, S., & Zhuang, X. (2001). Scalable, error-resilient, and high-performance video communications in mobile wireless environments. *IEEE Transactions on Circuits and Systems for Video Technology, 11*(7), 833–847. doi:10.1109/76.931110

Wang, Y., Wenger, S., Wen, J., & Katsaggelos, A. K. (2000). Error resilient video coding techniques. *IEEE Signal Processing Magazine, 17*(4), 61–82. doi:10.1109/79.855913

Watkinson, J. (2004). *The MPEG Handbook* (2nd ed.). Burlington, MA, USA: Focal Press.

Wu, D., Ci, S., & Wang, H. (2007). Cross-layer optimization for video summary transmission over wireless networks. *IEEE Journal on Selected Areas in Communications, 25*(4), 841–850. doi:10.1109/JSAC.2007.070519

Yan, B., & Ng, K. W. (2002). A survey on the techniques for the transport of MPEG-4 video over wireless networks. *IEEE Transactions on Consumer Electronics, 48*(4), 863–873. doi:10.1109/TCE.2003.1196414

Chapter 5
Reviewing HTTP and RTSP Work in Two Actual Commercial Media Delivery Platforms for Multimedia Services and Mobile Devices

Martin Zimmermann
University of Applied Sciences, Germany

Gilbert Seilheimer
University of Applied Sciences, Germany

ABSTRACT

The media streaming server technology have been developed in the recent past using free software and open source approaches. But commercial approaches are preferred by certain multimedia big enterprises. These enterprises dominate the mobile media delivery business. For this reason we think it is interesting to discover the main characteristics of the principal commercial platforms to deliver mobile video streaming to mobile devices and its main key features: Media and delivery components. We will give our vision on: Real Networks' Helix Media delivery components based on RTSP and Microsoft's IIS Media Services based on HTTP. The aim of this book chapter is to provide detailed insight into these different streaming approaches and how these platforms and protocols aid in building flexible streaming applications for mobile devices. A comparison among these approaches is also provided and also we present the delivery of video streaming to Apple mobile devices in the two commercial approaches.

DOI: 10.4018/978-1-61350-144-3.ch005

INTRODUCTION

The last ten years has seen an explosion of new mobile devices: from handhelds to smart telephones. All of them able to receive *High Definition* (*HD*) video from a video Server (live video or stored video). The techniques the media servers used to deliver information to the mobile devices is name *live*, *streaming* or *progressive* download. Each technique is implemented by a particular protocol and several delivery techniques in the Server. We center our attention in the protocol and Server technology for delivery information to mobile devices.

The most difficult task of streaming audio and video over a network is maintaining a continuous presentation to the user in a highly changeable environment. Buffering is caused when the client runs out of data in the buffer and must wait for more to arrive. The client will always run out of data if the bit rate of the incoming stream exceeds the current available bandwidth. The actual bandwidth at any given point is determined by network conditions and traffic, which are constantly fluctuating. The delivery of information to mobile devices is very hard due to the heterogeneity of these kinds of devices (Macias, E. & Suarez A., 2010). We first analyze the particular characteristics of mobile devices that differentiate them for other kind of devices:

They have reduced resources: display, processor and memory. That means the multimedia contents must be adapted to be delivered to them. Standards like *3rd Generation Partnership Project* (*3GPP*), *Moving Picture Experts Group* (*MPEG*), H.264 and codecs like *Advanced Audio Coding* (*AAC*), *Adaptive Multi-Rate Speech* (*AMR*) and *Flash* can be used for mobile devices. The information must be encapsulated in a determined container (for example, MPEG-2 transport stream container) compatible with the mobile device player.

They naturally use wireless technologies (for example, *Wireless Fidelity* (*WiFi*)) and mobile networks (for example, *Universal Mobile Telecommunications System* (*UMTS*)) to communicate. These technologies suffer from fluctuating bandwidth that may occur due to network congestion or the handover causing several problems to the delivery of media information (Macias, E. et al., 2009).

They use proprietary (Symbian, Apple iOS and Windows Phone ...) or open source operating systems (Android) and players.

The protocols traditionally used for video delivery are the following:

Live video. The video is sent using *User Datagram Protocol* (*UDP*) or *Real Time Protocol* (*RTP*).

Streaming. The *Real Time Streaming Protocol* (*RTSP*) is the first used (Schulzrinne, H. et al., 1998). A recent variation of this protocol is *Real Time Messaging Protocol* (*RTMP*) that is a proprietary solution of Adobe that runs over *Transmission Control Protocol* (*TCP*), for streaming audio, video and data over the Internet, between a Flash player and a Flash media server. A key benefit with RTSP / RTMP is its ability to provide multicast support. They are the only alternatives for streaming scenarios inside a corporate network where delivery to many customers is required without initiating new connections for each user.

In the last three years, adaptive *Hyper Text Transfer Protocol* (*HTTP*) streaming solutions for live and on demand content have appeared on the market mainly because HTTP does not have the same firewall and *Network Address Translator* (*NAT*) traversal problems as traditional streaming. Moreover, HTTP is not likely disallowed by routers and firewall settings. No ports need to be opened that are commonly closed by default. This is one of the main motives why Apple developed the protocol for its mobile internet devices. Important to note here is that YouTube chooses to offer HTTP Live Streams for iPhone visitors.

Because of this, the HTTP live streaming technology suddenly gained a huge installed base, as all the video's on YouTube are also available through this streaming technology.

One of the main advantages of HTTP streaming is the fact that all media objects that need to be transmitted are standard HTTP objects. These objects can be cached by proxy servers without any modification. *Content Distribution Networks* (*CDNs*), such as Amazon's Cloudfront service can distribute HTTP Streams just as they would distribute other HTTP objects such as images or *Hyper Text Markup Language* (*HTML*) pages. E.g., Akamai, one of the world's leading CDNs, has built a specific solution for media delivery to the iPhone platform using HTTP Live Streaming (Akamai, 2010). HTTP based adaptive streaming was used for broadcasting in the 2009 French Open (France Televisions Interactive), at the 2009 Wimbledon Championships (NBC Sports), and in the 2010 Winter Olympic games from Vancouver, British Columbia. A disadvantage of HTTP streaming is it does not provide broadcast.

In contrast to RTSP, the *progressive download* of 3GP audio/video clips over HTTP (Fielding, R. et al., 1999), available in 3GPP since Release 6, does not support client based adaptive streaming and has very limited support for live service. Therefore, 3GPP decided in January 2009 to open a work item to define an HTTP streaming standard that would allow for dynamic bit rate adaptation and improved live services with 3GP files (Gabin, 2010). Adaptive HTTP streaming was introduced in Release 9 of the *Packet Switch Streaming* (*PSS*) standard and was finalized in March 2010. Furthermore, extensions to the 3GPP file format were made related to the timing and synchronization of HTTP streaming content. 3GPP adaptive HTTP streaming provides an entire system for streaming including storage, transport, and media rate adaptation to available link bit rates.

The Apple iPhone provides a client-driven approach for simulating streaming audio and video objects. The iPhone client can use HTTP to download and play an entire QuickTime or MPEG file formatted for progressive download. Each file begins to play shortly after the download starts. Players can seek back and forth through the entire presentation. A segmented media presentation approximates on demand streaming and live broadcasting by encoding each presentation as a series of MPEG-2 transport stream (*.ts*) files. An index file lists the order of, and the URLs to, each segment file. After receiving the index file, the iPhone media client downloads segment files using HTTP, buffering and playing each segment of the media presentation in order. The viewer can seek backward and forward through media segments that have arrived.

Some recent works have been done to improve HTTP adaptive streaming. Fecheyr-Lippens, (2010) showed client implementations of the HTTP streaming protocol are the most efficient than RTSP, flash and Microsoft ones. He and Lee (2005) also showed that serving segments through a generic web server over HTTP is more efficient than using a RTP/RTSP based streaming server (the bottleneck is the network not the Server for HTTP, while the contrary is true for RTSP). Some new research projects propose new mechanisms to improve HTTP delivery. In (Kuschnig, 2010) a client-driven video transmission scheme is introduced which utilizes multiple HTTP/TCP streams. The scheme is largely insensitive to unanticipated packet loss and thereby reduces throughput fluctuations. Since it is based on HTTP, the scheme can easily be deployed in existing network infrastructures. It fosters scalability on the server side by shifting complexity from the server to the clients. Certain features of request-response schemes allow maintaining fairness, despite of using multiple HTTP streams. Ma et al., (2009) introduced an architecture for an HTTP streaming server using paced output. They discuss the scalability advantages of a HTTP streaming server architecture and compare it with traditional HTTP server response times and bandwidth usage. This approach also introduces an intelligent bursting mechanism and considers its effects on end user experience. Wang et al., (2009) suggested using multipath versus single path for TCP based streaming.

A brief comparison of characteristics of the above streaming protocols is presented in Table 1.

The above protocols can be used in several commercial platforms. We overview two of them: Real Networks' Helix Media delivery and Microsoft's *Internet Information System* (*IIS*) Media Services. For each of them we analyze several characteristics that allow them to deliver video to mobile devices (in concrete we will analyze the elements presented in Table 1):

Key features of the media platform to work with mobile clients: We first will briefly analyze the key features of the *Real Networks platform*: *Fast Channel Switching* (*FCS*), Server Side Playlist, custom statistic reporting and the encoder. Then we will analyze these characteristics for the *Microsoft Platform*: Cache/proxy management, Transform Manager, Expression Encoder and configuration of Publishing Points.

Key features of the Media delivery methods for mobile clients: We will first analyze the Real Networks features: Rate Adaptation, streaming live content, streaming pre-recorded content, multicast distribution and splitting technologies, and then the intelligent streaming, fast streaming, advance fast start (and *Forward* (*FF*)/*Rewind* (*RW*)), smooth streaming, bit rate throttling and publishing points streaming feature of Microsoft platform.

The aim of this overview is to compare the above key features and to analyze a comparison of both platforms specifying an example of how to deliver video to iPhone mobile devices for each of them.

In section 2 we present the key features of the both platforms. In Section 3 we present a comparison of the specification of mobile streaming for Apple devices. In Section 4 we present the future research directions analyzing possible evolution of the protocols and platforms and in Section 5 we present the main conclusions of this overview.

MEDIA STREAMING PLATFORMS FOR MOBILE DEVICES

In this Section we shortly review the principal benefits of Real Networks and Microsoft platforms are the following. First we resume those benefits

Table 1. Brief comparison among streaming protocols

	Using		
	HTTP	**RTSP/RTP**	**RTMP**
Protocol Type	Standard	Standard	Proprietary
Multicast support	No	Supported	Supported
Compatibility	Low	Using plug ins	Very high
Client Performance	Good	Good	High Central Processing Unit (CPU) in some platforms
Server performance	Good, network bounded	Good, CPU bounded	Good
Encryption	Built in	Custom module	Server version
Authentication	By protecting key files	Custom module	?
Built in Failover	Yes	Custom module	No
Seek Delay	Segments duration	< 2 s	4-10 s
Ease of Distribution	Easy (existing CDNs)	Custom modules (push, pull, splitting)	?

and then expose them from the server and delivery point of view.

The principal benefits of both platforms are:

Real Networks platform (RealNetworks, 2010b): Standards based multi-tiered architecture (3GPP) and RTSP compliant, support for any 3GPP compliant handset, RTP compliant, interoperable with standards based encoders on the back end, Encoder-server redundancy (forward error correction), Redundancy policies, Media delivery to mobile devices, e.g. Apple devices

Microsoft Platform: Adaptive media streaming (called "Smooth Streaming") media over HTTP, Live Smooth Streaming (for live adaptive streaming of broadcast events), Bit Rate Throttling (meters the speed that media is delivered to a player) Web Playlists (secure sequencing of media content), Transform Manager (acts as an extensible media transform engine for IIS Media Services)

Short Overview of Real Networks Platform Key Features

Real Networks Helix Mobile Server was designed for scalable and flexible delivery of different media types, such as audio, video, animations, images, and text, to a broad range of media players including: RealPlayer, Windows Media Player, Apple Quicktime Player, and 3GPP compliant players.

Real Networks Media Server Key Features

Helix Server supports the following media formats: AAC, AAC+, AMR, MP3, H.264, and MPEG-4.

Some key features of the Helix Media Server are:

Fast Channel Switching specially indicated for mobile clients. The idea is to allow the mobile user to change the channel it is viewing while it is connected to another channel (in this way latency of wireless networks is avoided). This switching capability eliminates the need to set up and tear down a RTSP session when the user selects a new stream. Clients can change channels without experiencing disruption in playback when a new channel begins.

The Server Side Playlists allows Helix to stream a sequence of media objects in a single RTSP session. Playlist management allows the Helix administrator to add, replace, or remove individual playlist files from the server's file system via HTTP commands.

The Helix server and the Client can report statistics that helps the customer can be appropriately billed for the service. The media players can relay back playback statistics to the media server, either at end of a session or on an interval basis. A client statistics template in the server periodically reports statistics from all media player connections. It also provides accurate information about startup time, playback time, re-buffering time, etc., so that the billing is as accurate as possible.

Helix Mobile Producer is the content encoding software component (encoder) for live and on-demand delivery over broadband networks. Encoding advancements include support for multiple source inputs, capability to simultaneously write multiple outputs per input source, enhanced monitoring and advanced features such as multi-rate delivery and on-demand file creation. With multiple outputs encoding, content is simultaneously encoded for desktop computers and mobile devices such as iPhone mobile telephone and iPad tablet. The encoding options include support for:

- Multiple outputs for a single input with a single pass on the source enables high-density simultaneous live and on-demand encoding, maximizes encoding capacity, throughput and supported devices.
- Creation of multi-rate files and live streams.
- Monitoring audio and video input signals and detection of signal acquisition issues such as no signal (color bars or solid color), noisy (static), frozen video signal,

and no audio (silence). This includes also a graphical audio/video preview, encoding and broadcast, statistics and session logging highlight bit rate, frame rate, latency, packets and quality.

- Automation of the encoding process is supported with a command-line interface and *eXtensible Markup Language* (*XML*) job file templates, e.g. for tailored recipes for mobile devices.

Real Networks Media Delivery Key Features

Helix Media delivery supports several: wireless standards: 3GPP, 3GPP2, Internet protocols: IP version 6 (*IPv6*), RTSP, RTSP under HTTP, and *Simple Network Message Protocol* (*SNMP*), and wireless technologies: *Evolution-Data Optimized* (*EV-DO*), *General Packet Radio Service* (*GPRS*), UMTS, *Enhanced Data rates for GSM Evolution* (*EDGE*), *Wideband Code Division Multiple Access* (*WCDMA*), CDMA 2000 and WiFi.

Some key features of the Helix Media delivery are:

Rate Adaptation for mobile technologies networks (live and on demand). The main idea is to sustain the quality of the user experience amidst the service gaps (RealNetworks, 2010c) based on the player's buffer status. Congestion control and rate adaptation are both supplied for live streams (some client dependencies may apply) and client behavior can be configured by type using the user agent profile.

Streaming *true live* video. Helix tries to reduce the *live latency* (difference of time between the delivery at the Server and the visualization in the player) managing packet loss and network jitter is handled in an end-to-end fashion. On certain networks live latency could be as low as 2 s.

Broadcasting methods. A live stream is delivered to one or more media servers, which then replicates the stream to each media player. The players allow the user to click a *mount point* inside a Web page. Before a player is allowed to receive any stream the IP of its mobile device is checked. Then it is checked the location of the media file is secure and finally the user is challenged for identification. If all checks are passed the broadcast starts. Helix implements three broadcasting methods:

Helix Push broadcast. The encoder maintains in addition to its data channel, an accounting channel that allows it to receive broadcast feedback and reestablish a dropped connection. The Figure

Figure 1. Helix push broadcast

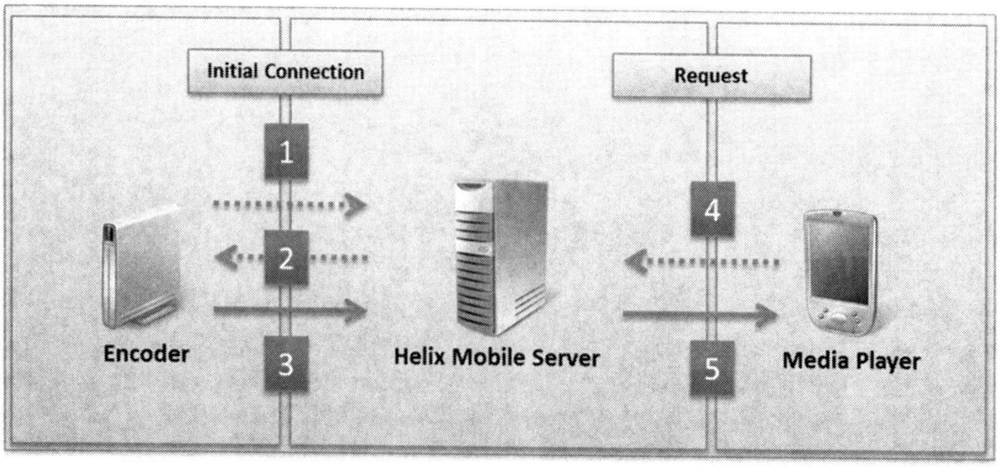

1 illustrates the interaction between an encoder, Helix Mobile Server, and a media player:

The encoder contacts Helix Mobile Server on its HTTP port and establishes a monitoring connection. The encoder sends a user name and password to authenticate its access attempt.

Helix Mobile Server authenticates the encoder's user name and password using its authentication database. It then informs the encoder how to make the broadcast connection, telling it which server ports to use, for example.

The encoder establishes the broadcast stream connection to Helix Mobile Server. It begins to send the encoded packets whether or not any media players have requested the broadcast yet.

A viewer requests the broadcast by clicking a link in a Web page.

Helix Mobile Server streams the broadcast data to the media player.

RTP Push broadcast is the pre-configuration of the Server for receiving MPEG or 3GPP media from standards-compliant, RTP based encoders. In Figure 2 is shown the interaction among server, decoder and player:

The RTP encoder begins to encode a stream and deliver it to the Helix Mobile Server address on a specific data port.

The RTP encoder creates an SDP file containing information about the encoded stream, such as the stream name and the Helix Mobile Server port where the packets are arriving.

The media player requests the live stream.

Helix Mobile Server delivers the replicated live stream to the media player.

Helix Pull broadcast is appropriated for services such as an online TV station that continuously encodes input. An encoder generates broadcast packets continuously but does not deliver the broadcast stream until the Server requests it. The stream is delivered only if a listener requests the stream from a server. If all listeners on the server stop the broadcast, Helix Mobile Server drops the stream, re-requesting it only if new listeners request the broadcast. Because a Helix Pull broadcast does not begin until the first media player requests the broadcast, the first viewer may experience a longer than normal delay as Helix Mobile Server contacts the encoder to acquire the broadcast stream. After the server acquires the stream, however, subsequent viewers experience no additional delay.

The streaming of pre-recorded content is adequate for communicating radio or TV programs based on playlists (name and locations of a set of media objects, title, author, and copyright informa-

Figure 2. RTP push broadcast

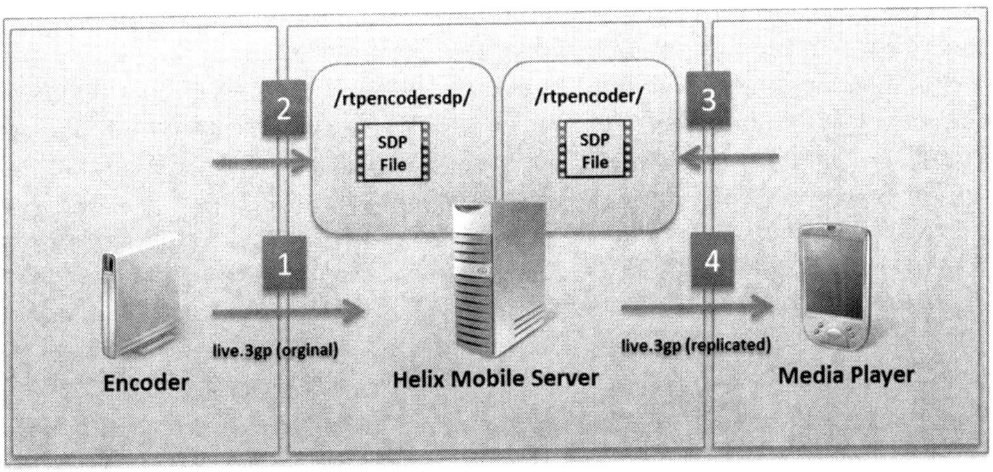

tion, as well as delivery options) that uses SMIL 2.0 (Bulterman D. & Rutlegde, L., 2004).

Multicasting requires a specially configured network (it is more suited for intranets than Internet delivery). The players connect to a stream (live or pre-recorded) rather than to a media server. Three multicasting methods are allowed:

Back Channel that works with all media players that support the RTSP. Each media player maintains a control channel to a media server that lets the server receive user identification and monitoring to track how many players are viewing the multicast.

Scalable multicast allows broadcasting to an unlimited number of media players because the transmission is one way. It works with any RTP based media player that complies with scalable multicasting standards, including the desktop version of Apple's QuickTime Player. In a typical scenario, viewers connect to scalable multicasts by clicking a link to a *Session Description Protocol* (*SDP*) file, i.e. all data is multicast once, and media players do not connect to a media server during the multicast.

Windows Media multicast permits to communicate Window Media encoders delivered media.

Optionally, back-channel and scalable multicasts can be publicized to anyone running a program that listens for the *Session Announcement Protocol* (*SAP*). These applications, such as SDR and ICAST Guide, display a list of all multicasts currently playing. Helix Mobile Server creates the SAP file automatically. Programs that listen for SAP announcements show the title, author, and copyright information encoded into the files.

Push splitting Technologies allows a transmitter (server or decoder) to deliver a live or pre-recorded stream to a receiver (server or decoder) and simultaneously broadcast or multicast the stream to players (only the server; the encoder only can deliver to server). The pull splitting works similarly but the delivery only starts when a receiver demands the stream. A combination of push and pull splitting can be implemented for optimal results in a particular installation.

Short Overview of Microsoft Platform Key Features

Microsoft offers two media delivery strategies: digital media objects can be delivered either from a *Windows Media Server* (*WMS*) or from an IIS 7. Clients use different methods to receive digital media content from WMS and IIS 7.

Microsoft Media Server Key Features

The WMS (Microsoft, 2010b) supports the following media formats: *Windows Media Audio* (*WMA*), *Windows Media Video* (*WMV*), and MP3. It also supports the use of JPEG files as media elements in server side playlists or as banner graphics to be displayed in *Windows Media Player* (*WMP*). WMS provides streaming over a wide range of bandwidths to WMP and to Web browsers that use the Windows Media Player 9 Series ActiveX control or the Microsoft *Silverlight* browser plug-in. If streaming on demand content was indexed during encoding, clients could use seek functions (e.g., FF/RW) on their player to jump directly to any part of the stream without waiting for a file download will finish.

Some key features of the WMS are:

It's *Cache/Proxy* plug in which can be used to configure it either as a cache/proxy server or as a reverse proxy server.

Expression Encoder. It offers enhancement, encoding, and publishing of media objects with *Silverlight*. It converts and compresses multiple video formats to *Society of Motion Picture and Television Engineers* (*SMPTE*) 421M video codec (VC-1) or H.264 for mobile, web, and HD video scenarios. It supports the following video formats: *Audio Video Interleave* (*AVI*), WMV, WMA, *QuickTime Movie* (*MOV*), MPEG (input), and VC-1, H.264 (output).

The IIS 7 supports the following media formats: WMA, WMV, and MP3. It also supports the use of JPEG files as media elements in server side playlists or as banner graphics to be displayed in WMP. WMS provides streaming over a wide range of bandwidths to Windows Media Player and to Web browsers that use the Windows Media Player 9 Series ActiveX control or the Microsoft *Silverlight* browser plug-in. If streaming on demand content was indexed during encoding, clients could use seek functions (e.g., FF/RW) on their player to jump directly to any part of the stream without waiting for a file download will finish.

Some key features of the IIS 7 are:

It's *Transform Manager* enables "watch folder" job submission, queuing, management, integrated media transcoding, and batch-encryption of on-demand audio and video files. Examples of such transforms include transcoding existing content in WMS or MP4 formats (or other encoder-supported formats), or transcoding smooth streaming files encoded with H.264 video and AAC-LC audio codecs into MPEG-2 Transport Stream (MPEG-2 TS) format for delivery to Apple mobile digital devices.

Configuration of *Publishing Points* is a fundamental concept to specify the delivery policy of CDN for media streams, especially for live streams. This configuration is stored as a file as part of IIS 7. This file contains a descriptive title (optional), a live source type (push delivery: they are posted to the publishing point from one or more live sources or pull media delivery), the estimated duration, and a maximum bit rate. The configuration can specify the following roles:

- *Client delivery.* Delivers the live streams from a live source to clients.
- *Server distribution.* Distributes the live streams to other servers (a push configuration) or gets the live streams from a publishing point on another server (a pull configuration).
- *Archiving.* Archives the live streams from a live source for later on demand playing.
- *Server cache.*
- *Client cache.*

Microsoft Media Delivery Key Features

WMS is designed specifically for streaming on-demand and live digital media to clients. It is especially useful for streaming large amounts of data over busy, congested networks and low-bandwidth connections. It sends data over the network only at the speed that is necessary for the client to render it correctly. This helps prevent the network from becoming overloaded and helps maintain system reliability. It uses UDP Resend: An intelligent UDP retransmission scheme that ensures that it only retransmits lost packets that can be sent to a player in time to be played, instead of the blind retransmission scheme employed by TCP. This smart resend feature conserves additional bandwidth on congested networks. It also uses *RTSP with UDP based transport* (*RTSPT*) to take advantage of UDP based streaming efficiencies. On networks that do not support UDP it uses a process called protocol *rollover* to try TCP based streaming, first by using RTSP, and if that does not work, it will uses its own version of HTTP for firewalls that allow Web traffic through port 80.

Some key features of the WMS Media delivery are:

Intelligent Streaming adjusts the bit rate of the content stream to counteract changes in available bandwidth. When a client first connects to a stream, intelligent streaming ensures that the client receives content at the bit rate that is most appropriate for the current bandwidth. As the stream plays, intelligent streaming dynamically and seamlessly adjusts the bit rate of the streaming content as the available bandwidth changes. Intelligent streaming progresses through a series of strategies to modify the bit rate of the stream so that it plays continuously on the client regardless of the current bandwidth. As bandwidth fluctuates between server and client, the server detects the changes and adopts the best strategy. When bandwidth is at its best, the server uses the first strategy. As conditions worsen, the server tries each strategy in the previous list of options in succession until the bit rate is optimized for the current bandwidth. The strategies are as follows:

Selects and delivers the stream with the most appropriate bit rate. The server and client automatically determine the current available bandwidth, and then the server selects and delivers the stream with the most appropriate bit rate. This requires that the content be encoded as a *Multiple Bit Rate* (*MBR*) stream.

Detect changes and switch to a stream with a lower bit rate. During transmission, if the bandwidth decreases, the server automatically detects the change and switches to a stream with a lower bit rate. If bandwidth improves, the server switches to a stream with a higher bit rate, but never higher than the original bit rate. The process of selecting the appropriate stream is transparent to the user.

Degrade image quality to avoid buffering. If the available bandwidth cannot support streaming video any longer, the client and server degrade image quality to avoid buffering. This method of decreasing the bit rate to accommodate the current bandwidth is called stream thinning. The server decreases the video frame rate first. If the bit rate is still too high, the server stops sending video frames.

Maintain a continuous audio stream. After the server stops sending video frames, it uses intelligent streaming to try to maintain a continuous audio stream. If audio quality starts to degrade, the client reconstructs parts of the stream to preserve quality.

Fast Streaming allow for a combination of streaming, downloading, and caching to improve media delivery:

Fast Cache. Enables Windows Media Services to use any additional bandwidth to send additional data to the player buffer, allowing the player to better withstand network bandwidth fluctuations.

Fast Start. Enables the beginning part of the content to be quickly downloaded to a player at the maximum available bandwidth, reducing the time required to fill the player's buffer requirement and the amount of time a user has to wait (also known as latency) before starting to receive the stream.

Fast Recovery. Enables the server to provide an uninterrupted viewing experience to users receiving content over high latency network connections, such as wireless and satellite networks, by using forward error correction to greatly reduce packet corruption and interruption.

Fast Reconnect. Enables the server to automatically restore client connections, including encoders, distribution servers, and players that are lost during a broadcast because of network conditions.

Advanced Fast Start. Adds to Fast Start capabilities by allowing the player to start to play content as soon as its buffer receives a minimum amount of data, further reducing the amount of latency before the user starts to receive the stream.

Fast Start delivers an instant-on playback experience by eliminating buffering time. When a player connects to a stream, the first few seconds of data are sent using the maximum available bandwidth so that playback can begin as soon as possible. *Advanced Fast Start* adds to these capabilities by allowing WMP to begin playing content as soon as its buffer receives a minimum amount of data, further reducing the amount of time a user must wait to begin receiving the stream. Improved FF and RW functionality for the video portion of content stabilizes network bandwidth availability by using separate files for each FF/RW speed. This results in a fixed bandwidth requirement per client, regardless of playback speed, and greatly smoothes the FF/RW experience. Potential server performance bottlenecks are reduced because the server must read less presentation data from the source content disk, while delivering a seamless experience to clients.

When a client requests a file from IIS 7, all available network bandwidth will be used to download the data to the client as quickly as possible.

Some key features of the IIS 7 Media delivery are:

Smooth Streaming is the Microsoft implementation of adaptive streaming technology, which is a form of Web based media content delivery that uses standard HTTP. Instead of delivering media as full file downloads, or as persistent (and thus stateful) streams, the content is delivered to clients as a series of MPEG-4 (MP4) fragments that can be cached at edge servers. Smooth Streaming compatible clients use special heuristics to dynamically monitor current network and local device conditions and seamlessly switch the video quality of the Smooth Streaming presentation that they receive. As clients play the fragments, network conditions may change (e.g., bandwidth may decrease) or video processing may be impacted by other applications that are running. Clients can immediately request that the next fragment come from a stream that is encoded at a different bit rate to accommodate the changing conditions. This enables clients to play the media without stuttering or freezing.

Bit Rate Throttling offers bandwidth-management and client-connection capabilities similar to Fast Start capability of WMS. Typically, this throttle rate is slightly greater than the encoded bit rate of the file, so that the client renders the throttled data as it is downloaded. When multiple clients request data from the Web server, they will consume the data at the throttled rate, instead of all at once. This frees additional bandwidth for servicing additional client requests. By throttling the content download rate so that clients receive just enough content to render it, content providers do not incur costs related to the bandwidth that is wasted in the unviewed parts of downloaded files. The Bit Rate Throttling extension offers advanced settings that allow more control of the cumulative network throughput.

Publishing Points streaming. Live Smooth Streaming encoders generate multiple streams, each containing MPEG-4 fragments encoded at various bit rates and delivers them directly to Live Smooth Streaming publishing points. When a client makes a request for a Live Smooth Streaming presentation, the publishing point will delivers an XML manifest that describes the available quality levels. The Live Smooth Streaming publishing point retrieves virtual fragments from the most appropriate stream in the live broadcast for the conditions at any given time in order to deliver the best content possible to the client. The client uses the Smooth Streaming heuristics to determine the best stream from which to request a media fragment at that instant. The Smooth Streaming heuristics continuously monitor the environment and at frequent intervals (typically 2 seconds), request the next contiguous MP4 fragment in the live stream (if available bandwidth and video-processing power have not changed) or a contiguous MP4 fragment from a different stream (if available bandwidth and video-processing power have changed). This process continues until the client stops playback or the end of the stream is reached, ensuring that the client has the best possible viewing experience without any buffering or stuttering. Publishing points are assigned various states to indicate the current status of media delivery. The Starting state means that the publishing point is ready to receive live streams. When the live source connects to the publishing point and begins pushing content to it, the Starting state changes to Started, meaning that the publishing point is receiving the live streams. In addition, the Stopped status changes to Started for Archive (the publishing point is archiving the live streams), Fragment, (the publishing point is delivering the live streaming fragments to clients), and Stream (the publishing point is distributing the streams to other servers on the network), if these publishing point roles are enabled. Finally let us remark that Publishing Points can be monitored.

MEDIA DELIVERY SUPPORT FOR APPLE IPHONE MOBILE DEVICES

The methods used for accessing media are very different than delivering content using RTSP or RTMP. Apple mobile devices use HTTP and those with *iPhone Operating system* (*iOS*) version 3.0 or greater supports the ability to switch to alternative versions of the content (if available) based on changing network throughput or content availability. They also enable the ability to request, decode and playback of AES-128 encrypted content which is used when a content owner would like to ensure their media is protected.

In this Section we present a case study that consists in to specify the delivery of multimedia information for iPhone mobile devices using the both platforms: Real Networks and Microsoft. The aim is to provide a new comparison between the both platforms that clarify the way they use the streaming protocols.

Real Networks Delivery to iPhone Mobile Devices

The Helix Media Delivery Platform supports progressive download and segmented download (or HTTP Live Streaming) for iOS devices. In progressive download, as the media file begins to progressively download, playback is initiated allowing an almost immediate viewing of the content; in the background, the media player continues to download the rest of the content. During such a playback process, users are able to seek back and forth through the entire media file. If the viewer seeks forward to a point in the timeline that has not yet downloaded, the media player will pause playback until the data arrives. In segmented download, live streaming is achieved in a similar manner as with on demand files though requires a live source. By taking the output from a live RTP encoder and producing a series of 10 second media files encapsulated in a MPEG-2 TS (*.ts*) container, these files are then delivered to the Apple. After creating the MPEG-2 TS segments, an index file (*.m3u8*) is created (or updated if the content is live) that lists the order of and URLs to each segmented media file. This index file is also known as a playlist and will also include some additional information about the content itself: Segment length, Decryption key and Multi-rate bandwidth information.

Real Networks recommends that all on demand content be delivered as a HTTP Live Stream. This also allows additional security over the content as the full media file is not available as a download for users to view offline and allows media file encryption for additional security. The Helix Universal Server offers further flexibility by allowing content to be started from a designated start point. This enables the content owner to choose where to view from, for example content previews. In addition Helix Universal Server incorporates bookmarking enabling a viewer to stop playback and then resume at a later time without having to start viewing from the beginning. The Helix bookmarking capability operates across all Helix Universal Server media clients from PCs to mobile devices.

The media formats for on demand video vary depending on the kind of download:

Progressive Download:

File. 3GPP (.3gp), MPEG-4 (.mp4,.m4v), MP3 (.mp3), QuickTime (.mov)

Video Codecs. MPEG-4, H.263, H.264

Audio Codecs. AAC, AAC+, AMR-NB, AMR-WB, MP3

HTTP Live Streaming:

File. 3GPP (.3gp), MPEG-4 (.mp4,.m4v,.f4v), QuickTime (.mov)

Video. H.264 Baseline Profile Level 3.0 (iPhone/iPod), Main Profile Level 3.1 (iPad)

Audio. HE-AAC or AAC-LC up to 48 kHz, stereo audio

Media files using these codecs are delivered by the Helix Universal Server to an Apple OS client either as a Progressive Download file or as a HTTP Live Stream without the content creator needing to do anything except place the media file in the relevant location on the Helix delivery server for the iPhone to access (step 1 in Figure 7).

If delivery on-demand files as a HTTP Live Stream are required, the Helix Server has the ability to on-the-fly segment (and encrypt if required for security) a correctly formatted on-demand media file into the components required by the Apple device (the TS media segments and the m3u8 playlist file as seen in step 3 in Figure 2). This process happens instantaneously on the first request by an Apple device (step 2) and the playlist is then delivered back to the device (step 4).

Once the device reads the playlist and works out the bandwidth available (if required for a multi-rate media object) the client then requests the relevant.TS media segment for playback to start (step 5). A Helix Server can also cache the resultant segmented streaming components to deliver to future requests for the same media object. This process is completely automated and as an additional benefit for both the content owner and the server administrator it means content management is simplified to the original on-demand file only.

The media formats for live video must be encoded using the formats supported by the HTTP Live Streaming specification and the Apple devices:

Video. H.264 Baseline Profile Level 3.0 (iPhone/iPod), Main Profile Level 3.1 (iPad)

Audio. HE-AAC or AAC-LC up to 48 kHz, stereo audio OR MP3 (MPEG-1 Audio Layer 3) 8 kHz to 48 kHz, stereo audio.

Live broadcasts using these codecs are delivered by the Helix Server to an Apple client as a HTTP Live Stream. A Helix Server will perform the following process:

- Accept the live broadcast feed from the encoder (step 1 in Figure 3)
- Create (and encrypt if required) the relevant TS media chunks and m3u8 playlist files (step 2)
- Generate an archive file if required (step 3)

When an Apple device requests the live broadcast (step 4), the Helix Server will redirect the client to the relevant playlist (step 5) where it will then make the request for the actual media on the Helix Server (step 6) and start playback of the live feed.

Figure 3. Helix segmentation of on-demand content

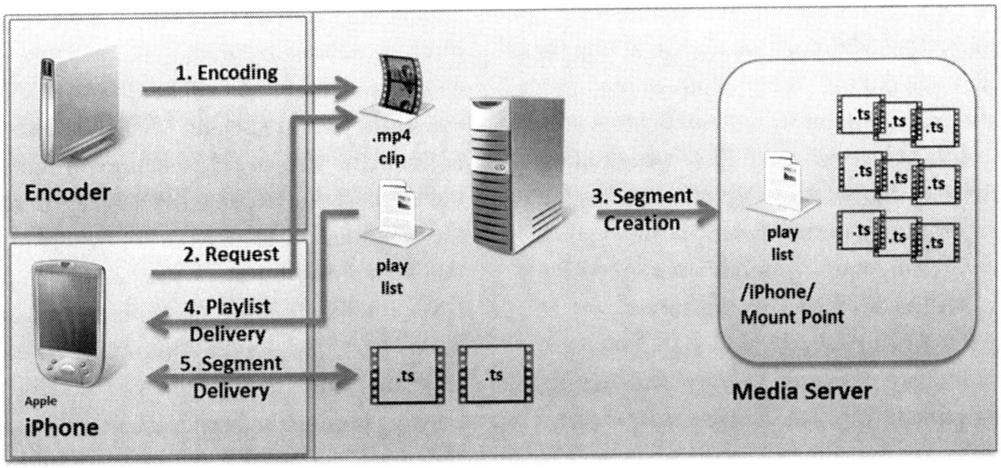

Microsoft Delivery to iPhone Mobile Devices

IIS Media Services 4.0 contains an option that enables delivery of Smooth Streaming content to Apple mobile devices using the Apple Live HTTP streaming format. Currently this new feature is supported for IIS Live Smooth Streaming, which transcodes the incoming Smooth Streaming fragmented-MP4 fragments into MPEG-2 TS segments and generates the appropriate *.m3u8* index for Apple mobile digital devices. In more detail, IIS Live Smooth Streaming converts incoming fragmented-MP4 streams that contain H.264 (AVC) video and AAC audio into one *.ts* file per stream in the server archive and then chunks the file on demand when a client requests it. This simplifies archive management, file movement, and assets management downstream. To stream content to an Apple mobile digital device, a Live Smooth Streaming capable encoder is required that supports encoding content using the H.264 video with AAC low complexity (AAC-LC) audio codecs.

The following video and audio formats are supported by the Apple HTTP Live Streaming feature in IIS Live Smooth Streaming 4.0:

Video. H.264 Baseline Profile Level 3.0 (iPhone / iPod / iPod touch), Main Profile Level 3.1 (iPad)

Audio. AAC-LC up to 48 kHz, stereo audio.

Video only stream cannot be transcoded.

To implement IIS Live Smooth Streaming for mobile devices the following steps are required:

Creation and configuration of one or more publishing points. It involves the following activities:

Define the output to Apple mobile devices, i.e. configuration of the publishing point to allow transcoding of incoming Smooth Streaming fragments to MPEG-2 TS files on the server

Define the size of the MPEG-2 TS fragments (chunks) that are delivered to the mobile device must be defined. Apple recommends 10 seconds (it should be not lower than 2 seconds). However, lowering this value will decrease the amount of buffering latency on the client site

Define the maximum bit rate (to the value of the highest bit rate stream for the MPEG-2 TS conversion). The default value is 1600 Kbps, which is also the maximum recommended bit rate for delivery to Apple iPhone and Apple iPad. Lowering this value reduces the number of streams that are transcoded to MPEG-2 TS, raising it allows more streams to be converted

Establishing a connection between an encoder and a web server

To establish a live stream Microsoft Expression Encoder 4 must enable Live Smooth Streaming and supports the bit rates and encoding requirements for the Apple HTTP Live Streaming for the target Apple mobile digital devices over a particular network. It provides presets for iPad and iPhone for both cellular and WiFi networks. The H.264 IIS Smooth Streaming iPhone WiFi preset uses the Apple recommended settings for delivering content to an iPhone device over a WiFi network. The profile defines the following properties:

- H.264 Baseline, Level 3.0
- Three different bit rates: 200 Kbps, 400 Kbps, and 700 Kbps
- 1-pass CBR encode with no B-frames

Figure 4 illustrates the encoding and media delivery process to Apple mobile devices. An encoder generates a live stream at a single rate or multiple rates using the supported codecs. As with an on demand media object the live stream can use a file format such as MPEG-4. The encoder delivers the stream (H.264 Smooth Stream) to a publishing point (step 1). To enable media delivery, the location, i.e. the URL of the publishing point that was created previously, has to be specified. IIS immediately begins to generate transport segments (*.ts*). The server continually deletes older segments and updates the playlist file so that viewers joining the broadcast start with the latest segment. The process for segmenting broadcast

Figure 4. Live broadcast to Apple device

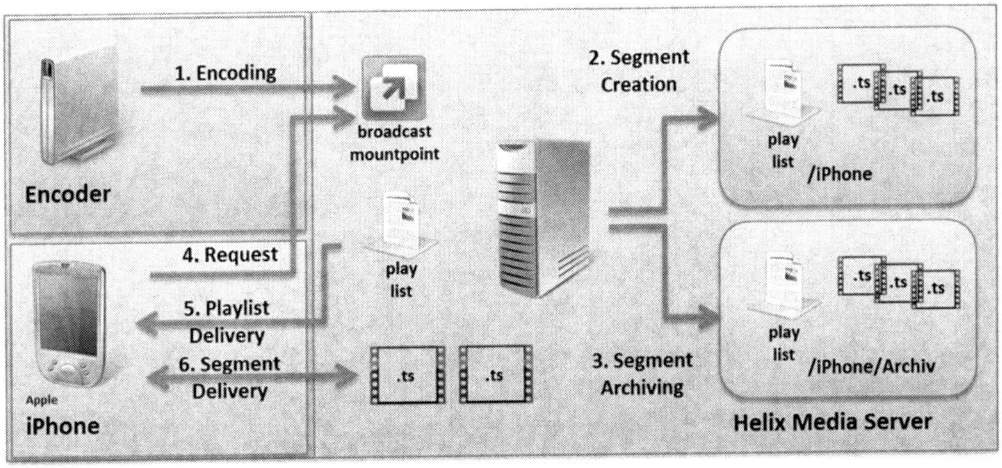

content is similar to that for on demand content, except that the playlist file is continually updated as new segments are generated. The watch folder chains tasks input and output together and uses an encoder to generate H.264 content and then transcode that output into MPEG-2 TS files for use with Apple HTTP streaming. This task also copies the output to the local Web server using two separate folder paths, one for Smooth Streaming, and one for MPEG-2 TS output files (step 2).

Live Smooth Streaming in IIS Media Services 4.0 currently does not support B-frames in H.264-encoded streams. Apple also recommends that B-frames should not be used when targeting iPhone and iPod devices. While the iPhone 3GS can play content that is at a level higher than Baseline 3.0, it is not recommended if older iPhone devices and iPod touch devices are used as clients.

The HTML page in Figure 5 illustrates a simple web page based on the HTML 5 video tag (Pilgrin, M., (2010). A new manifest request URL is also required to get access to the*.m3u8* file from the IIS Media Server. The important part is the section in the parenthesis, which specifies that the format of the manifest (*mpeg2ts*). Other HTML 5 video tags that support attributes such

Figure 5. Live broadcast to Apple devices based on IIS media services

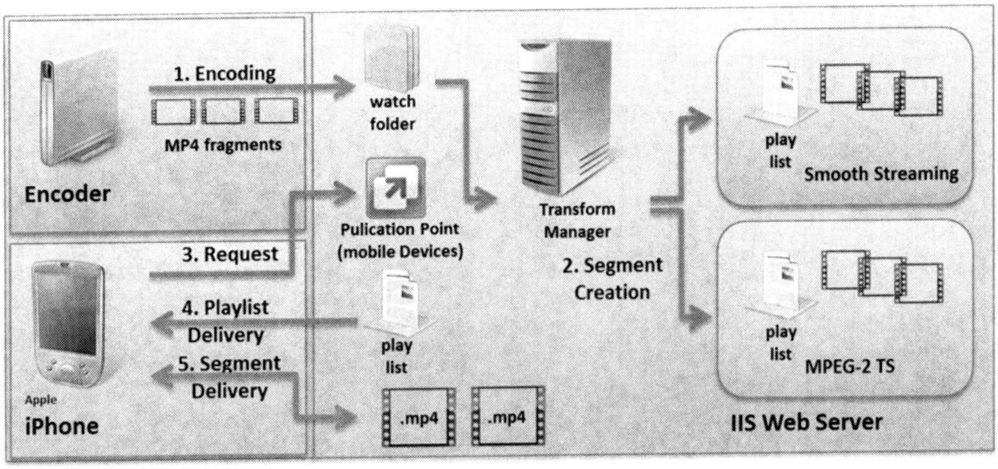

Figure 6. Live broadcast to Apple devices

```
<html>
   <head><title>iPhone live streaming page</title></head>
   <body>
      <h1>Live stream.</h1>
      <video width="640"
         height="480"
         src="smoothApple.isml/manifest(format=m3u8-aapl)"
         poster="MyPosterFrame.png"
         autoplay="true"
         controls="true" >Live
      </video>
   </body>
</html>
```

as autoplay, controls, and a poster frame can be added optionally.

Figure 6 shows the default HTML 5 video tag on an Apple device, which has a simple Play button on (Safari Web browser). When the Play button on the HTML 5 video element is clicked, the device will open the QuickTime player and begin loading the Live Smooth Stream (right part of the Figure 7).

FUTURE RESEARCH DIRECTIONS

One of the trends that have emerged in commercial media delivery platforms over the past years is that there has been a steady shift away from traditional media streaming protocols (RTSP, *Microsoft Media Streaming* (*MMS*), RTMP ...) back to the plain HTTP download, e.g. community video sharing sites that support only HTTP progressive download for media delivery. There are several strong reasons for this industry trend: Web download services have traditionally been less expensive than media streaming services offered by CDNs

Figure 7. Browser and player window on the mobile device

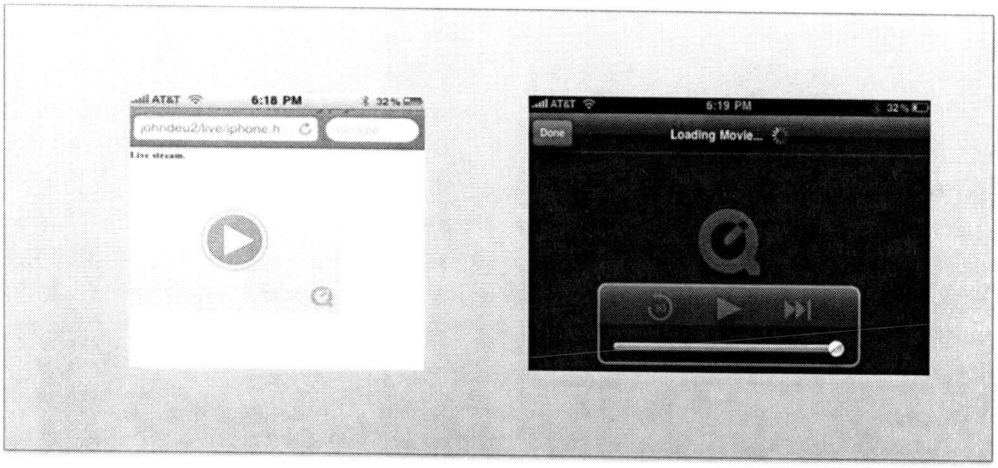

and hosting providers, media protocols often have difficulty getting around firewalls and routers because they are commonly based on UDP sockets over unusual port numbers. HTTP based media delivery has no such problems because firewalls and routers know how to pass HTTP downloads through port 80, and HTTP media delivery does not require special proxies or caches.

Future media delivery platforms have to solve the problem of dynamic adaptation of multimedia services in heterogeneous device contexts, e.g. alternative layouts within one service based on system properties, e.g. bit rate and screen size. In order to optimize the content based on the mobile device type, it is necessary to identify the type of device and browser. The necessary information includes screen resolution, markup language(s) supported, CSS capability... It will be necessary to (automatically) optimize content for the specific type of mobile device. This involves designing the mobile page flow and applying the appropriate content markup tags, such as *eXtensible Hypertext Markup Language* (*XHTML*), *XHTML Mobile Profile* (*XHTML-MP*), HTML, *Wireless Markup Language* (*WML*)... Appropriate stylesheets also need to be applied depending on the type of mobile device and its *Cascade Style Sheet* (*CSS*) support. When considering the range of mobile devices that must be supported in the future, it will be of utmost important to analyze the type of customer depending on the target demographic for the content and the type of application. Only supporting a limited range of high-end devices can severely restrict the market for media streaming services and potentially create customer resentment where their device is not fully supported.

New technology advancements are allowing innovative media applications and services to be delivered to mobile devices. Many of these applications and services will require security in protecting the confidentiality of media content and ensuring its authenticity. There is a demand from business scenarios which require secure media streaming mechanisms, i.e. support of media aware encryption, sender authentication and secure transcoding. E.g., Microsoft announced that the Apple HTTP streaming support will expand to include AES encryption.

CONCLUSION

Both media delivery platforms namely Real Networks's Helix platform and Microsoft's Media Services can easily and compliantly support the delivery of content to Apple and other mobile devices. The solution is fully interoperable with 3rd (third) party encoding components, which create standards compliant content compatible with mobile devices.

RealNetworks' approach is based on traditional streaming technologies, i.e. a streaming media system based on Helix media technologies typically consists of a system running a media encoder, one or more server running Helix Mobile Media Server, and a number of media players. The latest version supports native delivery of Flash content using the RTMP. It can also stream MPEG-4 or 3GPP content to Flash clients. Additionally, Helix Mobile Server can automatically segment on demand as well as live broadcast contents into MPEG-2 TS for delivery to Apple mobile devices. One of the key advantages of such traditional media streaming delivery platforms is the support of multicast distribution including scalable multicasting and back channel multicasting. Additionally, advanced functionality can be offered, for e.g., session management or advanced rate control strategies. Helix Mobile server support monitor, control, log and charge for the real time usage of media objects through the integration of additional middleware components. The Helix Rate Control subsystem is responsible for modulating the media transmission rate with a specific goal of sending data at the highest possible rate without over utilizing the capacity of the underlying network.

Microsoft offers basically two media delivery strategies: one can deliver digital media audio and video files and live streams either from a WMS (a server that is running Windows Media Services) or from a Web server (running IIS 7). Smooth Streaming is Microsoft's implementation of HTTP based adaptive streaming, which is a hybrid media delivery method. It acts like streaming, but is based on HTTP progressive download. HTTP downloads are performed in a series of small chunks, allowing the media to be easily and cheaply cached along the edge of the network, closer to clients. Providing multiple encoded bit rates of the same media source also allows clients to seamlessly and dynamically switch between bit rates depending on network conditions and CPU power. In a typical adaptive streaming implementation, video and audio objects are cut into many short segments (called "chunks" typically 2-to-4-s long) and encoded to the desired delivery format. In case of video objects, this typically means that each chunk is cut along video Group of Pictures boundaries (i.e. each chunk starts with a key frame) and has no dependencies on past or future chunks. This allows each chunk to later be decoded independently of other chunks. Clients request the chunks from a Web server in a linear fashion and downloads them using traditional HTTP progressive download. As the chunks are downloaded to the client, the client plays back the sequence of chunks in linear order.

Adaptive HTTP Streaming is a young technology that proves its value in embedded devices and is noteworthy for its simplicity. Deployment and distribution are remarkably easy and might open up video streaming to a much wider audience. Standardization and further adaption by other vendors and the open source community could boost the importance of this protocol in the upcoming HTML 5 specification. RTSP based media streaming is a traditional approach to deliver digital media content to large numbers of concurrent users, i.e. media servers send data at the same bit rate as the content, leaving more bandwidth available for servicing concurrent client requests for content and resulting in better quality audio and video for connected clients.

With its built-in bandwidth-management capabilities, a RTSP based media server is a traditional approach to deliver digital media content to large numbers of concurrent users. Such a media server sends data at the same bit rate as the content, leaving more bandwidth available for servicing concurrent client requests for content and resulting in better quality audio and video for connected clients. There is typically a delay between the time the stream is received by a player and the point at which it starts to play because the player must first buffer some data in case there are delays or gaps in the stream. This buffer allows the media to continue playing uninterrupted, even during periods of high network congestion.

RTMP is a good solution popularized by YouTube for embedding videos in a webpage when client compatibility plays a key role. Because of this broad compatibility it is also a good candidate as a fallback technology for any of the other two protocols. Nevertheless, the proprietary nature and the lack of adoption in the mobile market are two important disadvantages. RTSP remains a flexible solution for latency sensitive live or on-demand streaming. Tunneling over HTTP, as implemented in several streaming servers, has alleviated its biggest issue. The protocol allows streaming various media formats to a wide variety of (mobile) devices but is more of a hassle to setup and distribute.

REFERENCES

Akamai (2010). *Akamai HD for iPhone*, Retrieved December 04, 2010, from http:// www.akamai.com/ html/ solutions/ iphone.html

Bulterman, D., & Rutlegde, L. (2004). *SMIL 2.0: Flexible multimedia for Web, mobile devices and daisy talking books*. Heidelberg, Germany: Springer Verlag; X.Media.Publishing.

Fecheyr-Lippens, A. (2010). *A Review of HTTP Live Streaming.* Retrieved December 04, 2010, from http:// andrewsblog.org/ a_review_of_ http_live_ streaming.pdf

Fielding, R., Gettys, J., Mogul, J., Frystyk, H., Masinter, L., Leach, P., & Benners-Lee, T. (2009). Hypertext Transfer Protocol -- HTTP/1.1. [*Standard Track, IETF.*]. *Request For Comment, 2616*, 1999.

Gabin, F., Kampmann, M., Lohmar, T. & Priddle, C. (2010). *3GPP mobile multimedia streaming standards*, IEEE Signal Processing Magazine, 138, November 2010.

Kuschnig, R., Kofler, I., & Hellwagner, H. (2010). *Improving Internet video streaming performance by parallel TCP based request-response streams*, 7th IEEE Consumer Communications and Networking Conference (CCNC).

Lee, Y.-J., Min, O.-G., & Kim, H.-Y. (2005). Performance evaluation technique of the RTSP based streaming server, *Proceedings of the 4th Annual ACIS International Conference on Computer and Information Science (ICIS).*

Ma, K. J., Bartos, R., & Bhatia, S. (2009). *Scalability of HTTP pacing with intelligent bursting,* IEEE International Conference on Multimedia and Expo (pp. 798 – 801).

Macias, E. & Suarez, A. (2009). Proactive estimation of the video streaming reception quality in WiFi using a cross-layer technique, *IEEE Latin America transactions magazine,* 7(3), 383-389.

Macias, E., & Suárez, A. (2010). Video streaming based services over 4G networks: Challenges and solutions, in A. Mobasher, & Tofighbakhsh (Eds.), *4th generation wireless networks: Applications and innovations* (pp. 494-523). Hershey, PA: IGI Global.

Microsoft Corporation. (2010a). *Live and on-demand video with Silverlight and IIS Smooth Streaming*, Microsoft Corporation. Retrieved on August 02, 2010, from http:// download.microsoft.com/ download/ 3/ A/ 4/ 3A4A066C-6543-4BC1-A8BA965D0E55F296/ Live_and_On_Demand_Video_ with_ Silverlight_ and%20IIS_ Smooth_ Streaming_ FINAL.pdf

Microsoft Corporation. (2010b). *Windows Media services*. Retrieved August 02, 2010, from http:// www.microsoft.com/ windows/ windowsmedia/ forpros/ server/ version.aspx

Nelson & David. M. (2009). *Smooth streaming deployment guide*, Microsoft Corporation. Retrieved August 02, 2010, from download.microsoft.com/ download/ 4/ 1/ 0/ 4105D7BE- 1E3F- 4AEC-86B7- 76207DB829B8/ Smooth_ Streaming_ Deployment_ Guide.docx.

Pilgrin, M. (2010). *HMTL 5: Up and Running.* Sebastopol, CA, USA: O'Relily Media Inc.

RealNetworks, Inc. (2010a). *Delivering content to Apple iPhone, iPod Touch and iPad using RealNetworks Helix Solutions*. Retrieved August 02, 2010, from www.realnetworks.com/ uploaded-Files/ products-services/ Whitepaper-iPhone.pdf

RealNetworks, Inc. (2010b). *HELIX mobile systems integration guide*. Retrieved August 02, 2010, from docs.real.com/ docs/ server14/ mobile/ HelixMobileServerIntegrationGuide_ v14.pdf

RealNetworks, Inc. (2010c). *Minimize rebuffering and maximize media encoding rate delivered: Helix mobile server rate control for mobile networks*, Retrieved August 02, 2010, from http://www.realnetworks.com/ uploadedFiles/ products-services/ Whitepaper-Enhanced- Rate-Control.pdf

Schulzrinne, H., Rao, A., & Lanphier, R. (1998). Real Time Streaming Protocol (RTSP) [*Standard Track, IETF.*]. *Request For Comment, 2326*, 1998.

Wang, B., Wei, W. & Guo, Z. (2009). Multipath live streaming via TCP: Scheme, performance and benefits, *ACM Transactions on Multimedia Computing, Communications and Applications*, *5*(3), Article 25.

ADDITIONAL READING

Arregoces, M., & Portolani, M. (2003). *Data center fundamentals*. USA: Cisco Press (http://www.ciscopress.com).

Cameron, R., & Ghosh, J. (2010). *Silverlight recipes: A problem-solution approach, (2nd Edition)*. USA: Apress (http://www.apress.com).

Caputo, A. (2010). *Digital video surveillance and security*. UK: Butterworth-Heinemann (http://www.elsevierdirect.com).

Follansbee, J. (2006). *Hands-on guide to streaming media: an introduction to delivering on-demand media, (2nd Edition)*. USA: Focal Press (http://www.focalpress.com/).

Furht, B. (2009). *Encyclopedia of multimedia, (2nd Edition)*. USA: Springer (http://www.springer.com).

Huber, D. M., & Runstein, R. E. (2009). *Modern recording techniques, (7th Edition)*. USA: Focal Press (http://www.focalpress.com/).

Künkel, T., & Kunkel, T. (2002). *Streaming media: Technologies, standards, applications*. USA: (http://eu.wiley.com).

Minasi, M., Gibson, D., Finn, A., Henry, W., & Hynes, B. (2010). *Mastering Windows server 2008 R2*. USA: Sybex (http://eu.wiley.com).

Rayburn, D., & Hoch, M. (2005). *The business of streaming and digital media*. USA: Focal Press (http://www.focalpress.com/).

Ryu, H., & Parsons, D. (2009). *Innovative mobile learning: techniques and technologies*. Hershey, PA: IGI Global.

Simpson, W. (2008). *Video over IP: IPTV, Internet video, H.264, P2P, web RV, and streaming: A complete guide to understanding the technology*. USA: Focal Press (http://www.focalpress.com/).

Waggoner, B. (2009). *Compression for great video and audio, (2nd Edition): Master tips and common sense (DV Expert)*. USA: Focal Press (http://www.focalpress.com/).

Walker, M., Turnbull, R., & Sim, N. (2007). Future mobile devices — An overview of emerging device trends, and the impact on future converged services. *BT Technology Journal*, *25*(2), 120–125. doi:10.1007/s10550-007-0035-3

Chapter 6
Multimedia Services Offer Mixing Telco and Internet Assets

Luis Angel Galindo
Telefonica, Spain

Joaquín Salvachúa
Polytechnic University of Madrid, Spain

ABSTRACT

Traditionally, Internet has offered multimedia services independently of the services offered by companies associated to Telco (companies that traditionally offer telephony and associated services). In the last years, the technical community has assumed that the convergence of both worlds is a fact; however, there are many barriers to be removed. On one hand the emergence of cloud computing technologies will change the way of providing multimedia services. On the other hand economic issues are very important in this convergence process. In this chapter we describe how multimedia services are offered from Internet and the Telco. The convergence of Web 2.0 and Internet Protocol Multimedia Subsystem on the WIMS 2.0 initiative focuses on creating value for Internet and Telco and making multimedia services a reality for end users, independent of the underlying environment. We propose a novel strategy that uses cloud computing to facilitate the development and deployment of converged multimedia services. We also observe a sustainable business model accompanying the converged proposals to offer innovative multimedia services.

DOI: 10.4018/978-1-61350-144-3.ch006

INTRODUCTION

A *Service Delivery Platform* (*SDP*) usually refers to a set of components within a telecom operator service architecture that manages easing service creation, deployment and execution utilizing enablers (network capabilities) and service capabilities. The keystone objective of a SDP would be to reduce time-to-market of a service and the key characteristic for a SDP is thus the ability to ease the integration task of application logic with the enablers and capabilities in the telecom service architecture for the delivery of a service to users. The SDP enables thus the rapid transformation of a service concept into a full product including all the related service life cycle management and commercial interactions.

While SDP is being used by Telco operators to provide any service, and in particular, multimedia services, however this approach is different from the Internet world, where it is not used a logical entity like a SDP. The convergence between Internet and Telco World is a reality; nevertheless they are very different worlds with diverse assets. This represents some barriers in the converge process. In order to remove these barriers in the convergence of both Internet and Telco worlds, a new initiative, WIMS 2.0 (WIMS 2.0 initiative, 2010), was founded in 2007. The aim of WIMS 2.0 is to create, in a collaborative way, new innovative services and a sustainable business model and ecosystem in the merge of a common world from Internet and Telco. To do that, WIMS 2.0 has defined a general architecture to be used in this convergence, contributing with it to the technical community.

The cloud computing (Marston, Li, Bandyopadhyay, Zhang & Ghalsasi,, 2010) is the extension of services, traditionally hosted in operator data centres and offered to large enterprises, to other segments in the market, like *Small and Medium Enterprises* (*SME*) and mass market that, thank to high speed access networks (Berrocal, Alvarez, Gonzalez, Perez, Roman & Vazquez, 2009) and low cost storage infrastructures (Komorowski, 2010) are shaping a new way of offering multimedia services. Cloud computing services are strategic services for any Telco operator or Internet provider over the world; not surprisingly, these services are closely related to the interest of these actors in providing services from network. In this chapter we present some ideas about the introduction of a cloud computing model to offer multimedia services to the end user. This model will remove costly barriers for new entrants or for those companies which prefers to focus on their core business.

The theme of the chapter builds the motivation of the authors to provide a new proposal combining aspects of cloud computing and the SDP as a platform to offer multimedia services in a converged environment: the *SDPaaS* (that stand for *SDP as a Service*). This chapter does not end with the proposal of this new concept, but the authors apply it to the case of SDP defined in the WIMS 2.0 initiative, transforming this SDP in a 'Telco and Internet converged SDP in the cloud', providing a more practical proposal from the authors to the scientific community. In order to be an end to end perspective, it is commented a new economic model which is linked to the technical aspects, which has been defined and industrialized by one of the authors and currently is running in one of the biggest Telecom operators over the world. One of the most common errors is to misalign business and technical issues, so that, it is frequent to see good technologies failing due not to consider business aspects and vice versa. The business model described here is a sustainable business model based on the creation of partnerships activities.

The structure of this chapter is the following: In section 2 we show how multimedia services are supported in Internet and Telco world; section 3 covers the work did under the WIMS 2.0 initiative, while in section 4 we propose how to offer convergent multimedia services in the cloud;

finally, we introduce a sustainable business model applicable to the proposal explained.

MULTIMEDIA SERVICES: AN ANGULAR VISION

The aim of this Section is to present how multimedia services are supported not only from Internet, but also from the world of telecommunications.

The Way Internet Supports Multimedia Services

One of the key points for the success of Internet is the multimedia applications and services running on it. Most users got into Internet by the end of the XX century due to the different services, most of them using multimedia for streaming services, playing or video-conference systems. Nowadays, most of these services, initially tied to one single application, have merged and are "mash up" into a single service from the user perception. This enables end users to enjoy more complex services like online learning or online collaboration, without any knowledge of the underlying services and their working features.

Although, original Internet design was not intended to support multimedia protocols, several solutions have been deployed to support them. Normally, multimedia services are challenging for any kind of network: They impose soft real time constrains in the delivery of data to end users (the features of the networks must provide solutions to this constraints). In some cases, high bandwidth, low delay, low jitter, good errors control... Are demanded by these kinds of services (to the physical networks transporting them confederated in Internet). Although, a lot of *Quality of Service* (*QoS*) solutions have been defined and deployed in the networks, the most popular solution is to add more bandwidth to the physical level of the communication network in order to ensure the user's demands. This is not the best solution but it has been cost effective due to the low cost for adding bandwidth has been cheaper and cheaper each year.

We think the main problem to provide multimedia services in current Internet is the lack of true standards guided definition of multimedia services. Some facts show this thought:

1. The *Internet Engineering Task Force* (*IETF*) has provided the main standard solutions for multimedia communications in Internet (even some protocols initially used, like H.263, was dropped for the most flexible IETF *Session Initiation Protocol* (*SIP*) standard). This solution does not provide any framework for services, just define the low level protocols (usually *Real Time Transport Protocol* (*RTP*) over *User Datagram Protocol* (*UDP*)) and some session protocols, like SIP or *Real-Time Streaming Protocol* (*RTSP*), but does not provide a global application solution. Usually, application providers have worked providing solutions on these areas, but not tackling important aspects like interoperability and, sometimes, using proprietary extra protocols or, even more, black box solutions. The IETF proposal implies working with a not easy solution to grow up a multimedia application in a standardized way. On the other hand, the Telco world, normally isolated from the usual Internet way of doing things, has been developed based always on standards. The fast development of prototypes and solutions in Internet is the engine that does not take into account the behavior of the underlying network, so that each developer just use the best approach in order to get the QoS needed for the application.
2. Since there is not a single solution for the optimal communication of multimedia over Internet there has been several applications, sometimes with proprietary solutions that has been used to provide these services, for

example some media streaming solutions like Real Player (RealPlayer, 2010) or Adobe Flash media Server (FMS, 2010) This has been slowly changing to open solutions that may be deployed on any platform without any limitation:

i. Web browser based solutions has been arising continuously. This has been a good solution due to several facts: a) The easiest way of having the client deployed in all of the different operating systems and devices, b) Just develop a solution on a single environment, like some of the available frameworks, for example, using java. These solutions have not been available till lately, mainly due to the problems requiring a lot of *Central Processing Unit* (*CPU*) power and memory to properly handle the multimedia. Lately this limitation has been removed due to the evolution of CPU and the usage of GPU included on most computers and now, on high-level smart phones.
ii. Current Internet has become balkanized into several domains where direct connection is not possible for multimedia applications due to the Firewall/*Network Address Translation* (*NAT*) solutions deployed on most organizations. This implies that extra protocols like *Inter-Client Exchange* (*ICE*) or *Simple Transversal of UDP over NATs* (*STUN*) must be used in order to open the connections or that these protocols may be pigged back over non multimedia protocols that are able to cross these organisational borders, like *Hypertext Transfer Protocol* (*HTTP*). This is not a good solution that has been widely used, for example by youtube, to be able to reach any computer on the world. Most of these solutions use the Adobe Flash proprietary technology that has been given for free to all the final users. Flash Technology provides some severs and toolkits that helps developers to create new applications, but it is not included a common framework to where build upon it. This great limitation, jointly with the additional problems to work with it in smart phones platforms, is moving developers to look for some other open solutions...
iii. Some proprietary solutions using a P2P approach, like for example Skype, include intermediary nodes acting as a communication overlay on top of *Transport Control Protocol* (*TCP*)/*Internet Protocol* (IP). These are non standard solutions that may not be even used as a service by other computer components, contributing little to the evolution of the global Internet multimedia architecture.
iv. It seems that the open standard for video support into the applications is the *HyperText Markup Language* (*HTML*) 5 (HTML5, 2010), which is ongoing developed by *World Wide Web Consortium* (*W3C*) standard body. Even supposing it would be available around 2012; most of the Web browsers implement it right now, so it is emerging a sound solution especially for video streaming. The supported video encodings are H.264, Theodora Ogg (Theodora, 2009) and Webm/V8 (WebM, 2009). The state of the open specifications for these codecs is still been work out to get them patent free, so anyone may implement them without any licensing. This is a key to get an exponential usage by the users and, so, allowing all the actual internet multimedia applications to move to this. This issue is quite important for

mobile terminals since it provides an easy way to get a cross platform implementation. It uses JavaScript (ECMA, 2009) and a set of predefined new tags for controlling video. Even is not as mature as the flash based solution, it is slowly growing as more sound solution for streaming. Even still it has to be proven that can be used for videoconference as it promises. Since it is an ongoing standard, for sure, it will provide facilities on the end, but this is not available currently. In the current move to "as a Service" (Zhou, Zhang, Zeng& Qian, 2010) still there is not a global solution. Most developer uses proprietary solutions deployed as an infrastructure even some of the Amazon object cloud storage facilities now are providing streaming capabilities, even limited to flash and HTML 5 de facto standards.

The above thought shows that over internet there is a still a long way to go and that may, in this time, learn something by the way the Telco world of services has been evolving.

The Way Telco Networks Support Multimedia Services

Telco networks were traditionally focused on voice networks with high availability and a set of basic and supplementary services that complement it. The advent of the intelligent network, led to the emergence of a number of value added services, but always related to voice, where a set of operations were real-time performed, triggering to Telco servers.

The arrival of the video-call to the Telco may be considered the first multimedia service, to the extent that there is a flow of voice and video, but are transmitted together through traditional circuit switched communication channels.

The success of packet-switched technology in the Internet resulted in the proliferation of multimedia services, which triggered the demand for higher bandwidths. The mobile networking environment is currently experiencing a similar phenomenon. End-user's demand for higher bandwidths is driven by their need to access multimedia services and Internet applications on the mobile devices connected to mobile access technologies. The so called Mobile Internet is driven the evolution of originally circuit-switched mobile networks to the current all-IP network architecture.

The emergence of mobile data networks associated with traditional voice networks opened a new vista in the multimedia concept in Telco networks. However, both networks were acting independently, carrying different streams without any synchronization between these flows. The usage of mobile data services was originally limited due to a lack of available bandwidths. The low bit rates inherent in *Global System for Mobile Communications* (*GSM*) technology made the data applications over *Global System for Mobile Communications* (*GPRS*) not widely used, and the bandwidths available for the end user become a key factor. The introduction of 3G radio technology enabled the launch of data services from a mobile terminal, enabling a step forward to mobile multimedia services.

The telecommunication networks are evolving to an all-IP-based core network that has enabled the convergence of different access technologies. Speaking of multimedia Telco networks involves talking to *Internet Protocol Multimedia Subsystem* (*IMS*) (3GPP TS 23.228, 2009) and SIP. IMS was originally conceived to make the mobile Internet paradigm come true by offering multimedia and Internet services anywhere and anytime on the cellular access technologies. The first release of IMS was tailored to the GPRS/*Universal Mobile Telecommunications System* (*UMTS*) cellular network. However, later releases of IMS aimed at true access independence by incorporating other access networks: *Wireless Local Area Net-*

work (*WLAN*), *Digital Subscriber Line* (*xDSL*), cable ... (3GPP TS 23.228, 2009). Prior to IMS, services were tightly coupled with a specific transport network and signalling protocol. With the introduction of IMS, the communication network architecture paradigm has transitioned from being stove-pipe vertically integrated to a horizontally integrated, layered architecture. The layered architectural design ensures that the transport and bearer services are separated from the signalling and session management services. It aims for minimum dependency between layers. It can be inferred that IMS is designed to separate service-related functions from the underlying transport-related technologies. This independence among the service, session and transport layers makes the underlying technology invisible to the user regardless of where in a multiservice, multiprotocol, multivendor environment the user resides. This nomadic behaviour enables seamless communication between fixed and mobile users. IMS architecture provides many common functions at the control layer and service layer that can be reused for fast service creation and delivery.

A high level of abstraction at the service layer enables the service developers to develop applications quickly and focus more on the logic of the application rather than lower layer complexities. This also ensures a large developer base for IMS services, yielding to more scope for innovation. Finally, the end-users are catered for by a variety of blended multimedia services (blend of telecom and Internet services) which otherwise would not have been possible. IMS has introduced new capabilities into the Telco architecture through the addition of new network elements called the *Call State Control Functions* (*CSCFs*), which are added to allow delivery of basic IP multimedia services, such as initiation of multimedia sessions between two subscribers and which has complete control of the sessions. The CSCFs are limited only to the session control layer and are in charge of the signalling routing function. The session control capabilities of the CSCFs are abstracted and hooked on to the application plane in the IMS. The application plane stimulates new and innovative IP multimedia services, taking advantage of basic multimedia session control and services capabilities provided by the core network.

Since the IMS uses only the packet-switched domain for transport and local mobility management, it can be deployed without a circuit-switched domain, building an all-IP network.

In order to understand the service capabilities of IMS and its ability to act as a comprehensive service delivery framework, it is essential to study the IMS core network components, protocols and their functional operation (Galindo, Moro, Balakrishna & Al Begain, 2009).

THE CONVERGENCE OF BOTH WORLDS

Once we have presented the both worlds: Internet and Telco, in this section we present our contribution to the convergence of both worlds: WIMS 2.0. Specifically, we first present the main idea and then its strategic lines of actuation. We continue with the WIMS 2.0 SDP, where we explain how it can be deployed in the real world and the web technologies used.

WIMS 2.0 is an open initiative founded by Luis Angel Galindo and David Moro. WIMS 2.0 seeks convergence between the Web 2.0 and the telecom new generation services based on IMS (IP Multimedia Subsystem), to create innovative, appealing and user-centric services and applications. These services will combine features from both Web 2.0 and Telco worlds. On the one hand, relevant features like interactivity, ubiquity, social-orientation, user participation and content generation will be adopted from Web 2.0 Services. On the other hand, IMS will enhance WIMS 2.0 services features with a collection of capabilities like multimedia telephony, media sharing, push-to-talk, presence and context, online address book... These capabilities are all of them applicable to mobile, fixed or convergent telecom networks. In addition, in order to provide more versatile and enriched services and a short-term implementation, WIMS 2.0 also considers the utilization of pre-IMS Telco capabilities like *Short Message Service* (*SMS*)/*Multimedia Message Service* (*MMS*) messaging, circuit-switched video and voice calls, networked address books, etc.

This initiative comprises the strategies, technologies and service platforms that will allow telecom operators to achieve this convergence with innovative Web 2.0 services in All-IP networks. Following the Web 2.0 revolution and the Telco 2.0 initiative (Telco 2.0, 2010), a change in Telco traditional business model is recommended following the adoption of a user-centric and open-garden philosophy consisting in the enhancement of the openness, flexibility and freedom in the fashion services are provided. IMS, along with its associated set of service enablers, has been identified as the base platform to support this convergence and business model change.

The two main goals of the WIMS 2.0 initiative are to create innovative services mixing the Web 2.0 and Telco World and to create a new fair ecosystem: the WIMS 2.0 ecosystem (Galindo & Moro, 2008): a fair ecosystem, where we all the actors can play a sustainable role. Currently, we are competing not only, as usual, between operators but also with those new Internet players. In this new ecosystem we are speaking about co-opetition instead of competition.

In order to make WIMS 2.0 to work in actual networks, we have defined two strategic approaches:

1. *Incorporation of Telco capabilities into Web 2.0 services through open Web Application Programming Interfaces* (*APIs*) allowing the integration of Telco into Web 2.0 mashups. This potentially applies to any Telco capability, which would therefore be usable by/from any Web 2.0 service. Two different, but interrelated, strategies are proposed here:
   i. *Web 2.0 mashups using Telco assets*. Here it can be divided in two possible strategies:
      a. Mash-ups based on widgets or *Portable Service Elements* (*PSEs*). Telco applications can be incrusted into Web 2.0 services in the form of web-widgets, which would be the technology-specific implementation of a PSE. The PSEs provided by the operator (or a Third Party) can be easily integrated by end users into Web 2.0 sites and, once incrusted, they are able to handle the interaction with the operator's open Web APIs, thus, enabling the use of Telco communication capabilities from Web 2.0 sites. Specific application examples are a Call-Me button for Blogger (Blogger, 2010) or a more sophisticated Facebook (Facebook, 2010) application providing voice and presence services for the members of a group.
      b. Mash-ups based on APIs. Once Telco capabilities are exposed through open Web APIs, they can be directly incorporated into any Web 2.0 service in order to provide complementary functionalities, just like any other regular mash-up. The final outcome is a complete integration of telecom capabilities in the resulting service. A specific service example could be "Real-time sharing of youtube (youtube, 2010) videos to mobile users", while maintaining an Instant Messaging session among the involved users.
   ii. *Telco-enabled user-generated content publication and generation*. This con-

vergence guideline aims to obtain new ways for content publication through the usage of Telco multimedia transmission capabilities. Operators can provide new solutions to receive the multimedia content from the user and automatically upload it to Web 2.0 sites. An example application could be "youtube real-time video generation", where users video call to youtube in order to create a spontaneous clip stored in youtube servers.

2. *Exploiting Web 2.0 for telecom services*. In this approach, the operator keeps the role of final service provider. Accordingly, two different guidelines are considered for the enrichment of telecom services:
   i. *Inclusion of Web 2.0 contents and events in the operator services*. The introduction, integration and distribution of videos, advertisements, podcasts, news, etc from the Web 2.0 within Telco services is of great interest for the end-user. As example services, we may consider special feed readers for mobile handset, the inclusion of social networks events into Telco Presence information or news, video and music distribution.
   ii. *IMS on-line applications*. Through the use of *Asynchronous JavaScript And XML (AJAX)* and other technologies, Web 2.0 applications have achieved major advances in the field of user interfaces. WIMS 2.0 favours for the usage of these technologies in order to build IMS on-line applications. The benefits of such kind of IMS applications are very important: ubiquity and a great simplification of service development and deployment, since service logic resides on the network and not on the terminal.

The WIMS 2.0 SDP

The WIMS 2.0 SDP is an intermediary for the adaptation between both worlds, enables the convergence in the desired terms. The proposed reference model is reflected in Figure 1. This reference model defines a framework that identifies the required logical entities, as well as the relationships existing among them. Two main types of entities are identified in the reference model: adaptation entities and entities for the provision of IMS-online applications. Within the adaptation entities, we also find two further sub-types: *entities for the exposure of IMS capabilities* and entities *for the exchange of multimedia content and events*. As a relevant difference, for the interaction with the Web 2.0 world, the entities of the first group use APIs defined by the operator, while the entities of the second group use the APIs defined by the Web 2.0 services to be contacted. In fact, since the entities of the latter group are actually using functionalities of Web 2.0 services along with Telco capabilities, they can be considered as platforms for the development and execution of service mashups by the operator itself.

The entities for the exposure and exploitation of IMS capabilities, and their functionalities, are the following:

1. *IMS Exposure Layer*. This entity exposes IMS capabilities to the outer world through open Web APIs. On the outer side, it manages HTTP messages associated to the execution of a specific procedure, while, on the inner side, it interacts with IMS capabilities using the appropriate protocol (SIP or *XML Configuration Access Protocol (XCAP)*). Obviously, the interactions on each side must be properly interrelated. It can be clearly seen that it is actually a gateway between HTTP and IMS protocols. The WIMS 2.0 holds the IMS Exposure Layer as a major axis for the architecture.

Figure 1. The WIMS 2.0 SDP reference model

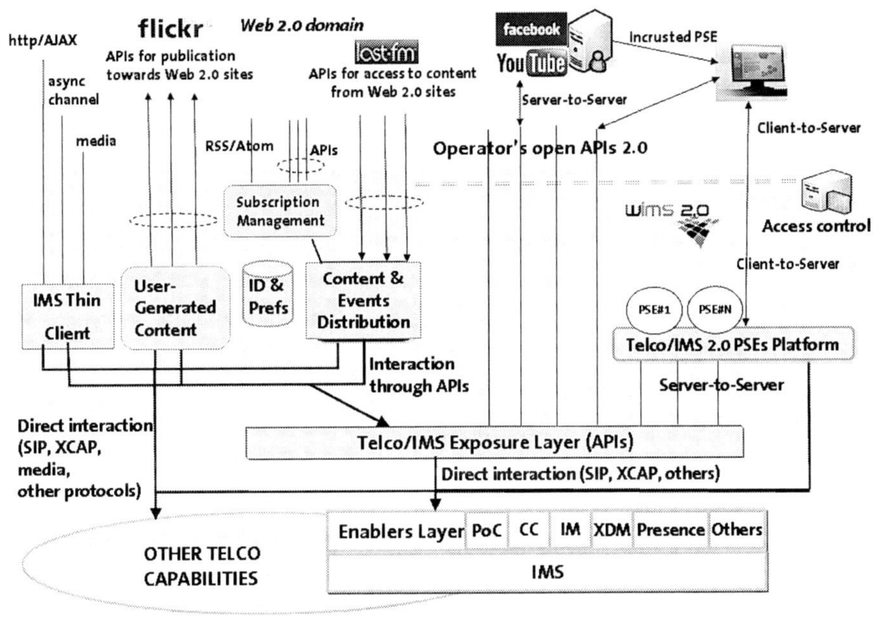

2. *IMS 2.0 PSEs Platform.* This entity holds and servers the operator's PSEs to be incrusted into the Web 2.0 sites. PSEs implement the required logic to interact with a specific group of open IMS APIs.
3. *Access control entity.* This entity performs the required access control mechanisms to secure the use of the operator's APIs,

The entities for the exchange of multimedia content and events, and their functionalities, are the following:

1. *Subscription Management enabler.* It subscribes to content and events of any kind generated in Web 2.0 services. Subscriptions may be made on behalf of the operator or directly on behalf of the final user. This entity talks with the Content & Events Distribution enabler to inform about the existence of new content/events to be obtained.
2. *Content & Events Distribution enabler.* It downloads, adapts and transmits content and events from Web 2.0 services to Telco users.
3. *User-Generated Content enabler.* It receives content from terminals and, after adapting its format, it uploads this content to Web 2.0 services.
4. *IDs & Preferences.* Database that stores the relationships between Telco identities and the identities used by Web 2.0 services. The other entities of this group consult this database to perform the conversion of identities when using the Web 2.0 services APIs.

Finally, we find an individual entity for the provision of Telco on-line applications. On one side, this entity interacts with *IMS& Telco* capabilities, acting as the final Telco client. On the other side, it shows a rich multimedia Web interface to the user, according to the service being provided. This entity must also handle a continuous media interface towards the user, as well as an asynchronous interface to signal the need to refresh the Web interface.

Web 2.0 Technologies in the WIMS 2.0 SDP

The WIMS 2.0 platform eases the technology transition from the Telco technology domain towards the pure Internet technology domain aiming at facilitating the dialogue between a telecom world and the web 2.0, from the operator perspective. In Figure 2 we depict a summarised view of the entire web technologies considered, representing the domain of scope of each one of them.

Before describing each of the different groups identified in Figure 2 and the relationship with the implementation of the WIMS 2.0 service platform, it is important to stress the principles that organise this technology map:

1. Technologies have been located following the client-server paradigm that governs the web world. Within the server and the client sides, the different groups comprise technologies for the implementation of similar functionalities which might be complementary like e.g. the *AJAX* set of technologies, or represent an alternate choice to provide the same functionality like e.g. JavaScript and ActionScript (Schnier, 2008)
2. *HTTP* is the base for all interactions between remote elements, independently of the nature of those elements. These interactions are retrieval and manipulation of web page contents and the usage of web-ready APIs.
3. As a featured concept in Web 2.0, remote APIs are of paramount importance to support the mash-up of functionality within a given service. In the map, there are two variants of the usage of web-ready APIs:
   i. Client-server scenario, where a browser directly invokes the API by sending HTTP request to a remote server.
   ii. Server-server scenario, where a server is the actor invoking the APIs provided by another server. The invoking server is acting as a web client as far as the

Figure 2. The WIMS 2.0 SDP technology map

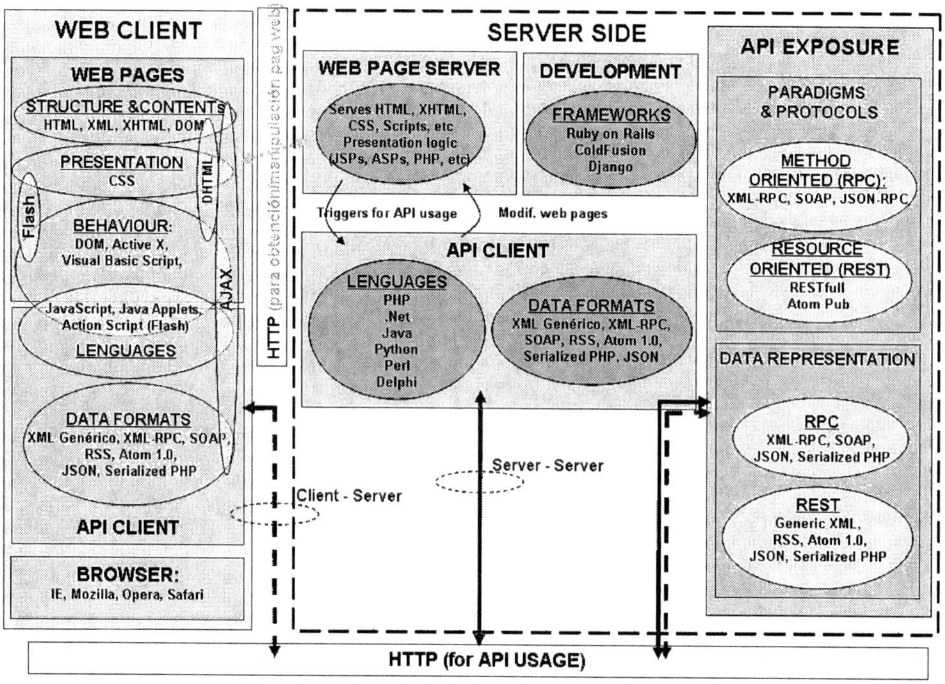

usage of API is concerned. Normally this scenario is used to prevent the web browser to realise this kind of actions, and thus, all the service logic to handle the remote API is managed by the server.

Once principles are explained, technologies are grouped as follows:

1. *Client Side*. The client side comprises all technologies that typically are used by web browsers. The following groups and subgroups are considered:
   i. Web pages technologies related to presentation and processing of web pages. There are three technological levels:
      a. Structure, comprising all technologies to support information, both contents and semantics, and the web page structure. HTML, *eXtensible HyperText Markup Language* (*XHTML*), *Extensible Markup Language* (*XML*) and *Document Object Model* (*DOM*) are technologies within this level.
      b. Presentation, comprising technologies to control how the information is presented. *Cascading Style Sheets* (*CSS*) is the most representative one in this level.
      c. Behaviour, which gather all technologies used to implement the logic that control local actions and dynamic behaviour of the web page. This dynamic activity can result into modifications over the presentation and structure levels. VisualBasic Script, Javascript, ActionScript, Active X are some of the ones considered.
   ii. Technologies for web API clients in the web browser, includes the technologies in charge of process, form and send HTTP request towards remote APIs and to receive and interpret responses according to specific data formats. Since these actions can be considered as dynamic behaviour of the web page, languages which rule this level are the same than those utilised in the web page behaviour level.
   iii. Browsers are those alternative options as client platforms like Internet Explorer, Mozilla Firefox, Opera, Safari and Google Chrome. Even if functionalities presented by these platforms are similar, some browses can present specific features that may affect the usage of certain technologies considered for the client side.
2. *Server Side*. It comprises technologies typically utilised in network infrastructures. Even though they are mostly associated to a web server, the role of a web API client shall also be considered. The following categories apply:
   i. Web page server technologies, which groups technologies to serve web pages such as software containers, servlets, *Java Server Pages* (*JSPs*), *Active Server Pages* (*ASPs*), etc. These are mature technologies which does not represent a key differential aspect of the web – telecom convergence.
   ii. API technologies. The exposure of telecom capabilities in the form of web APIs enables the offering of distributed functionalities within the web world. In this group, technologies for implementing the serving API logic are considered, and then, they are valid for implementing both the client-server and server-server

scenarios. There are different options when considering how the HTTP protocol is employed and how the functionality behind the API is modelled. In the WIMS 2.0 technology map, the *Remote Procedure Call* (*RPC*) and *Representational State Transfer* (*REST*) approaches are considered. The map also considers options to encode the data exchanged between the API client and the API server, like XML, *JavaScript Object Notation* (*JSON*), Atom, etc.

iii. API client technologies for the server side. As previously mentioned, in the server-server scenario, web servers act are API clients, and in this particular case of the web loom, they act as web clients in order to invoke remote web APIs. Technologies are those languages at the server side to rule API invocation like PHP, Ruby, Java, Python, etc. as well as the data formats that the specific API is utilising to encode the data exchange. As mentioned before, this is a feature of the API server and there are a variety of possibilities, most of them based on different variants of XML.

iv. Development frameworks, which includes environments geared at easing the web service creation. These frameworks feature a high level view on the development process and a rapid deployment over different server platforms like Java or.NET. A popular framework for web 2.0 application is Ruby on Rails.

The Key Entity in the WIMS 2.0 SDP

The axial concept for convergence within the WIMS 2.0 SDP is the exposure of capabilities through web-friendly APIs. The entity in charge is the IMS Exposure Layer. There are two aspects for the technical realisation of this entity: on one side, the web-friendly APIs that exposes IMS service capabilities towards web applications and, on the other side, the internal logic to interwork and interact with the underlying IMS infrastructure in order to execute the functionalities exposed via the API.

The IMS Exposure Layer shall implement a gateway adapting the HTTP-based API primitives into SIP and XCAP operations to enable communication between the web world and the IMS. This mission affects two levels of communication in the telecom side: the signalling plane and the media plane. Wherever possible, both planes need to be adapted from a pure technical perspective, although continuous real-time media is not supposed to be adapted by means of API approaches. This means, that audiovisual media exchange between a Telco endpoint and a web component is not processed through an API and thus, not handled by the IMS Exposure Layer.

The following IMS service capabilities constitute a non exhaustive list of capabilities that can be exposed via the IMS Exposure Layer, and thus, the underlying gateway shall be able to communicate with the enabling servers implementing these service capabilities:

1. Presence: an API client might be able at least to subscribe, consult or modify Presence information for a given user.
2. *XML Document Management* (*XDM*): to manage lists and contact items in the IMS domain.
3. Instant Messaging both in pager mode as well as session mode.
4. Multimedia Telephony, to provide third party call ability to web applications and

call control features of sessions established between two endpoints.

5. Multimedia conferencing, to provide a web application with the possibility to schedule and create multimedia multi-participant calls and manage them.
6. Rich CS calls, to provide features related with media sharing combined with circuit switched calls.
7. SIP Push, to provide web applications the ability to push events and contents to IMS terminals.
8. Customized Alerting Tone enables a web application the control over ring back tones and personalised alerting tones (ring forward tones).
9. Multimedia inbox control.

To expose capabilities, the APIs of the IMS Exposure Layer will implement a resource-oriented model as a basis. That is, the REST philosophy will be utilised to model the APIs for IMS capabilities. Since most IMS services inherit a session orientation, some RPC techniques might be added to the basic REST representation in order to obtain a similar result as when modelling telecom services as Web Services. However, REST provides a simpler approach which reduces the entry barrier for web developers to integrate and utilise telecom service features in web applications. Most internet players in the Web 2.0 ecosystem provide APIs utilising this resource-orientation philosophy.

Besides, since REST is naturally stateless, scalability increases in comparison with pure RPC alternatives like *Simple Object Access Protocol* (*SOAP*) and Web Services. Beyond that, the processing for a lightweight option like REST is not comparable to the SOAP computational cost.

For the technical realisation of the REST APIs of the IMS Exposure Layer, there are several protocol alternatives. Among all them, the *Atom Publishing Protocol* or AtomPub is one of the most suitable, besides it is a standardised mechanism by the IETF.

In Figure 3 we present an example of the IMS Exposure Layer exposing Presence by means of REST API. In this examples, the Presence APIs relies on some of the intrinsic features of REST i.e. it works with resources e.g. a presentity. The resources can be freely defined and can keep states (e.g. presence status). The client acts on the resources: get, modify (e.g. state), delete, etc. and these resources are always univocally identified by URIs. The actions on the resources are performed through standard HTTP verbs: GET, POST, PUT and DELETE, which has a specific meaning in the context of each API. For instance, in this presence example HTTP GET is utilised to retrieve a presentity's status; HTTP POST can be used for publishing presence information; HTTP PUT for refreshing soft state presence information; and HTTP DELETE to erase this soft state information. The application of REST principles to IMS Presence is thus immediate.

Aside from the protocol aspects, the API exchange information between the API client and the API server. To accommodate the expectations of a bigger number of developers, APIs shall provide different data formats so the user can select the most suitable one according to the programming language and development tools the user utilises. The IMS Exposure Layer at least shall provide the following data formats:

1. Client-Server scenario: XML, Atom 1.0, RSS and JSON
2. Server-server scenario: XML, Atom 1.0, RSS and serialised PHP.

This allows utilising the remote APIs and the exchanged data from most of the languages being utilised in the Internet.

Furthermore, the Telecom Exposure Layer might be simpler to utilise if the developers are provided with language libraries that handle the API client behaviour. That means that the developer does not need to deal with HTTP request and response construction and processing. Most inter-

Figure 3. Example REST API for Presence

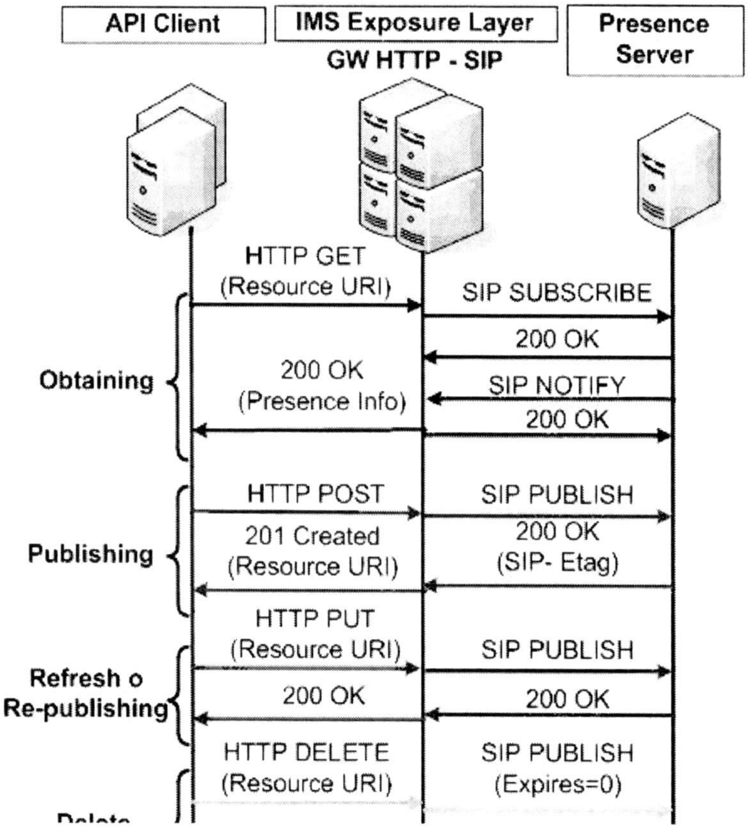

nct players providing functionalities via API are adopting this best practice, offering client libraries in a limited number of programming languages and also the bare REST interface with different data formats available for those programmers utilising other programming languages besides those considered by the client libraries.

For instance, the Telecom Exposure Layer should offer client libraries in JavaScript and ActionScript languages for client-server scenarios, and Java, PHP, Python and Ruby for server-server scenarios. For developers utilizing other languages, they still can access the functionality by handling the REST APIs directly.

Furthermore, the Telecom Exposure Layer might be simpler to utilise if the developers are provided with language libraries that handle the API client behaviour. That means that the developer does not need to deal with HTTP request and response construction and processing. Most internet players providing functionalities via API are adopting this best practice, offering client libraries in a limited number of programming languages and also the bare REST interface with different data formats available for those programmers utilising other programming languages besides those considered by the client libraries.

For instance, the Telecom Exposure Layer should offer client libraries in JavaScript and ActionScript languages for client-server scenarios, and Java, PHP, Python and Ruby for server-server scenarios. For developers utilizing other languages, they still can access the functionality by handling the REST APIs directly.

The WIMS 2.0 SDP Deployment

Previous sections have shown the reference model of the WIMS 2.0 SDP. In order to deploy onto the service delivery domain of an operator is required to analyze how the different entities of the WIMS 2.0 reference model can be either implement by existing entities in the telecom Service Architecture or how new ones –specific for WIMS 2.0- fits into the service delivery domain. In Figure 4 is shown the functional entities of the WIMS 2.0 reference model that are new in the service delivery domain of an operator are coloured in strong grey. For instance, a case for the later would be the "access control entity" in the WIMS 2.0 reference model; here it is mapped onto a general gateway for securing the access of third parties to the service delivery domain i.e. the integrator towards third parties of the SDP of the layered architecture.

Figure 4. WIMS 2.0 SDP deployment

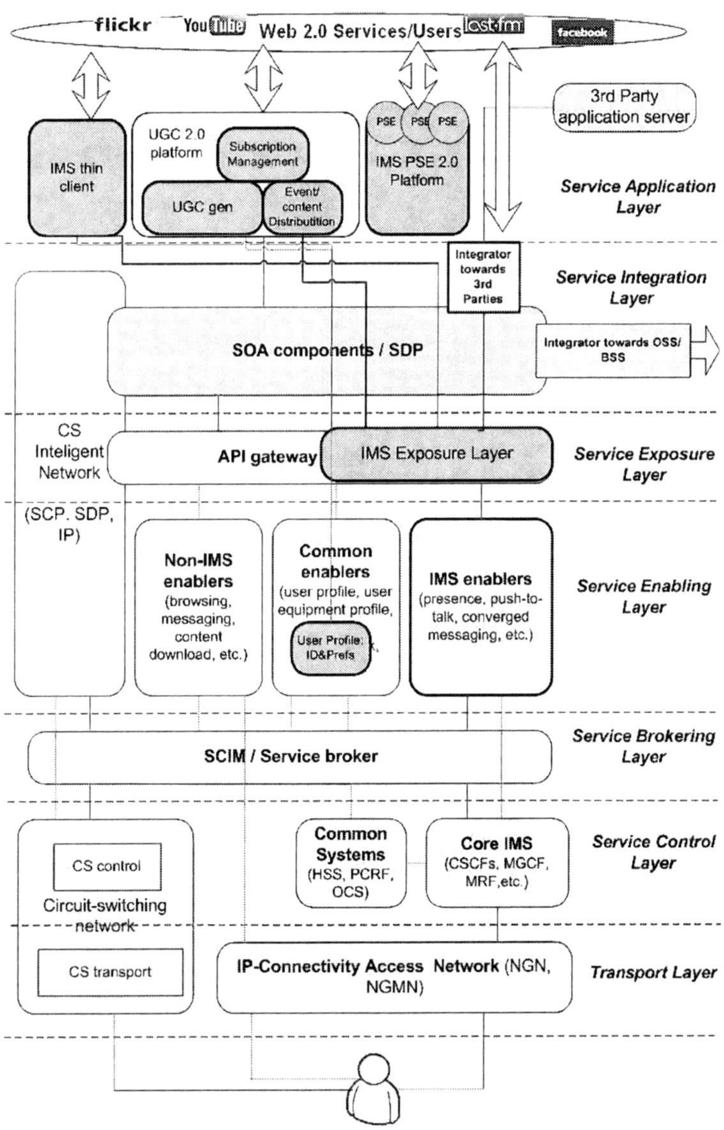

As for WIMS 2.0 specific entities, the IMS Exposure Layer is located at the Service Exposure Layer of the layered architecture, while the rest of WIMS 2.0 functional entities are considered service-platform and thus, they are represented in the Service Application Layer. That would be the case for the IMS Thin Client, containing service logic for providing end-user application online representations of an IMS phone; the functional entities of WIMS 2.0 related with content and media delivery or generation are gathered under a service platform called *User Generating Content* (*UGC*) 2.0 at the Service Application Layer, except for the database containing the user IDs and preferences, which is stored together with other user data at the User Profile directory (a common enabler) at the Service Enabling Layer; and finally, the IMS 2.0 PSE platform is regarded as a container for end user applications, in fact, the Portable Service Elements.

MULTIMEDIA SERVICES IN THE CLOUD

One of the most important transitions from the Grid Infrastructure (Foster, 2002) to the cloud infrastructure has been the change of the W3C Web Services (WS-*) for the REST way in designing services (Olsen, 2008). This makes a small philosophical change from the plain *Service Oriented Architecture* (*SOA*) to a *Resource Oriented Architecture* (*ROA*) where the different services emerge as a set of Resources to be reused and composed by the different players; usually as a mash-up of resources. This approach has been proven quite reasonably and sound over the different Web 2.0 start-ups and all the collaborative services that have shown up all along Internet.

Cloud computing is based on the virtualization of the resources, mainly computing power and storage. It allows using the resources in a very flexible way and being dynamically able to add new resource with a fine grain measure. It provides an elastic way of paying as you grow for the resources needed. This have been using into private clouds, where it is virtualized all the resources available in the computer department in order to allow users to gain access exactly with the resources they need. Also there have appeared public clouds where users may be able to buy, quite cheap, the resources needed on demand. This solution is useful for end-users, helping them in the implementation of new services in an important issue: how to dimension your computing resources. Most of the technologies in Cloud computing are based on using virtual operating system images into the cloud server and instantiate, with the proper parameters, an image of the operating system and services needed.

This image is quite closer to the one provided by the SDP layer into IMS architecture. In the SDP, there are different services available, usually by an API or enablers, without need to care where are deployed or the real details for the configuration.

Cloud computing has been deployed into 3 layers:

1. *Infrastructure as a Service* (*IaaS*).
2. *Platform as a Service* (*PaaS*).
3. *Software as a Service* (*SaaS*).

Most of the Internet web 2.0 applications fall into the *SaaS* layer. Applying this model to the IMS architecture, it is possible to find similarities and, in the case of SDP, it can be considered as a Platform as a Service or a Software as a Service, depending if the SDP services are taking into account like a final service. This approach may be quite convenient from the architectural perspective and provides a convergence of the Telco and Internet world. This proposal currently does not consider that any operator may take into account the outsourcing of their capabilities to a public cloud; nevertheless this may be used to organize Telco resources in a virtualised way. Following this way, it is possible to have a fine grain tuning of the resources in use and the exact cost of them, clarifying if the revenues from the services are in the guidance of the forecast business plan. From the point of view of *Operational Expenditures* (*OPEX*), it also may be a great cost reduction, since it unifies procedures and maintenance interfaces, which means fewer people to do these tasks.

One of the most powerful skills from the Telco is the ability to manage non-stop services. In that way, the five 9's paradigm is a Telco characteristic. This know-how could be a great contribution from Telco in order to operate Cloud services. So, it can foresee new business lines with a hybrid approach, where Telco can offer cloud infrastructure to their end users also deploying the different services.

One of the current missing capabilities for the cloud is *Communications as a Service* (*CaaS*). This is an Achilles-heel for it on its usage into the Telco world, although it can not be ensured any Quality of Service on the communication. Most of the current providers just bypass the situation declaring that their communication infrastructure is just enough for any application. Also in this case, the IMS services have special needs in the areas of the communication and interconnection with the operator infrastructure. This means that the interconnection of actual infrastructure may not be as easy as it is in other Internet services, most of them based on asynchronous interactions over HTTP. One remarkable aspects is that into Internet a wide development of Quality of Services technologies have fail up to now, mainly due that it has not been required by end-users. Perhaps with a convergence of Internet with IMS, Internet networks may gain capabilities that guarantee some of the communication skills without loosing the freshness of Internet.

The convergence of both worlds implies to consider the characteristics of the SIP protocol: synchronous interactions and a protocol that is not just question-response like HTTP, which implies that most of the cache infrastructure will not be reusable. Under this consideration it may not be able to deploy real SIP services integrated into an operator infrastructure into a public or hybrid (mixing public and private infrastructure) clouds. This does not mean that there is no a place for the cloud computing paradigm and technology into the IMS architecture. The Service Brokering Layer may be the right interface where to interconnect a cloud service with the in-company SIP communication infrastructure.

Under this approach, it is possible to offer SDP capability to other operators using the Cloud computing paradigm and business model. For example, for a virtual operator, it can be offered a "pay as you grow" model or even not paying for the IaaS, but for "SDP as a Service" where virtual operators or those which have still not deployed the SDP may pay for the capability of implementing the services provided in the SDP.

In this line of work is quite important to enable the creation of mash-ups by third parties with the virtual services offered. This is why now WIMS 2.0 proposal is even more interesting. IT will provide a standard way of joining both worlds with the right set of APIs coming from the Telco world, but also allowing implementing the web 2.0 needs.

The WIMS 2.0 SDP in the Cloud

Previous sections have shown how a traditional SDP can be migrated to a cloud computing architecture. This section provides the strategy to follow in the way of moving the real WIMS 2.0 SDP to the cloud.

As it has been exposed in section 3.3, the different entities that compound the WIMS 2.0 SDP has been placed in existing or new entities under the traditional SDP architecture. Analyzing the proposal for migrating a traditional SDP to a cloud computing architecture done in section 3, it is understandable that the migration for the WIMS 2.0 SDP can be easily performed. This migration can follow two paths:

1. Considering that both SDPs, the traditional and WIMS 2.0, are moved to a cloud computing architecture. In this path, the point of contact with the in-company infrastructure of the operator is the Service Broker. Under this consideration, some access control functionalities should be included in this entity that ensures that *Service Level Agreement* (*SLA*) is fulfilled.
2. Considering that the traditional SDP remains under the infrastructure of the operator and the WIMS 2.0 SDP is moved to the cloud. This way requires a multilevel integration and it is more complex. Entities like IMS thin client, the UGC 2.0 platform and the IMS PSE 2.0 platform interacts with the in-company SDP in the same way, but other like the IMS exposure layer or the User profile entity requires an adaptation. The IMS Exposure Layeris a platform where the IMS APIs are exposed to third parties. In this case, this platform requires a connection with the IMS and common enablers. The User Profile entity is a database that stores the relationships between Telco identities and the identities used by Web 2.0 services and it requires or connecting both databases or to have a replica of the Telco identities in the WIMS 2.0 SDP. Under this consideration, the access control functionalities should be included in the SDP framework.

Any of the two options mentioned are possible and will provide a collection of mixed Telco-internet services in a cloud computing strategy, enabling a scalable solution for those operators which are not in the position to tackle huge costs in the deployment of a SDP allowing the convergence of both worlds, the Telco and Internet.

A NEW BUSINESS MODEL FOR MULTIMEDIA SERVICES

With the current economic situation (Krugman, 2009), operators have dealt with various processes for efficiency with the clear objective of improving operations. With this situation, operators have reduced investments in all lines except those that are considered core business.

With this situation, investment in innovation in value-added services is very limited. However, the success of the IP world, and thanks to technologies for development simpler and focused on the developer, has produced a ferment of ideas and companies, which offer services and solutions from the ICT sector to, practically, any sector. The opening of APIs that enable to execute operations more easily, without having to know the technology or the underlying protocols, represents a significant step in expanding the number of developers. Latest techniques for syndicating content and creating mash-ups reduces development complexity and allows reusing previously developed code, as if we were using building blocks, in this case based on software. The number of start-ups is growing steadily in an environment that facilitates this situation, not on the economic side, but the technical part.

Attempts to start-ups working with Telco operators normally have ended in failure. Operators are used to provide high availability services, as happened in the traditional circuit-switched environments in its offer of basic services or value added around the voice, and with some very good, but also very costly, levels of customer care and O&M. These demands of the operators to start-ups made it very difficult to participate, providing a decoupling of both worlds.

In order to remedy this situation, on one side a lack of investment of the operators, on the other the interest of the start-up in entering in the operators' sales channel, Luis Angel Galindo has leaded the industrialization of an open innovation model, called Innovation 2.0 which has been deployed in

Telefonica and it is based on the model defined by Win Win Consultores (Win Win Consultores, 2009), a leading consultancy company in the open innovation arena for Telco operators.

This model is an open innovation model for the twenty-first century and is based on specialization of various actors that shape it in its work. Figure 5 shows how the operation is, where what has been done is put in front of the SDP of an operator of the adapter.

In the proposed model, each agent focuses on the tasks where they have more experience:

1. Operator. It focuses on marketing services under their brand or a co-branding, using its various sales channels and customer base that operator has.
2. Adapter. This agent handles the more complex task to make by a start-up: second level O&M and customer care. Both tasks, cost-intensive, involve a drag on the margins of the start-up.
3. Start-ups. They are the owners of the idea and the service providers. It performs the third level *O&M*. With their bright ideas bring freshness to the traditional Telco environments.

This is a win-win sustainable model based on the shared success of a service and where revenues and risks are shared among all. Because the adapter makes the work more expensive for start-up, the operator can benefit from changing its image, having a most innovative portfolio of services tailored to the demands of their customers.

The model is organized around the definition of a model framework that orchestrates the actions of the operator and the adapter; eliminating repetitive tasks and hiding the complexity of the operator to start-ups.

It is not intended to explain the model in depth, but the interested reader can find more information in a book under publication (Galindo, 2010).

FUTURE RESEARCH DIRECTIONS

The WIMS 2.0 initiative has been working from the end of 2007 and within its framework various services have been developed. However the group of companies under the initiative continues to investigate new technologies as they emerge.

With the current investment restrictions that telecommunications operators are suffering, proposals like SDPaaS form a very interesting alternative to the Telco business sustainability. Our

Figure 5. Innovation 2.0 Model

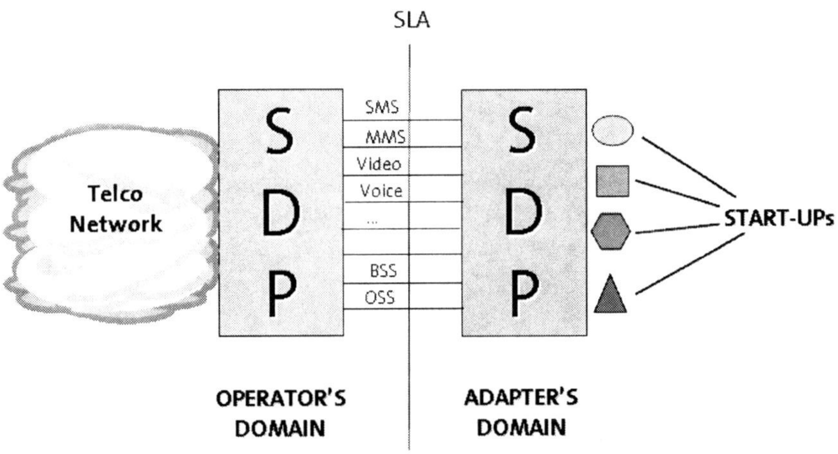

proposal for SDPaaS needs to be tested against the market, as has been proven technologically possible.

A deployment of the SDPaaS to be offered to other operators would demonstrate business viability. This requires research in the field of marketing to define the right strategy where to consolidate the model.

Currently the team that has defined this proposal is working with specialists in the areas of marketing and strategy to define an approach that allows the development of converged, fixed-mobile and internet, value-added services. These services must establish new revenue streams to offset decreases in revenues from traditional services.

CONCLUSION

The way to offer multimedia services is different in the two converging worlds, Internet and Telco. While in Internet, multimedia services are growing in a not very structured way, with many proprietary solutions, in the Telco, multimedia services are based on the SDP concept. SDP is a key component in order to deploy commercial Telco multimedia services especially when the operator is planning to launch services using advanced enablers like Presence, XDM, PoC, etc.

The WIMS 2.0 initiative has defined its architecture for converging Internet and Telco assets based on the WIMS 2.0 SDP, a set of entities which enables the creation of innovative services mixing features coming from both worlds, which, at the end, is demanded by end users. The WIMS 2.0 initiative is an open initiative providing the results of its investigations and developments to the international community with the goal to create a sustainable scenario for all of the players in this convergent world.

A deployment of the SDP architecture is expensive in CAPEX and requires expertise during all the phases of the project. After deployment, OPEX could represent a huge amount, overall in a Telco-Internet convergent scenario where operators are demanding the same quality of service levels that normally they offer in a pure Telco service. Thus, moving this infrastructure to a cloud computing architecture mitigates the expenses in both terms, capital and operational.

Other important issue to remark is the reality for most of the operators in the world: they spend more than nine moths in *Time To Market* (*TTM*) for a new service. Having a deployed 'Cloud SDP', scalable, with a pre-defined O&M SLA, connected with the operator's provisioning systems ensures that the TTM decreases thank to a part of the operator's processes are pre-validated or removed, allowing a quicker commercialization of services without forgetting that it enables to create a richer service portfolio.

With users demanding convergent Telco and Web 2.0 services, it seems that capabilities of both worlds should merge in a common platform. Thus, the Cloud WIMS 2.0 SDP is a proposal following this line of convergence. With this approach, a quick launch of new innovative and convergence services is ensured, but moreover with the Cloud computing functionalities, the WIMS 2.0 SDP becomes a multioperator platform, where new operators using this SDP could offer most of the services running in it, avoid costly deployments of platforms and software.

Users are demanding more customization and remixability in order to satisfy their necessities (WWC, 2009). As it has been shown, the technical solution is available, now it is needed to be introduced in the operator's strategy. For doing this, it has been included a reference to a business model proposed, which is successfully deployed in Telefonica.

REFERENCES

Berrocal, J., Alvarez, M., Gonzalez, F., Perez, R., Roman, I., & Vazquez, E. (Eds.). (2009). *Tecnologías de banda ancha y convergencia de redes*. Centro de Publicaciones del Ministerio de Industria, Turismo y Comercio.

Blogger (2010). Retrieved August 15, 2010, from http://www.blogger.com

ECMA-262. (2009). *Ecmscript language specification 2009* (ISO/IEC 16262). Retrieved December, 31, 2010, from http://www.ecma-international.org/publications/files/ECMA-ST/ECMA-262.pdf

Facebook (2010). Retrieved August 15, 2010, from http://www.facebook.com

FMS. (2010). Retrieved December, 31, 2010 from *http://www.adobe.com/devnet/flashmediaserver.html*

Foster, I. (2002). *What is the grid? A three point checklist*. Argonne National Laboratory & University of Chicago.

Galindo, L. A. (2011). *Cases on SMEs and open innovation: Applications and investigations. Innovation 2.0: Creating a sustainable business model and a win-win ecosystem*. Hershey, PA, USA: IGI Global.

Galindo, L. A., & Moro, D. (2008). WIMS 2.0 ecosystem. *IIR IMS global congress*. Retrieved 12 January, 2009, from http://www.iir-events.com/IIR-conf/Telecoms/DocumentView.aspx?EventID=1800sDocumentID=2691.

Galindo, L. A., Moro, D., Balakrishna, C., & Al Begain, K. (2009). *IMS. A development and deployment perspective*. Hoboken, NJ, USA: John Wiley & Sons.

3GPP TS 23.228 v 10.2.0 (2009). *IMS: IP Multimedia Subsystem, Stage 2*. Retrieved December 19, 2009, from http://www.3gpp.org.

HTML5. (2010). *W3C ongoing standard*. Retrieved December, 31, 2010 from http://www.w3.org/TR/html5/

Komorowski, M. (2010). *A history of storage cost*. Retrieved December 28, 2010, from http://www.mkomo.com/cost-per-gigabyte

Krugman, P. (in press) (2009). How did economist get it so wrong? *New York Times*.

Marston, S., Li, Z., Bandyopadhyay, S., Zhang, J., & Ghalsasi, A. (2010). (in press). Cloud computing - The business perspective. *Decision Support Systems*.

Olsen, D. (2008). *Putting the Web services specification to REST*. Thesis, Brigham Young University, Utah.

Realplayer (2010). *RealPlayer company*. Retrieved December, 31, 2010 from www.real.com/realplayer

Schnier, J. (2008). *ActionScript 3: Basic tutorial, FlashXML applications*. Boston, MA: Focal Press.

Telco 2.0 Manifesto (2010). Retrieved August 15, 2010, from http://www.Telco2.net/manifesto

Theodora (2009). *Xiph.Org Foundation. Theodora Specification*. Retrieved December, 31, 2010 from http://theora.org/doc/Theora.pdf

Web, M. (2010). *WebM open web media project*. Retrieved December, 31, 2010 from http://www.webmproject.org/

WIMS 2.0 initiative (2010). Retrieved August 15, 2010, from http://www.wims20.org

Win Win Consultores. (2010). Retrieved August 15, 2010, from *http://www.winwinconsultores.com*

youtube (2010). Retrieved August 15, 2010, from http://www.youtube.com

Zhou, M., Zhang, R., Zeng, D., & Qian, W. (2010). *Services in the cloud computing era: A survey.* Software engineering institute, East China Normal University, Shanghai 200062, China Universal Communication Symposium (IUCS), 2010 4th International.

ADDITIONAL READING

Blum, N., Magedanz, T. & Schreiner, F. (2009). *Management of SOA based NGN service exposure, service discovery and service composition.* Integrated Network Management 2009.

Changwoo Yoon & Hyunwoo Lee (2010). *Service delivery platform for convergence service creation and management.* Advanced Communication Technology (ICACT), 2010.

Galindo, L. A. (2010). *Cases on SMEs and Open Innovation: Applications and Investigations. Innovation 2.0: Creating a sustainable business model and a win-win ecosystem.* Hershey, PA, USA: IGI Global.

Galindo, L. A., Lozano, D. & Gonzalez, D. (2008). *WIMS 2.0: Convergencia de la Web 2.0 con IMS. Implementación de un API REST de Presencia en una arquitectura WIMS 2.0.* JITEL 2008.

Galindo, L. A., Lozano, D. & Moro, D. (2008). *WIMS 2.0: converging IMS and Web 2.0- New Multimedia Services for Telecom Networks.* ICWMC 2008.

Galindo, L. A., Lozano, D., Moro, D., & Garcia, L. (2008). *WIMS 2.0: Converging IMS and Web 2.0. Designing REST API for the exposure of session-based IMS capabilities.* NGMAST 2008, Cardiff.

Galindo, L. A., Lozano, D., Moro, D. & Gonzalez, D. (2008). *WIMS 2.0: converging IMS and Web 2.0.* ICT Mobile Summit 2008.

Galindo, L. A., Moro, D., Balakrishna, C., & Al Begain, K. (2009). *IMS. A development and deployment perspective.* Hoboken, NJ: John Wiley & Sons.

Galindo, L. A. & Salvachua, J. (2007). *El Web 2.0 móvil: Camino del Teleco 2.0.* Mundo Internet 2007.

Galindo, L. A., & Salvachua, J. (2008). *User centered mobile applications.* NGMAST 2008, Cardiff.

Khajeh-Hosseini, A., Greenwood, D., & Sommerville, I. (2010). *Cloud Migration: A Case Study of Migrating an Enterprise IT System to IaaS.* 3rd International Conference on Cloud Computing (CLOUD), 2010.

Liu, F., Li, L., & Chou, W. (2009). *Communications enablement of software-as-a-service (SaaS) applications.* Global Telecommunications Conference (GLOBECOM), 2009.

Hanhua Lu, Yong Zheng & Yanfei Sun (2008). *The Next Generation SDP Architecture: Based on SOA and Integrated with IMS.* IITA 2008.

Lv, C., Li, Q., Lei, Z., Peng, J., Zhang, W., & Wang, T. (2010). *PaaS: A revolution for Information Technology platforms.* International Conference on Educational and Network Technology (ICENT), 2010.

Rossi, D., Mellia, M., & Meo, M. (2009). *Evidences Behind Skype Outage.* International Conference on Communications (ICC), 2009.

Dongyan Zhang, Chao Zheng, Hongli Zhang & Hongliang Yu (2010). *Identification and Analysis of Skype Peer-to-Peer Traffic.* Internet and Web Applications and Services (ICIW), 2010.

Chapter 7
Media Resource Adaptation for Multimedia Services and Streaming Media for Mobile Telephones

Tibor Szkaliczki
Hungarian Academy of Sciences, Hungary

ABSTRACT

Nowadays an intensive diversification can be experienced among the devices that can display multimedia content to the user. Typically, clients are unable to receive large audio-visual data volumes in original quality because of resource limitations and deficiency of terminal capabilities. The aim of media resource adaptation is to provide multimedia items for users in a format that is most appropriate with respect to the current terminal capabilities, network characteristics, server capacities and user preferences. The execution time of adaptation is crucial in order to shorten startup delay. The chapter is intended to give a survey on media resource adaptation. Our aim is to introduce the related standards and specifications, show an adaptation framework. We put special emphasis on the algorithmic aspects of adaptation decision taking and present some optimization methods to efficiently select adaptation parameters.

DOI: 10.4018/978-1-61350-144-3.ch007

INTRODUCTION

Recently the mobile telephones have increase their computing power and displays. But they are still limited in the visualization of media information (Chittaro, 2006; Loizides and Buchanan, 2010). On the other hand, mobile networks have increased their bandwidth but still they cannot transport efficiently high definition and 3D multimedia information (Wortham, 2009; Väätäjä and Männistö, 2010). Moreover, mobile devices are not able to receive always and everywhere multimedia content in high quality. This makes that, the access to online multimedia content, especially streaming videos, from the own mobile device provokes disappoint. Still, in many cases the requested video can be displayed only on a high performance *Personal Computer* (*PC*) and efficiently communicated through a high bandwidth communication network. Sometimes the multimedia stream is jerky because it overloads the display device and/or the network. Sometimes the service providers are too cautious and the quality of the media stream is poorer compared to the quality accepted by the device and the network.

Nowadays, there is a high diversity (Chukwu, 2009; Higgins et al., 2010) of: a) devices that can display multimedia content (the capabilities of devices are different in terms of graphical resolution, storage capacity, processor speed, audio quality), b) supported media formats, c) networks and d) content providers. Indeed, mobile device can be used as a source of multimedia to other devices. This makes very challenging the delivery of multimedia contents to any current mobile device. To solve this media objects or media resources (content) must be adapted. The goal of content adaptation (Chang and Vetro, 2005; Hu, 2010) is to present it in a format that is appropriate for the current mobile devices capabilities, network characteristics, server capacities, user rights and user preferences. The replacement of the content itself is not considered content adaptation. A simplest way to do resource adaptation is to store media objects of different qualities. An advanced way is to store only one instance of the high-quality content which can be adapted to the different constraints of the mobile devices at delivery time. Media resource adaptation can be applied to standalone information (e.g., multimedia streaming) or to structured content (e.g., *HiperText Markup Language* (*HTML*) objects).

Figure 1 gives an overview of resource adaptation components and concepts (that will be explained in detail in next sections). Adaptation needs information both on the original content (resource metadata) and the destination consuming the media (usage environment). The media content consists of the multimedia resources and the related metadata containing their properties. *Usage Environment Description* (*UED*) provides information on the entities as follows:

- The terminal capabilities.
- The network characteristics.
- The special requirements and preferences of the person requesting the content.
- The characteristics of the natural environment.

The *Adaptation Decision Taking Engines* (*ADTE*) controls the adaptation process; its task is to determine the parameters of the adapted multimedia resources. Good selection of the adaptation parameters may highly increase the satisfaction of the users. *Adaptation Engine* (*AE*) executes the transformation in accordance with the adaptation parameters coming from the output of the adaptation decision. ADTE should be aware of the capabilities of AE (which parameters can be adapted, what is the range of their possible values, what kinds of transformations are possible, what is the performance of the AE). A general adaptation framework is interoperable with different usage environments and in principle, it can process media resources with arbitrary feature values.

Moving Picture Experts Group (*MPEG*) developed MPEG-21 standard aiming at creating a

Figure 1. Overview of adaptation framework

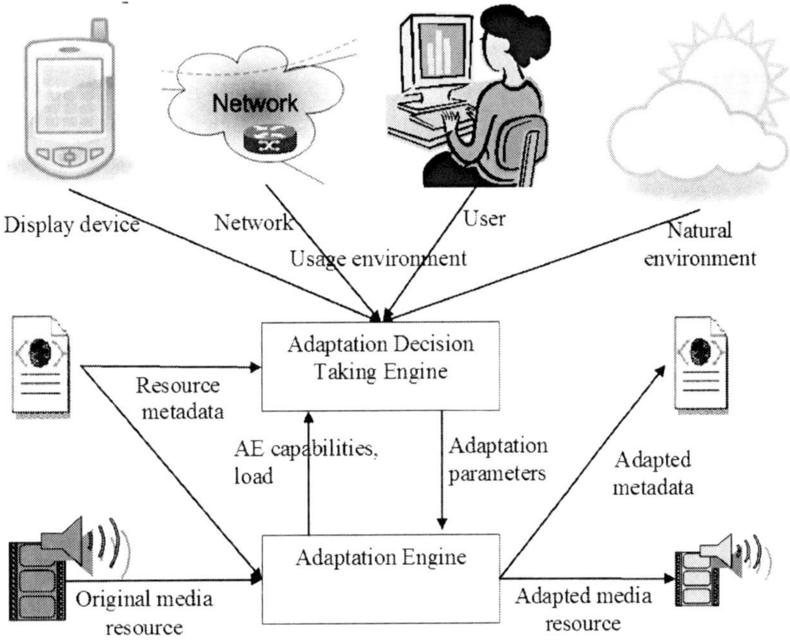

multimedia adaptation framework (Burnett et al., 2003; Bormans & Hill, 2002). Since many elements are already standardized (e.g., codecs), the standard intends to integrate the already existing components, discover the parts where standardization is still needed and specify additional tools to complete the adaptation framework. It specifies the syntax and semantics of tools supporting device and coding format independence (usage environment description, universal constraints description, bitstream syntax description...) without standardizing the methods for adaptation decision and the resource adaptation.

Media Resources

We use the term media resource for text, image, audio and video. We enumerate the physical attributes of the different kinds of media as follows:

- *Spatial resolution* (screen width and height). The resolution may vary significantly on different display terminals, and usually, several different resolutions are available on the same mobile telephone.
- *Color resolution.* If the number of possible color codes is larger than that of the used ones, palettes can make possible to use short color codes even in case of high color resolution.
- *Frame rate.* It is a very important parameter in case of video and animation defining the smoothness of the movements.
- *Audio sample rate.* Audio sample rate influences decisively the quality of the played audio.
- *Audio bits per sample.* It determines in the number of discrete levels in which the intensity values can be represented.
- Number of audio channels.
- *Bitrate.* It gives the bandwidth needed for the continuous streaming of the audio or the video.
- *Format.* Different devices are able to recognize different image formats, audio and video codecs.

Resource metadata facilitates not only multimedia searching, content-based retrieval and personalization but content adaptation as well (Van Deursen et al., 2010). The metadata can define the values of the physical attributes. Ontology for Media Resource developed by *World Wide Web Consortium* (*W3C*) defines basic media resource properties. It gives a mapping from the defined properties to the elements of existing metadata formats which is very useful for transformation between different metadata formats as well. Furthermore, it is a valuable summary on multimedia metadata: It compares the ontology with 17 other media description schemes, CableLabs 1.1, TV-Anytime, TXFeed, XMP, YouTube... (Lee et al, 2010). A brief introduction of some standards and specification on the metadata of multimedia objects follows:

MPEG-7 (Martinez, 2004) specifies *Extensible Markup Language* (*XML*) Metadata for Multimedia Content Description and facilitates multimedia searching, content-based retrieval, personalization and content adaptation. It can describe the media information (such as resolution, frame rate, audio codec, video codec, bitrate...), semantics, structure and other content information (author, creation date...).

In Table 1 we show MPEG-7 description of a video resource. It starts with standard header information with references to XML namespace of MPEG-7. In the code there are bold key words related to the properties of the video (content type, identifier, bit rate (average and maximum bit rates are 525 kbps and 3,147 kbps, respectively), visual coding format (MPEG-1), frame height of 272 pixels, width of 480 pixels and frame rate of 25, audio coding format (*MPEG-1 or MPEG-2 Audio*

Table 1. Sample media resource description in MPEG-7

```
<Mpeg7 xmlns="urn:mpeg:mpeg7:schema:2001" xmlns:xsi=http://www.w3.org/2001/XMLSchema-instance
xmlns:mpeg7="urn:mpeg:mpeg7:schema:2001" xmlns:xml="http://www.w3.org/XML/1998/namespace"
xsi:schemaLocation="urn:mpeg:mpeg7:schema:2001 Mpeg7-2001.xsd">
<Description xsi:type="ContentEntityType">
<MultimediaContent xsi:type="VideoType">
<Video>
<MediaInformation id="test_1231418911_MobyPorcelain.mpg">
<MediaProfile>
<MediaFormat>
<Content href="MPEG7ContentCS:2001"> <Name>audiovisual</Name> </Content>
<BitRate average="525" maximum="3147">2000</BitRate>
<VisualCoding>
<Format href="urn:mpeg:mpeg7:cs:VisualCodingFormatCS:2001:2" colorDomain="color">
<Name xml:lang="en">MPEG1</Name> </Format>
<Frame height="272" width="480" rate="25" />
</VisualCoding>
<AudioCoding >
<Format href="urn:mpeg:mpeg7:cs:AudioCodingFormatCS:2001:3.2">
<Name xml:lang="en">MP3</Name> </Format>
<AudioChannels front="0" side="2" rear="0" lfe="0" track="0">2</AudioChannels>
<Sample rate="48000" bitsPer="16" />
</AudioCoding>
</MediaFormat>
</MediaProfile>
</MediaInformation>
<MediaTime> <MediaDuration>0H4M1S</MediaDuration> </MediaTime>
</Video>
</MultimediaContent>
</Description>
</Mpeg7>
```

Layer 3, MP3), number of audio channels (stereo), audio sample rate of 48 kHz, 16 bits per sample and duration of 4 minutes 1 seconds).

MPEG-21. Digital Item Declaration (*DID*) part of MPEG-21 gives a syntax describing digital items. It is able to define structured current and several variants together with the conditions for selecting a variant (Van de Walle et al., 2005). It embeds MPEG-7 schemes for resource metadata if it is possible.

Let us see the example below (Table 2) for selection of media resource based on the available bandwidth in the *Digital Item Declaration Language* (*DIDL*). It is in XML format starting with reference to DIDL namespace. It contains the identifier *(MOVIE_Item1)* and description of the media resource. The choice structure selects exactly one case from the possibilities. Script named *speed* is used to determine the connection speed. In the *Selection* tags, the speed is classified as high if it is at least 2000 kbit/sec, otherwise it is considered as low. Two components are defined where proper resource variants *movie-v1.mpg* and *movie-v2.mpg* are assigned to high and low connection speeds, respectively.

Resource Description Framework (*RDF*) is a specification of W3C (Hayes, 2004). It provides a general framework for representing information in the Web. It supports a wide variety of applications where knowledge exchange on the Web is important. The RDF Specifications are built on *Uniform Resource Identifier* (*URI*) and XML technologies.

Dublin Core is a basic metadata standard supporting broad range of purposes and business models (dublincore.org, 2010). Dublin Core Metadata Element Set is standardized by *International Organization for Standardization* (*ISO*).

Learning Object Metadata (*LOM*) is the first eLearning standard as a result of the *Institute of Electrical and Electronics Engineers* (*IEEE*) standardization process developed by the *Learning Technology Standards Committee* (*LTSC*) (ltsc.ieee.org, 2002). It contains 64 metadata elements for the description of learning resources. It is compatible with the metadata of digital online libraries. *Sharable Content Object Reference Model* (*SCORM*), the most common eLearning specification (Rustici, 2010) uses the LOM standard for metadata specification.

Table 2. Sample media resource description in DIDL

```
<DIDL xmlns="urn:mpeg:mpeg21:2002:02-DIDL-NS">
<Item id="MOVIE_Item1">
<Descriptor> <Statement mimeType="text/plain"> My Movie. </Statement> </Descriptor>
<Choice minSelections="1" maxSelections="1" choice_id="SPEED">
<Descriptor> <Statement mimeType="text/plain"> Select a proper media resource variant. </Statement> </Descriptor>
<Descriptor> <Statement mimeType="application/x-script"> var speed = system.connectionSpeed(); </Statement> </Descriptor>
<Selection select_id="HIGH_SPEED">
<Descriptor> <Statement mimeType="text/plain"> HIGH SPEED </Statement> </Descriptor>
<Descriptor> <Statement mimeType="application/x-script "> (speed &gt;= 2000 ? 1: 0); </Statement> </Descriptor>
</Selection>
<Selection select_id="LOW_SPEED">
<Descriptor> <Statement mimeType="text/plain"> LOW SPEED </Statement> </Descriptor>
<Descriptor> <Statement mimeType="application/x-script "> (speed &lt; 2000)) ? 1: 0); </Statement> </Descriptor>
</Selection>
</Choice>
<Component id="MyMovie1"> <Condition require="HIGH_SPEED "/> <Resource ref="movie-v1.mpg" mimeType="video/
mpeg"/> </Component>
<Component id="MyMovie2 "> <Condition require="LOW_SPEED "/> <Resource ref="movie-v2.mpg" mimeType="video/mpeg"/>
</Component>
</Item>
</DIDL>
```

The scalable media formats such as scalable image and video coding standards facilitates media resource adaptation. The scalable coding divides the media into layers in such a way that the base layer gives poor quality that can be improved by subsequent enhancement layers. Advantage of the scalable coding technique is that the adaptation can be easily solved by dropping layers (Zhang et al., 2010). Several methods belong to the scalable image coding and they are usually based on a wavelet coding. *Embedded Zerotrees of Wavelet Transforms* (*EZW*) is an example for an early method on which several other methods were built later (Shapiro, 1993). The image with small bit rate is embedded into an image with high bit rate. Scaling can be done both in the terms of *Peak Signal-to-Noise Ratio* (*PSNR*) and resolution. *Joint Photographic Experts Group* (*JPEG*) developed the most known scalable image coding standard named as JPEG2000 (Marcellin et al., 2000). It contains the image in several resolutions. After downloading a small part of the image, the image can be seen blurred with bad contrast and the quality is gradually improving as more and more parts are downloaded from the picture.

The *Scalable Video Coding* (*SVC*) standard is an extension of H.264/MPEG-4 *Advanced Video Coding* (*AVC*) standard of Telecommunication Standardization Sector (ITU-T) of *International Telecommunication Union* (*ITU-T*) and *Joint Technical Committee 1 of the International Organization for Standardization and the International Electrotechnical Commission* (*ISO/IEC JTC 1*) organizations (Schwartz et al., 2007). A scalable video stream contains several sub-streams that can be decoded standalone representing the same content but in reduced quality. The sub-streams can be gained by simply dropping layers from the original stream. There are four kinds of scalabilities:

- *Temporal.* Whole frames can be dropped. Widespread methods are known for this case. Due to its simplicity, frame dropping is the most common method in lack of bandwidth or *Central Processing Unit* (*CPU*) power. It is part of H.264/MPEG-4 AVC standard as well.
- *Spatial.* The video is encoded in several spatial resolutions.
- *Quality.* The spatial and temporal resolutions are the same in the substreams but their qualities are different in terms of fidelity or *Signal to Noise Ratio* (*SNR*).
- *Combination of the above methods.* We speak about fully scalable video stream when the stream can be scaled in terms of all adaptation parameters.

Figure 2 represents a fully scalable video stream in a three dimensional space. The dimensions represent the different domains of scalability (e.g., temporal, spatial, and quality). The spheres represent the packets of the media stream. All packets form the stream with full quality. The dark packets represent the adapted stream obtained by dropping the white ones.

The main problem with most of the SVC types, except scaling in time, was that they required quite complex decoder at the client side and decoding overloaded the display device. Nowadays,

Figure 2. SVC. Dark and white spheres represent the retained and dropped packets, respectively, during adaptation

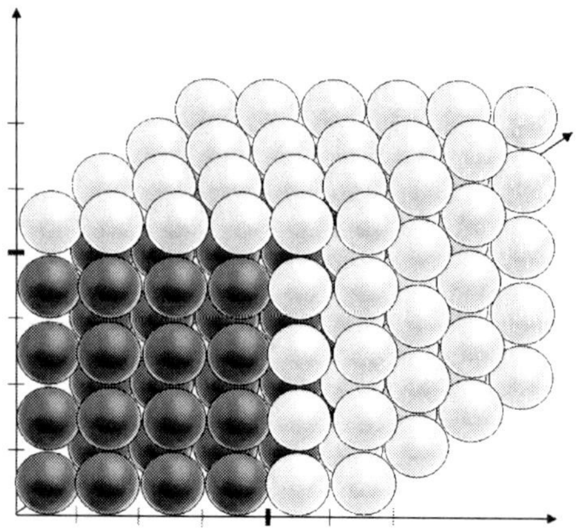

SVC is spreading in real applications, due to the improvement of the performance of the devices and to sophisticated coding techniques. Profiles are defined in the standard for specific class of applications and target devices. The extension of the standard to *Scalable Baseline Profile (SBP)* specifies scalable coding especially suitable for the mobile applications.

Device Description

Besides media resources the properties of the display devices should also be defined. Multimedia and Web experts meet common problems as new kinds of devices such as *Personal Digital Assistants* (*PDAs*) and mobiles appeared (Glover and Davies, 2005). Both communities developed new standards and recommendations in order to describe multimedia data and device capabilities. Let us see the most important standards and recommendations on the existing metadata sets describing the capabilities and features of devices.

W3C Mobile Web Initiative (*MWI*) *Device Description Working Group* (*DDWG*) provides ways to describe devices in order to facilitate adapting Web content to the capabilities of mobile devices. Among other specifications (e.g., recommendation on *Application Programming Interface* (*API*) to the Device Description Repository), they defined a core vocabulary identifying the most important properties of devices (vendor, model and version of the browser, display width and height, display color depth, input devices, markup, stylesheet, image format support, input mode, cookie and script support), see Hanrahan et al. (2008).

User Agent Profile (*UAProf*). The aim of UAProf is to describe the capabilities of the mobile devices (Butler, 2002). It specifies the properties of the hardware and software and contains information on the network to where the device is connected. It is based on the RDF specification of the W3C: It contains an RDF scheme and vocabulary that can be used for media description. The UAProf information on a device can be queried from the Web.

Let us see an example for a UAProf code fragment describing a hardware platform (Table 3): It specifies a device model supporting Bluetooth headsets, the screen resolution is 121 x 87, the model supports three kinds of character sets, the screen is monochrome with grayscale resolution

Table 3. Sample device description in UAProf

```
<rdf:Description ID="HardwarePlatform">
<rdf:type resource="http://www.wapforum.org/profiles/UAPROF/ ccppschema -20010430#HardwarePlatform" />
<prf:BluetoothProfile> <rdf:Bag> <rdf:li>headset</rdf:li> <rdf:li>dialup</rdf:li> </rdf:Bag> </prf:BluetoothProfile>
<prf:ScreenSize>121x87</prf:ScreenSize>
<prf:Model>SampleModel</prf:Model>
<prf:InputCharSet> <rdf:Bag> <rdf:li>ISO-8859-1</rdf:li> <rdf:li>US-ASCII</rdf:li> <rdf:li>UTF-8</rdf:li> </rdf:Bag> </
prf:InputCharSet>
<prf:ScreenSizeChar>15x6</prf:ScreenSizeChar>
<prf:BitsPerPixel>2</prf:BitsPerPixel>
<prf:ColorCapable>No</prf:ColorCapable>
<prf:TextInputCapable>Yes</prf:TextInputCapable>
<prf:ImageCapable>Yes</prf:ImageCapable>
<prf:Keyboard>PhoneKeypad</prf:Keyboard>
<prf:NumberOfSoftKeys>0</prf:NumberOfSoftKeys>
<prf:Vendor>myprofileprovider</prf:Vendor>
<prf:OutputCharSet> <rdf:Bag> <rdf:li>ISO-8859-1</rdf:li> <rdf:li>US-ASCII</rdf:li> <rdf:li>UTF-8</rdf:li> </rdf:Bag> </
prf:OutputCharSet>
<prf:SoundOutputCapable>Yes</prf:SoundOutputCapable>
<prf:StandardFontProportional>Yes</prf:StandardFontProportional>
</rdf:Description>
```

of 2 bits, text characters can be input, the device is able to display images, etc.

Wireless Universal Resource File (*WURFL*). Similarly to UAProf, WURFL specifies the capabilities and features of the mobile devices. WURFL is the result of the effort of *Free and Open Source Software* (*FOSS*) community in order to make the content displayable on a broad range of wireless devices. It is a single XML configuration file containing information on all known wireless devices. The developers provide information on the devices from all over the world and WURFL is regularly updated with the data of the new wireless devices appearing on the market. Unfortunately, the data provided by the vendors are not always exact but the mistakes are continuously corrected. WURFL can be stored locally and in contrast to UAProf, it can be used even without Web access.

The possible capabilities refer to the areas as follows: general product information, browsers, markup languages, display, image formats, downloadable objects, *Wireless Application Protocol* (*WAP*), *Multimedia Messaging Service* (*MMS*), *Short Message Service* (*SMS*), audio formats, streaming video, flash... The display information refers to the width and height. The number of displayable colors can be found in the group of capabilities referring to the image formats. The downloadable objects can be ringtone, wallpaper, screen saver, image and video. Besides the format, the streaming video can be characterized by the maximal width and height, bit rate, frame rate, size in bytes, support of H263 codec... The complete list of capabilities used by WURFL can be found in (Passani, 2010).

Let us see an example for a device description (Table 4). It starts with a general description of the device (device label) followed by groups (group label) of individual capabilities (capability label) and their values. The device part specifies a User Agent String and an identifier. Property *fall_back* points to a general device family in the hierarchy of devices. The sample description contains groups providing product info such as model name and specifying *eXtensible HyperText Markup Language* (*XHTML*) and markup language capabilities.

On one hand, the aim of the chapter is to help the reader to better understand the problem and give an overview on the techniques related to adaptation. We focus on the simplest way and standalone content. We also present our contribution that is choosing the best adaptation decision considering resource limitations on client and server sides as well as network characteristics in order to find appropriate adaptation parameters in strict time limits.

Table 4. Sample device description in WURFL

```
<device user_agent="VendorXXXX" actual_device_root="true"
fall_back="vendor_generic_series40" id="vendor_XXXX_ver1">
<group id="product_info"> <capability name="model_name" value="XXXX" /> </group>
<group id="xhtml_ui">
<capability name="xhtml_format_as_css_property" value="true" />
<capability name="xhtml_supports_table_for_layout" value="true" />
<capability name="xhtml_supports_css_cell_table_coloring" value="true" />
<capability name="xhtml_readable_background_color1" value="#99CCFF" />
<capability name="xhtml_readable_background_color2" value="#FFFFFF" />
</group>
<group id="markup">
<capability name="preferred_markup" value="html_wi_oma_xhtmlmp_1_0" />
<capability name="html_wi_w3_xhtmlbasic" value="true" />
<capability name="html_wi_oma_xhtmlmp_1_0" value="true" />
</group>
</device>
```

The rest of the book chapter is organized as follows. The main adaptation techniques and the adaptation decision taking, i.e. the procedure how to select the adaptation parameter values are examined in Section 2. Optimization methods are presented to find as good adaptation parameters as possible in Section 3. Future research directions are shown in Section 4. Finally, the chapter is summarized in Section 5.

ADAPTATION ENGINE

The AE executes the adaptation on the original content and it produces a variation of the media resource with given adaptation parameters. It can be built from several building blocks for adapting the different types of media resources with different adaptation parameters. It is not standardized; it is the developers' responsibility to decide what kinds of adaptation techniques are needed. We have to remark that the term AE can be used in the literature in a sense designating a module including both the adaptation execution and the adaptation decision taking, but we mean the executing part only.

The three main AE are:

- Adaptation by selection. It is the simplest way where content developers create different content for each device type (for mobile phones, PDAs, net PCs, desktop PCs...). In this case, several versions of the same content are stored and when the content is updated, each version has to be modified. It can handle a limited number of variants but can be performed in a short time. It is usually the only solution for adapting text length.
- Transcoding. The content is stored only in one high-quality instance which can be adapted to different devices. The drawback of transcoding is that the CPU load on the serving node is quite high. The media resource is typically decoded into an uncompressed format, and then encoded with the given adaptation parameters. Typical task in adaptation is to re-sample images (or frames) on a regular grid in order to change the spatial resolution. Color quantization reduces the number of colors on an image; the resulting image may well represent the original one even with limited number of colors if the color palette will be appropriately selected. Vetro et al. (2003b) gave an overview on transcoding architectures.
- Transformation needs scalable format that makes the adaptation easy by dropping enhancement layers (Thang et al., 2009). The price of the simple adaptation is the increased complexity in encoding and decoding using a scalable format: the adaptation is easy in case of layered content but the decoding process has high computation need. Therefore, it was not a preferred method for weak display devices (low memory and small CPU) although scalable decoders have become more efficient and they are available nowadays for mobile devices as well.

The above techniques can be combined. For example, as we mentioned, transcoding is a computation intensive task. Therefore, caching can help in accelerating the adaptation process by storing the frequently requested media resource variants and only the variants not found in the cache should be transcoded. Caching technique can be regarded as a combination of transcoding and adaptation by selection techniques.

The AE can run in the node assembling and delivering the content (i.e., server node in case of client-server architecture), in the client node or somewhere between them. Adapting on the serving node has several advantages: typically, the server nodes are powerful enough to perform the time-consuming transcoding process and the network load is decreased if only the adapted variant is sent to the network. Adaptation at the client is an option if the serving node does not care with the adaptation or cannot consider all circumstances. Some recent solutions are known where the adaptation is performed in the network elements: network routers or gateways can react fast on changes in bandwidth by dropping enhancement layers of scalable video streams. Dick et al. (2005) developed a system with distributed adaptation.

Furthermore, the adaptation can be performed in a distributed way where several nodes may participate in the adaptation by executing different steps of the adaptation (Hellwagner et al., 2009). The distributed architecture may improve the performance of the adaptation. It is possible that the server node sends a degraded media resource variant that is further adapted in the network or at the client node, knowing the current state of the usage environment. For example, the server creates a video stream that the client is able to play in normal circumstances, but if the client is overloaded by other applications, it can skip frames.

The AE not only receives adaptation parameters but also sends data to the adaptation decision taking engine: AE capabilities (e.g., codec types, quantization levels) and current AE resources (e.g., available server CPU capacity).

Although wide diversity can be experienced also in the coding formats it is possible to realize a coding format independent adaptation. MPEG-21 multimedia framework offers tools for this as *Bitstream Syntax Description* (*BSD*) and BSDLink for describing bitstream structure and adapting bitstreams (Vetro and Timmerer, 2005).

Several open-source tools are available for performing adaptation e.g., *mencoder* for video (mplayerhq.hu, 2010), *ImageMagic* for images (Still, 2005), Lame for Audio (Cheng, 1998) and *iconv* and International Components for Unicode for character encoding translations (Haible, 2000).

Adaptation Decision Taking

The task of the ADTE is to choose the appropriate adaptation parameters. The aim is to find the adaptation parameters which fit in the given constraints typically referring to resource limitations and terminal capabilities, and which are optimum, e.g. in the sense of utility value for the consuming user. *Adaptation Decision Taking* (*ADT*) in general is a quite complex task and is not standardized; except input descriptions. It is the responsibility of the developers of the adaptation module how to make balance between the accuracy of the decision and the complexity of the problem model. They have freedom in selecting parameters to be considered in the decision process, defining constraints on the parameters and giving optimization goals.

The ADT engine can run in different times. In case of static adaptation, the decision is taken at the start of the media resource delivery. This is a good approach when only the device capabilities are considered or when the available resources are assumed to be fixed while the media is being served. These assumptions are not valid in multimedia streaming when a large amount of data is served for a long time. In this case dynamic adaptation is performed when the decision is repeated periodically during the streaming in order to react on the changes in the usage environment (e.g. fluctuations in the available bandwidth or CPU load).

The general form of the problem model is a mixed-variable multi-criteria optimization with general constraints. The *Digital Item Adaptation* (*DIA*) Part of MPEG-21 specifies ADTE in a general way, and even the constraints and the optimization goals can be given as inputs and offers several tools for supporting ADTE (Vetro et al., 2007). Next we define the inputs, outputs, constraints and the goal function of the optimization problem.

The inputs of ADT are the metadata on the resource, the description of the usage environment and the capabilities and load of the AE (Figure 1). Instead of the media resource, the metadata describing it is enough for the decision. The usage environment includes, first of all, the terminal. The network is also an important part of the usage environment because it is often a bottleneck in the service. User characteristics and the natural environment may also influence the final decision. It can be avoided that the ADTE should select a variation that the AE cannot produce because the ADTE has information about the AE capabilities as well. The above information would be enough if the problem model of the adaptation decision taking was fixed.

We have observed that the possible input parameters considered by the final decision can be quiet wide. Using more and more input parameters, the decision can be more adequate but some problems may be caused as well by giving exact input data. For example, if the current network characteristics and node resources on the client and server side are considered, their monitoring should be realized in the network as an additional functionality and the collected data should be sent to the ADTE.

The outputs of the ADTE are the set of adaptation parameters specifying the target media resource. Naturally, it depends on the AE which parameters can be changed. The output values should fall within the range of the adaptation parameters fitting into the terminal and AE capabilities. Remember that the range usually cannot be given separately for each adaptation parameter since the possible output values of a parameter may depend on the values of other parameters. Therefore, the range of the possible output values can be imagined as an arbitrary vector set in the multidimensional space of the adaptation parameters.

Multimodal adaptation is also possible where, e.g. audio is replaced by text or video by an image. In this case the output feature space should be able to describe not just the original kind of media resource but the reduced media type as well.

Optimization may have several different contradicting goals (e.g. maximizing the quality of the served stream, minimizing the cost). Adaptation decision can be a complex multi-criteria optimization problem. In case of multi-criteria optimization, usually, does not exist a unique global optimum but a set of Pareto optimal solutions (a solution is Pareto optimal if it will be not dominated by any other solutions i.e., there is no other solution that is better from all the optimization aspects.) Here we give some typical optimization goals as follows:

- *Adaptation parameters*. Typically, the possible values of the adaptation parameters are comparable and can be ordered. For example, users desire as high spatial and color resolution and as high frame rate as possible on their terminal. Therefore, the adaptation problem may be optimized for the adaptation parameters. This goal can be well followed if the available resources do not limit the service, but the device capabilities. However, if the resources cause bottleneck in the service, optimizing for all adaptation parameters is a complex task where one adaptation parameter can be improved on the price of degrading the other. The metrics enlisted below offer a possibility to give an overall evaluation on the goodness of the media and to compare different media variants.
- *Perceptual Quality* (*PQ*) is a metric based on how a user perceives the audio visual media. For example, the user prefers a picture with sharp contours to a blurred one, or continuous video is preferred to a jerking one. The PSNR and *Perceptual Evaluation of Audio Quality* (*PEAQ*) (Thiede et al., 2000) metric can be used for video and audio PQ estimation, respectively, that can be calculated by using numeric methods for the adapted media resource variants.

- *Semantic Quality* (*SQ*) is a more subjective metrics than the perceptual one and it tries to measure in what extent the user receives the semantic content of the adapted variant of the media resource. For example, it can express how well the client receives the information brought by the media or how well a delivered audio stream containing music can entertain listener (Thang et al., 2005). It is not a trivial task to give a numerical formula for it. Setting up a semantic quality model often requires expensive and time consuming subjective tests in order to gather information on the users' opinion. As a subjective term, semantic quality may depend on the user preferences and the content itself as well.

Prangl et al. (2007) defines the uni-modal semantic quality as a weighted sum of the relative values of the adaptation parameters. The weights express the relative importance of each adaptation parameter. The weights can be used for customizing the model, taking the user preferences and the content type into account. For example, if users prefer color to the spatial resolution, they can give higher weight on the color resolution. Let us see an example how the quality of the media resource can depend on the content type (or genre): in case of a sport program (e.g., soccer) the fast movements are more important than the high colors resolution, therefore, frame rate should get an increased weight, while the weight of the color resolution is decreased. On the other hand, the color resolution is significant in a natural scene.

- *Utility.* This metrics is used in general term and is closely related to the perceptual and semantic quality. It expresses the overall satisfaction of the user consuming the media resource. It can be the combination of the other metrics. There exist cross-modal utility models that estimate the total utility of a media stream consisting of two or more modalities, e.g., video and audio as a function of the utilities of the substreams (Winkler & Faller, 2005).
- *Distortion.* In many cases, instead of maximizing the quality or utility of a media resource, a minimization problem is solved, namely, the distortion of the media resource is minimized i.e., the degradation of the adapted variant relative to the original media resource should be kept as small as possible.

Regarding the input data and the output parameters we have several constraints. An output satisfying all the constraints should specify a variant of the media resource that can be produced by AE, transported to the client and played by the terminal. The constraint may refer to the network and node resource limits, a range of the output parameters. It is a challenge for modeling how to match the physical parameters of the media and the device parameters (e.g., how the CPU need on the client side depends on the parameters of the video). Functions can be given in different formats: e.g., look-up tables, lists or formulas can be used.

Some of the constraints are quite simple and can be easily managed: For example, the capabilities of the display device (e.g., spatial and color resolution) can specify a range for a single adaptation parameter from which a value has to be selected during the adaptation decision taking. The adaptation is different upon reducing the resource usage of media content delivery and display, which is a critical task in a nearly overloaded system. In this case several options are possible to reduce the parameters of the overloaded system.

Rate-distortion optimization appeared in media (image, audio, video) compression but it is also useful in adaptation as well in order to find the best quality stream by using limited amount of resources (Berger, 1971). Although rate-distortion optimization does not consider the static device capabilities but it can be well used in dynamic bit rate adaptation that is a crucial process in case of changeable network characteristics.

Many solutions use rate-distortion model. Kim et al. (2005) present a general framework formulating adaptation as constrained optimization problem considering feasible adaptation operations, constraints and utilities. Formulations for both the resource-constrained utility maximization and the utility-constrained resource minimization are presented. Rate adaptation problem is defined where the distortion of the video should be minimized while not exceeding the target bitrate. They apply the problem model to rate adaptation of videos where frame dropping and discrete cosine transformation coefficient dropping are allowed as adaptation operations.

In the model of Prangl et al. (2007) the possible variants of the adapted multimedia streams can be characterized by the combination of video and audio adaptation parameters. The input data provide information on the terminal capabilities and current resource limits of the usage environment. Furthermore, ADTE is aware of the capabilities of the AE and the resource limits of AE as well. The output of the decision taking is an adapted variation of the multimedia stream consisting of the video and the audio streams. They use the cross-model utility model for goal function where the total utility of a multimedia stream can be given as the linear combination of the utilities of the particular modalities. One of the general constraints claims that a variation can be selected if AE is able to produce it and the terminal is able to play it. There exist general constraints on the resource limits as well. In order to make the model complete, the utility and resource needs (CPU clock cycles in AE and terminal, memory of the terminal, bandwidth over the network) of the variation should be given: they can be part of the input, or additional constraints can specify functions regarding how they can be calculated from the adaptation parameters.

ALGORITHMS

The ADT has to run fast in order to keep the startup delay of the requested session in the acceptable range. The decision has to be taken in real-time especially when the decision is continuously re-evaluated because of fluctuations in the available resources (e.g. bandwidth, CPU capacity...). Finding the optimal combination for the individual user within strict time constraints is a challenging optimization problem. In many practical cases the time-consuming algorithms aiming at the exact solution are not applicable. In this case approximation or heuristic methods can be applied that find a feasible solution close to the optimum.

It is possible to process requests for media resources one by one or consider several requests together. The reason of the first approach is that the adaptation decision should be performed promptly when a new request arrives, while the second approach can better utilize the resources if several requests can be managed together. We consider the decision on a single media resource, because otherwise the problem leads to the more complicated one of integrated resource management.

The selection of proper algorithm depends highly on the complexity of the problem model. Let us see some possible searching and optimization techniques as follows:

- *Adaptation rules*. In the simplest case, the constraints define adaptation rules in a switch-case structure for the adaptation parameters defining the adaptation steps taken in certain cases. Look-up tables are also appropriate to describe the cases and the corresponding outputs. That offers a fast and straightforward method to solve the problem but it can be applied if the number of adaptation options is limited.
- *All Combinations*. The total enumeration is possible for the case when the possible adaptation parameter values are discrete

(Mukherjee et al., 2005). This simple approach takes all combinations of the adaptation parameters one after the other and selects the optimum one(s) from the combinations that comply with the constraints. This method can be used for ADT if the total number of the combinations is limited, and it is useful for validating the results gained by other algorithms.

- *Dynamic programming.* Thang et al. (2008) applied the Viterbi algorithm of dynamic programming to find optimal transformation of several simultaneous SVC streams.
- *Merging Video and Audio Variations.* Merging can be used for finding the maximum of a separable nonlinear global optimization problem. We call an optimization problem separable into two groups if the goal function and the resource constraints will be written as weighted sums of functions depending on two separate variable groups. This method can be used for ADT on multimedia stream if the constraints consider the range of the adaptation parameters and the resource needs only. A further requirement of applicability is that the common goal function and overall resource needs could be defined as the weighted sum of the goal functions and resource needs of the audio and video streams. The method can be efficiently used if the number of limiting resources will be at most two (e.g. bandwidth and CPU capacity).

The enumeration of all the combinations of video and audio variants can be avoided by exploiting the fact that it is enough to combine a video variation only with the audio variation of the highest utility among those whose resource needs are less than the available resources reduced by the video resource need (Prangl et al., 2007). A merging algorithm proceeds on the video and audio variations simultaneously, according to the increasing and decreasing order of bandwidth demand, respectively. In order to efficiently manage the ordered list of candidate audio variations, it is implemented in a special balanced tree, namely red-black tree.

- *Border Scan.* This approach can be used if the goal function and the resource needs are monotonic in the function of adaptation parameters, which is typical. There exist several methods for solving monotonic optimization problems (Tuy, 2000). The monotony of the resource needs and utilities implies that a point representing the optimum solution in the space of adaptation parameters is located directly below the surface (or border) defined by the resource constraints. This fact implies that, instead of all points, it is enough to enumerate the points located directly below the border for finding the optimum solution (Prangl et al., 2007). The algorithm generates the border starting from a single initial value. Then the algorithm takes the different adaptation parameters one after the other and extends the border in the subspace of the parameters already examined, while the values of the not yet selected parameters are fixed. After including the last parameter, all border points are generated and evaluated for optimality.
- *Hill Climbing.* Heuristic search methods, e.g., steepest-ascent hill climbing (Rich & Knight, 1991), offer an efficient approach for real-time application at hand. Monotonicity in the resource needs and goal functions is advantageous from the aspect of the convergence to the optimum. In each iteration step, the adaptation parameter is increased in the direction where the improvement in the goal function is the highest and the new variation satisfies the constraints. Unfortunately, this algorithm may run into a local optimum but it is very

fast. The accuracy of the algorithm can be enhanced by running it repeatedly with different initial points (Prangl et al., 2007).

- *Langrangian relaxation* can be well used to constrained optimization problems. It transforms the constrained problem into the series of unconstrained ones. In this case the goal function is penalized by the product of the Lagrange multiplier and the degree of the violation of the constraints if any. The Lagrange multiplier is gradually increased for each execution of the unconstrained optimization in order to move the found optimum into the region of feasible solutions. Ortega and Ramchandran (1998) apply it to convex rate-distortion optimization.
- *Generalized Pattern Search (GPS)*. In most cases it would be enough to restrict ADT to discrete optimization, since even continuous parameters (audio bit rate, sample rate) can be well approximated with integer values or the range of the possible values is discrete. However, we may get more accurate solutions if the continuous parameters are managed really as continuous variables instead of sampling them to discrete values. Therefore, let us consider the methods dealing with continuous adaptation parameters.

Direct search methods (Kolda et al., 2003) can be well applied to continuous unconstrained optimization. Mukherjee et al. (2005) recommended the Generalized Pattern Search (Audet & Dennis, 2003) for ADT, which is one of the most promising direct search approaches proving strong convergence properties. Like other search methods, GPS tries to improve a solution in each iteration step by moving from a point to another one in the parameter space. The algorithm tries to step from the current point in the directions defined by a lattice. If an improvement is found, the step will be really taken and the search continues from the more optimal point, using the same or increased lattice. If no improvement is found using the lattice, its scale will be reduced in order to decrease the lengths of the trial steps. Compass or co-ordinate search is a special version of the GPS. In this case, the adaptation parameters of the current point are taken one after the other in each iteration step, and each parameter is decreased and increased at a given value in order to find an improved solution.

The above method is used for the unconstrained case. The constraints can be introduced in different ways into the search methods. Lagrangian relaxation is a good possibility as we have seen. There is another simple option for including constraints to check the constraints at each trial step and to skip the steps that lead to points violating the constraints. *Mesh Adaptive Direct Search (MADS)* (Audet, C., Dennis, J. E., 2006) can help in this case to ensure convergence to the optimum.

If a variant can be characterized by both discrete and continuous parameters, the continuous search methods should be extended to deal with discrete values as well. In this case while the changes in the continuous parameters are defined by the lattice, the discrete parameters can jump to neighboring discrete values in each step.

Artificial Intelligence (AI) methods. Jannach et al. (2006) recommended AI approach for ADT. In their model the output consists of not just adaptation parameters, but a sequence of adaptation steps (e.g., first adaptation in spatial domain and then in temporal domain). The ADTE finds the proper sequence of adaptation steps for the multimedia resource. This problem can be considered as a state space planning problem and be solved by using Prolog.

Performance Evaluation

We implemented four algorithms from the above methods, namely, generating all combinations, merging video and audio variations, border scan and hill climbing in the adaptation framework of

Prangl et al. (2005). Generating all combinations was selected in order to validate the other methods. Merging and border scan exploits the monotonicity of utility and resource need functions therefore we expected better performance from them than from general optimization methods. Among the implemented methods, the hill climbing has the best time complexity but it may fail to find the optimum.

We tested the implemented methods on real multimedia data. Table 5 shows the running times of the methods as a function of the number of possible multimedia variants. You can see that the heuristic hill climbing method was clearly the fastest. The first three methods always found the optimum while the average relative error of hill climbing was 1% but in the worst case the error was even 15%. Repeated run of hill climbing with different starting points reduced the error to 0.2% and 1.3% in the average and worst case, respectively. Among the exact methods, the border scan had the shortest running time. The results show that the running time of ADT can be significantly reduced in comparison to generating all combinations that is a common method in the case when each parameter has discrete values.

FUTURE RESEARCH DIRECTIONS

MPEG-21 standard gives tools for an interoperable multimedia framework. However, components can be found in this standard with free competition in the development (methods applied in AE and ADTE) as indicated by López et al. (2010).

The adaptation of interactive components seems to be a hard task; however, they are indispensable part of most multimedia services.

Different kinds of media resources have different kinds of *Quality of Service* (*QoS*) requirements during delivering through network (e.g., text and video have totally different QoS requirements). Therefore, media-aware network elements with media resource adaptation capabilities could improve the performance of the systems (Zahariadis et al., 2009).

There exist many approaches to formulate the utility of a multimedia content, but it is quite hard to express it numerically. Besides automatic ADTs, the solutions where the user can participate in the decision are an interesting research topic as well. In this case clients could adjust the adaptation parameters of a stream for themselves, customizing it for their current preferences by requesting a smoother play, higher resolution, more colors, better audio quality… (of course, at the expense of degrading the other parameters).

A gap can be observed between the implementation of the adaptation in commercial products and the standard methods. We expect the spreading of standard solutions in the systems used in practice.

Although several optimization methods are known for the ADT, it is still a challenge to find optimal or nearly optimal methods with short running time. In the future, we could perform a comparison study of ADT including more methods than the evaluated ones in this chapter.

Table 5. Running time of implemented algorithms in seconds

Number of variants	*500*	*1000*	*1500*	*2000*	*2500*	*3000*	*3500*	*4000*
All combinations	3.095	5.516	7.171	11.972	13.699	17.314	23.893	27.299
Merging	3.175	7.11	7.51	14.464	17.083	17.351	27.745	31.835
Border scan	1.321	0.918	1.326	2.078	4.616	6.918	8.008	8.369
Hill climbing	0.003	0.003	0.003	0.026	0.042	0.035	0.13	0.089

CONCLUSION

In the chapter an overview on the media resource adaptation was presented and the main components related to this area were shown. Mobile multimedia services and streaming have increasing popularity nowadays and the need for standard solution in content adaptation is increased. We suppose that the current constraints would allow better quality multimedia services and streams than the users receive them in many implemented services. Sophisticated adaptation techniques can better exploit the available resources and yield a better multimedia experience for the client. Metadata play a relevant role in providing media and device format independence. We have observed that there are plenty of standards and specifications that support different areas related to resource adaptation (e.g., metadata schemes, device descriptors, coding techniques). Since the quality of the media resources to be played develop quickly and vendors keep on appearing with new products, a general multimedia framework is needed that is able to manage the wide range of inputs and is ready to manage future display devices and media formats. MPEG-21 is the most recent standard defining an interoperable multimedia framework for universal multimedia access.

The scalable media coding is a technique supporting quick adaptation. The improvement of the performance of this technique made it available on mobile devices as well. Finding resource media stream variations that maximize the media stream's utility for the user under given constraints, represents a challenging complex optimization problem in the multimedia area. Providing short running time both in the adaptation execution and ADT is important for avoiding long startup delay and proper quality of service.

Finding media stream variations that optimize the utility (or the media experience) for the user under given resource constraints represents a complex optimization problem. We presented and implemented algorithms to find optimal video and audio variations for multimedia content adaptation. The adaptation methods yield a better multimedia experience for the client.

REFERENCES

Audet, C., & Dennis, J. E. (2003). Analysis of generalized pattern searches. *SIAM Journal on Optimization*, *13*(3), 889–903. doi:10.1137/S1052623400378742doi:10.1137/S1052623400378742

Audet, C., & Dennis, J. E. (2006). Mesh adaptive direct search algorithms for constrained optimization. *SIAM Journal on Optimization*, *17*(1), 188–217. doi:10.1137/040603371doi:10.1137/040603371

Berger, T. (1971). *Rate distortion theory-A mathematical basis for data compression*. Englewood Cliffs, NJ: Prentice-Hall.

Bormans, J., & Hill, K. (Eds.). (2002). *MPEG-21 Overview* v.5, ISO/IEC JTC1/SC29/WG11/N5231

Burnett, I., Van de Walle, R., Hill, K., Bormans, J., & Pereira, F. (2003). MPEG-21: Goals and achievements, tutorial, *IEEE Multimedia.*

Butler, M. H. (2002). CC/PP and UAProf: issues, Improvements and future directions, in *W3C Workshop on Delivery Context,* 4-5 March 2002, Sophia-Antipolis, France.

Chang, S.-F., & Vetro, A. (2005). Video adaptation: Concepts, technologies, and open issues, in *Proceedings of the IEEE, 93*(1), 148-158. Piscataway, NJ, USA: IEEE Press.

Cheng, M. (1998). *Rationale for LAME development*. Retrieved on January 5, 2011, from http://lame.sourceforge.net/rationale.php

Chittaro, L. (2006). Visualizing information on mobile devices. *IEEE Computer*, *39*(3), 40–45.

Chukwu, M. (2009). Media adaptation for connected mobile devices, in *Proceedings of the 6th IEEE Conference on Consumer Communications and Networking Conference (CCNC '09)* (pp. 168-169). Piscataway, NJ, USA: IEEE Press.

Dick, M., Brandt, J., Kahmann, V., & Wolf, L. (2005). *ICIP 2005. IEEE International Conference on Image Processing, 3*, pp. III - 700-3. Digital Object Identifier: 10.1109/ICIP.2005.1530488

dublincore.org. (2010). *DCMI metadata terms*. Retrieved on January 05, 2011, from http://dublincore.org/documents/dcmi-terms/

Glover, T., & Davies, J. (2005). Integrating device independence and user profiles on the Web. *BT Technology Journal, 23*(3), 239–248. doi:10.1007/s10550-005-0045-ydoi:10.1007/s10550-005-0045-y

Haible, B. (2001). *The Unicode HOWTO*. Retrieved January 5, 2011, from http://tldp.org/HOWTO/Unicode-HOWTO.html

Hanrahan, R., Trasatti, A., & Rabin, J. (2008). *Device description repository core vocabulary*. Retrieved August 20, 2010, from http://www.w3.org/TR/2008/NOTE-ddr-core-vocabulary-20080414/

Hayes, P. (Ed.). (2004). *RDF semantics*, W3C recommendation, February 10, 2004. Retrieved August 20, 2010, from http://www.w3.org/TR/2004/REC-rdf-mt-20040210/

Hellwagner, H., Kuschnig, R., Stütz, Th. & Uhl, Andreas. (2009). Efficient in-network adaptation of encrypted H.264/SVC content. *Signal Processing Image Communication, 24*, 740–758. doi:10.1016/j.image.2009.07.002doi:10.1016/j.image.2009.07.002

Higgins, B. D., Reda, A., Alperovich, T., Flinn, J., Giuli, T. J., Noble, B., & Watson, D. (2010). Intentional networking: Opportunistic exploitation of mobile network diversity, In *Proceedings of the 16th Annual International ACM Conference on Mobile Computing and Networking (MobiCom)* (pp. 73-84). New York, NY, USA: ACM.

Hu, S. (2010). Personalized content adaptation using multimodal highlights of soccer video, in *Proceedings of 11th Pacific Rim Conference on Multimedia* (LNCS 6297, Issue PART 1, pp. 537-548). Berlin, Heidelberg: Springer Verlag.

Jannach, D., Leopold, K., Timmerer, C., & Hellwagner, H. (2006). A knowledge-based Framework for multimedia adaptation, *Applied Intelligence - Special Issue. Innovations in Applied Artificial Intelligence, 24*(2), 109–125.

Kim, J.-G., Wang, Y., Chang, S.-F., & Kim, H.-M. (2005). An optimal framework of video adaptation and its application to rate adaptation transcoding. *ETRI Journal, 27*, 341–354. doi:10.4218/etrij.05.0105.0019doi:10.4218/etrij.05.0105.0019

Kolda, T. G., Lewis, R. M., & Torczon, V. (2003). Optimization by direct search: New perspectives on some classical and modern methods. *SIAM Review, 45*, 385–482. doi:10.1137/S003614450242889doi:10.1137/S003614450242889

Lee, W. S., Bürger, T., Sasaki, F., Malaisé, V., Stegmaier, F., & Söderberg, J. (Eds.). (2010). *Ontology for media resource 1.0, W3C working draft 2010*. Retrieved August 20, 2010, from http://www.w3.org/TR/mediaont-10

Loizides, F., & Buchanan, G. (2010). Performing document triage on small screen devices. Part 1: Structured documents, In *IIiX 2010 - Proceedings of the 2010 Information Interaction in Context Symposium* (pp. 341-344). New Brunswick, NJ: ACM.

López, F., Martínez, J. M., & García, N. (2010). *Towards a fully MPEG-21 compliant adaptation engine: Complementary description tools and architectural models* (including subseries lecture Notes in artificial intelligence and lecture notes in bioinformatics) (. *LNCS*, *5811*, 155–169.

ltsc.ieee.org. (2002). *Draft standard for learning object metadata*. IEEE Standard 1484.12.1. New York, NY: Institute of Electrical and Electronics Engineers. Retrieved August 20, 2010, from http://ltsc.ieee.org/wg12/files/LOM_1484_12_1_v1_Final_Draft.pdf

Marcellin, M. W., Gormish, M., Bilgin, A., & Boliek, M. (2000). An overview of JPEG 2000, in *Proc. of Data Compression Conference*, Snowbird, Utah, March 2000.

Martínez, J. M. (Ed.) (2004). *MPEG-7 overview* (version 10), ISO/IEC JTC1/SC29/WG11N6828; International Organisation for Standardisation; Palma de Mallorca. Retrieved August 20, 2010, from http://www.chiariglione.org/mpeg/standards/mpeg-7/mpeg-7.htm

mplayerhq.hu. (2010). *MPlayer - The movie player*. Retrieved January 5, 2011, from http://www.mplayerhq.hu/DOCS/HTML/en/index.html

Mukherjee, D., Delfosse, E., Kim, J.-G., & Wang, Y. (2005). Optimal adaptation decision-Taking for terminal and network quality of service. *IEEE Transactions on Multimedia*, *7*(3), 454–462. doi:10.1109/TMM.2005.846798doi:10.1109/TMM.2005.846798

Ortega, A., & Ramchandran, K. (1998). Rate-distortion methods for image and video compression. *IEEE Signal Processing Magazine*, *15*(6), 23–50. doi:10.1109/79.733495doi:10.1109/79.733495

Passani, L. (2010). *WURFL devices and WURFL capabilities*. Retrieved August 20, 2010, from http://wurfl.sourceforge.net/help_doc.php

Prangl, M., Szkaliczki, T., & Hellwagner, H. (2007). A framework for utility-based multimedia adaptation. *IEEE Transactions on Circuits and Systems for Video Technology*, *17*(6), 719–728. doi:10.1109/TCSVT.2007.896650doi:10.1109/TCSVT.2007.896650

Rich, E., & Knight, K. (1991). *Artifical intelligence*. New York, USA: MacGraw-Hill.

Rustici, M. (2009). *SCORM 2004 Overview for Developers*. Retrieved January 5, 2011, from http://scorm.com/scorm-explained/technical-scorm/scorm-2004-overview-for-developers/

Schwarz, H., Marpe, D., & Wiegand, T. (2007). Overview of the scalable video coding extension of the H.264/AVC standard. *IEEE Transactions on Circuits and Systems for Video Technology*, *17*(9). doi:10.1109/TCSVT.2007.905532doi:10.1109/TCSVT.2007.905532

Shapiro, J. M. (1993). Embedded image coding using zerotrees of wavelet coefficients. *IEEE Transactions on Signal Processing*, *41*(12), 3445–3462. doi:10.1109/78.258085doi:10.1109/78.258085

Still, M. (2005). *The definitive guide to ImageMagick*. New York, NY, USA: Apress.

Thang, T. C., Jung, Y. J., & Ro, Y. M. (2005). Semantic wuality for content-aware video adaptation, *Proceedings of the IEEE International Workshop on Multimedia Signal Processing.*

Thang, T. C., Kang, J. W., Choi, H., & Yoo, J.-J. (2008). MPEG-21 enabled SVC adaptation for multi-user QoS-adaptive streaming, in *Proceedings of the ICCE2008.*

Thang, T. C., Kim, J. G., Kang, J. W., & Yoo, J. J. (2009). SVC adaptation: Standard tools and supporting methods. *Signal Processing Image Communication*, *24*, 214–228. doi:10.1016/j.image.2008.12.006doi:10.1016/j.image.2008.12.006

Thiede, T., Treurniet, W. C., Bitto, R., Schmidmer, C., Sporer, T., Beerends, J. G., & Feiten, B. (2000). PEAQ - The ITU standard for objective measurement of perceived audio quality [AES]. *Journal of the Audio Engineering Society. Audio Engineering Society*, *48*(1-2), 3–29.

Tuy, H. (2000). Monotonic optimization: Problems and solution approaches. *SIAM Journal on Optimization*, *11*(2). doi:10.1137/S1052623499359828doi:10.1137/S1052623499359828

Väätäjä, H., & Männistö, A. A. (2010). Bottlenecks, usability issues and development needs in creating and delivering news videos with smart phones, in *Proceedings of the 3rd workshop on Mobile video delivery*, ACM Multimedia Conference (pp. 45-50). New York, NY, USA: ACM.

Van de Walle, R., Burnett, I., & Drury, G. (Eds.). (2005). *ISO/IEC 21000-2 DID 2nd Edition FDIS*, ISO/IEC JTC1/SC29/WG11/N6927, January 2005, Hong Kong.

Van Deursen, D., Van Lancker, W., De Bruyne, S., De Neve, W., Mannens, E., & Van De Walle, R. (2010). Format-independent and metadata-driven media resource adaptation using Semantic Web technologies. *Multimedia Systems*, *16*(2), 85–104. doi:10.1007/s00530-009-0178-9doi:10.1007/s00530-009-0178-9

Vetro, A., Christopoulos, C., & Ebrahimi, T. (Eds.). (2003a). Special Issue on Universal Multimedia Access. *IEEE Signal Processing Magazine*, *20*(2). doi:10.1109/MSP.2003.1184335doi:10.1109/MSP.2003.1184335

Vetro, A., Christopoulos, C., & Sun, H. (2003b). Video transcoding architectures and techniques an overview. *IEEE Signal Processing Magazine*, *20*(2), 18–29. doi:10.1109/MSP.2003.1184336doi:10.1109/MSP.2003.1184336

Winkler, S., & Faller, C. (2005). Audiovisual quality evaluation of low-bitrate video, *Proc. SPIE/IS&T Human Vision and Electronic Imaging, 5666*, 139-148.

Wortham, J. (2009). AT&T to urge customers to use less wireless data, New York Times. Dec. 2009. Retrieved January 04, 2011, from http://www.nytimes.com/2009/12/10/technology/companies/10iphone.html

Zahariadis, T., Lamy-Bergot, C., Schierl, T., Grüneberg, K., Celetto, L., & Timmerer, C. (2009). Content adaptation issues in the future Internet. In G. Tselentis, J. Domingue, A. Galis, A. Gavras, D. Hausheer, S. Krco,…T. Zahariadis (Eds.), *Towards the future Internet - A European research perspective* (pp. 283-292). IOS Press.

Zhang, K., Chen, L., Wu, Z., Xie, S., & Chen, X. (2010). Video broadcast system based on scalable video coding (SVC), in *Proceedings - 2010 3rd IEEE International Conference on Computer Science and Information Technology (ICCSIT 2010)* (vol. 4, pp. 368-371). Piscataway, NJ, USA: IEEE Press.

ADDITIONAL READING

Angelides, M. C., Sofokleous, A. A., & Schizas, C. (2005). Mobile computing with MPEG-21. In: Lecture notes in computer science 3823: *proceedings of the IFIP International Conference on Embedded and Ubiquitous Computing* (EUC 2005) Workshops: 2nd International Symposium on Ubiquitous Intelligence and Smart Worlds (UISW2005), Nagasaki, Japan (pp. 556–565).

Burnett, I. S., Pereira, F., de Walle, R. V., & Koenen, R. (Eds.). (2006). *The MPEG-21 Book*. John Wiley & Sons. doi:10.1002/0470010134doi:10.1002/0470010134

Kasutani, E. (2004). *New frontiers in universal multimedia access*. Tech Rep ITS Report 04.22.

Pereira, F., & Burnett, I. (2003). Universal multimedia experiences for tomorrow. *IEEE Signal Processing Magazine*, *20*(2), 63–73. doi:10.1109/MSP.2003.1184340doi:10.1109/MSP.2003.1184340

Perkis, A., Zhang, J., Halvorsen, T., Kjode, J. O., & Rivas, F. (2002). A streaming media engine using digital item adaptation. *IEEE Workshop on Multimedia Signal Processing* (pp. 73 – 76).

Poynton, Ch. (2003). *Digital video and HDTV: Algorithms and interfaces* (p. 430). Waltham, MA, USA: Morgan Kaufmann.

Sun, H., Vetro, A., & Asai, K. (2003). Resource adaptation based on MPEG-21 usage environment descriptions. In *Proceedings of the IEEE International Symposium on Circuits and Systems* (*ISCAS*), Bangkok, Thailand (pp. 536–539).

Timmerer, C. (2008). Generic adaptation of scalable multimedia resources: Interoperable device and coding format independent multimedia communication for universal access. *VDM.* Berlin-Heidelberg, Germany: Verlag.

Van Deursen, D., De Schrijver, D., De Bruyne, S., & Van de Walle, R. (2007). Fully format agnostic media resource adaptation using an abstract model for scalable Bitstreams, *IEEE International Conference on Multimedia and Expo* (pp. 240 – 243).

Vetro, A. & Timmerer, C. (2005). Digital item adaptation: Overview of standardization and research activities, *IEEE Trans. on Multimedia, 7*(3).

Vetro, A. (2004). MPEG-21 digital item adaptation: Enabling universal multimedia access. *IEEE MultiMedia*, *11*(1), 84–87. doi:10.1109/MMUL.2004.1261111doi:10.1109/MMUL.2004.1261111

Chapter 8
High-Performance Solutions for Adaptive and Customizable Streaming of Interactive Content to Mobile Devices

Gianluca Paravati
Politecnico di Torino, Italy

Fabrizio Lamberti
Politecnico di Torino, Italy

Andrea Sanna
Politecnico di Torino, Italy

Cesare Celozzi
Politecnico di Torino, Italy

ABSTRACT

Recent improvements in technology have opened new, intriguing, and challenging scenarios for the latest generation of mobile devices, and users are asking for an ever larger spectrum of applications. In particular, implementation of interactive applications is an exciting task. In spite of the continuous improvements in the hardware components of mobile devices, several applications are still based on the remote visualization paradigm that adopts streaming based solutions. Nevertheless, some interactive applications introduce hard constraints to be met when low-delay systems have to be designed. Moreover, these systems have to cope with unstable network bandwidth and limited device capabilities. This Chapter first reviews classic strategies to design and implement remote visualization architectures and then presents recent developments regarding high-performance solutions for streaming interactive and customizable contents to mobile devices. In particular, the newest techniques that are specifically able to efficiently cope with bandwidth fluctuations are discussed, and a comparison between optimization based and control based approaches is addressed.

DOI: 10.4018/978-1-61350-144-3.ch008

INTRODUCTION

During the first decade of this millennium, mobile devices have pervaded and changed the way of life of many people. While many types of devices, from cellular phones to laptops, may be considered mobile, we are concerned specifically with portable handheld devices, particularly focusing on the latest generation, which includes smart phones, *Personal Digital Assistants* (*PDAs*), Ultra-Mobile PCs, and e-Readers.

One of the key factors behind the success of latest generation of mobile networks is the availability of useful, attractive, and low cost services for the end-user. Hardware improvements have led to design of increasingly more powerful handheld devices and to the integration of digital imaging and wireless networking technologies, therefore enhancing the range of applications that they can support. At the same time, technological advances in wireless communications now make ubiquitous access to the network possible, allowing more users to complete their tasks while on the move. Moreover, increasingly capacious and fast wireless networks allow researchers to design and implement a wide range of multimedia services specifically tailored for the mobility scenario. For these reasons, mobile devices are the key factors for the growth of pervasive computing. Indeed, the number of handheld devices sold (mainly smart phones) is continuously increasing (Elkin, 2010), at a rate much faster than that of desktop systems, as people use handheld devices primarily for data communication and internet connections.

On the other hand, although today's mobile devices are equipped with hardware resources that were unthinkable until a few years ago in terms of their computational power, storage, and network connectivity; they still present some limitations that prevent people from using them for intensive processing tasks, such as computer graphics applications. To overcome this issue, several applications are based on the remote visualization paradigm, where streaming based techniques are used to visualize remote multimedia content on user devices, thus moving the need for performance from client (mobile) devices to specialized, remote hardware machines.

Streaming applications are widely adopted today; they are subjected to demanding requirements and are expected to perform as so called "killer applications" on future wireless networks because of their high bandwidth usage. In particular, video streaming applications have been successful thanks to the number of higher bandwidths available after the improvement of wireless networks.

The delivery of video content may be realized in two ways: progressive download and real-time streaming. The progressive download method is based on a direct download onto the storage unit of a client from a server; it can be used whenever the applications do not require real-time visualization, and the length of the content is short. On the other hand, real-time streaming is used whenever the content must be consumed by the client device as soon as it has been received. It is pertinent to note that "real-time" in this context does not necessarily mean that the application is mission critical, and the total correctness of an operation depends on the time in which it is performed. Rather, traditional streaming systems can tolerate some delays and may result in decreased *Quality of Service* (*QoS*) while maintaining an active service. For example, during a video streaming session, some frames can be omitted while displaying a video because of network congestion (i.e., the packets that encapsulate an image frame might follow a different path and arrive after some packets from a frame sent later) or limited computational capabilities on the client device (i.e., image frames might be delivered faster than the device is capable of processing). For this reason, streaming systems belong to the so-called "soft real-time systems".

Soft real-time streaming can be performed using two main types of delivery options: live and on-demand. If the media content concerns live events, this type of streaming is called live video streaming. Instead, if the media content is delivered at the user's request, it is called *Video-on-Demand* (*VoD*). Moreover, streaming based applications can be interactive and non-interactive. Both types of delivery options live and on-demand, can present different interactivity requirements. For example, mobile TV streaming applications are non-interactive because the user cannot have the control over visualization. In on-demand video streaming, users have a limited amount of control over what they see (e.g., user controls are usually limited to simple start, stop, rewind, and fast forward functionalities). Other kinds of live video streaming solutions are highly interactive because feedback from users is frequently requested to control the application, for instance for online interactive gaming, remote control, and tele-surgery.

In this Chapter, the focus is on interactive applications that need remote visualization solutions to deliver multimedia content to mobile devices. These solutions rely on a remote server that is responsible for computing the visual representation of the data whereas the mobile device acts as a client in charge both of displaying data representation and receiving user inputs to be forwarded to the server. In particular, the VoD scenario will be addressed, where highly interactive, closed loop systems impose strict requirements in terms of *Round Trip Time* (*RTT*) and need very low delay between user inputs and their visible effect on the screen.

This chapter first introduces and reviews classic strategies to design and implement the media streaming based solutions covered in Section 2, which also discusses the technological challenges. Section 3 reviews the adaptive, media streaming solutions that aim to cope with unstable mobile network conditions; some bandwidth estimation techniques useful in video streaming scenarios are also analyzed. Then, we present recent developments and future trends regarding high-performance solutions for streaming interactive and customizable content to mobile devices. In particular, Section 4 focuses on remote visualization architectures and Section 5 on the drawbacks of streaming solutions based on video coding schemes. The newest techniques are specifically able to efficiently cope with bandwidth fluctuations, which can be caused by network link congestion and by poor signal strength in fringe coverage areas and will be discussed in Section 6. Finally, further research developments are outlined in Section 7 and conclusions are drawn in Section 8.

MEDIA STREAMING ON MOBILE DEVICES

There exists a wide range of applications based on the transmission of visual content over wired and wireless networks; they mainly differ according to timing requirements, rather than the degree of interactivity. It can be claimed that the timing constraints tighten as the degree of interactivity increases.

In this context, latency is the time delay between an event occurring on the server side and seeing its effect on the client side. More specifically, it is the delay between the content generation on the server and the content fruition on a mobile device. In live video-streaming applications, the main factors that cause latency are the following (Endoh et al., 2008): capturing/rendering delay; encoding delay; transmission delay; decoding delay; display delay; buffering delay.

The capturing/rendering delay depends on the source of the content to be streamed. If the source is a camera, this delay depends on its performance. If the source is a rendered three-dimensional scene, the delay depends on the graphic card. Most of the latency is introduced both by the encoding/decoding process (encoding and decoding delays), thus it depends on the equipment used, and on the time required by data packets to traverse the wired and wireless networks (transmission delay). There is little the equipment can usually do to change the network latency as it depends on the network access of the mobile terminal. Moreover, the network latency presents fluctuating characteristics that are unavoidable over best-effort packet networks, such as the Internet, and might cause annoying interruptions during visualization. This is the main reason many systems introduce buffering as a solution. Buffering notably reduces the chance to run the risk of playback breaks and to experience jerky visualizations; however, it noticeably increases playback delay. Lastly, the display delay comes into play when the received image has been decoded and needs to be transferred to the screen frame buffer in order to be displayed. Moreover, the display delay can be protracted by an additional time due to an eventual scaling up process aimed to fit the image frame to the available screen resolution. Indeed, mobile users almost always watch videos at full screen resolution, because mobile terminals have small displays. The display delay depends on the equipment used.

Video Streaming Applications

Latency-sensitive video applications require that the time lapse between source and decoded video is extremely small, but how small depends on the application. Users of mobile TV streaming and, in general, non-interactive streaming applications can tolerate very high delays in the order of magnitude of several seconds or even minutes. For these kinds of users, it is more important to receive smooth visualization, i.e., playback without interruptions, than to experience a faster, but of poorer quality, visualization. The requirements of mobile television include high buffer management (and therefore higher memory requirements) and, in general, the management of a large number of receiver devices. Indeed, streaming content must address a wide range of mobile devices with different features and characteristics; for example, video transcoding mechanisms are used to deliver visual contents to a variety of terminals with different network conditions and different display capabilities in terms of resolution and color depth (Kim et al., 2006). These traditional spatial transcoding schemes are usually based on spatial downsizing by a power of two factors. As an alternative, because displays of mobile terminals are significantly small, an automatic modeling approach has been proposed to crop the most informative region within the original image and scale the cropped region according to a certain display size (Fan et al., 2003).

Most streaming systems are used for recreational use, as for VoD services. This kind of multimedia service presents, as seen before, limited interaction requirements (playback controls) and, as a consequence, more important timing constraints with respect to mobile TV streaming. On the other hand, these timing constraints are more relaxed compared to those needed for highly interactive systems. Indeed, in these systems, the image quality often takes priority over the playback. Video-on-demand users can tolerate several seconds delay, mainly due to buffering time. In order to reduce the start-up service latency for mobile users, proactive video-on-demand broadcasting protocols have been proposed (Hua et al., 2004). These techniques divide a video into several data segments that are repeatedly broadcasted with different frequencies.

Emerging real-time video surveillance applications focus on ubiquitous video acquisition and ubiquitous video processing and viewing, addressing both computer based and human based surveillance. Computer based surveillance is used for people, detection and tracking, whereas human based surveillance refers to human operators controlling areas of interest. Some examples of this scenario are police officers viewing live images coming from a camera installed inside a building on his/her handheld device, a camera mounted on a police car, or a robot with an on-board camera to monitor disaster areas. These systems provide live content and must guarantee efficient video streaming by keeping low latencies and low frame skipping. For example, a latency-care system implemented by Gualdi et al. (2008) was able to provide during streaming tests an average latency that in the best case was in the order of 1 s. The maximum experienced delay was in the order of 6 s.

More strict requirements are needed when highly interactive applications are designed. Video conferencing may take place on a data network without any QoS standard, such as the Internet. Excessive latency increases the chances of people talking over one another because they don't realize the person at the other end has started speaking too. Another problem is that the latency for the audio and video may be different, and hence lip movements do not appear synchronized with the audio. This is a function of both the terminal and the network, and it can vary dramatically. Similarly, high latencies in online interactive gaming can make it very difficult to properly play the game (Nave et al., 2008; Eisert et al., 2007). In the last years, streaming game technologies started investigating the possibility of providing high-quality gaming on low end machines, letting people pick up and play a new game on just about any Internet-connected device, including mobile devices such as smart phones and PDAs. These streaming technologies promise to completely free people from the need to download and install games in the more traditional sense and instead stream them from a remote server cluster far away from where the game is played. The success of this technology will certainly depend on the interactivity, i.e., the latency that will be experienced by players during the game. Other applications that require near real-time performance may concern remote control (Kim et al., 2002; Hamasaki et al., 2008; Zanaty et al., 2008), tele-surgery, telemedicine (Natarajan & Ganz, 2008), and military applications. Real-time image transmission is also important in new services on camera-equipped mobile phones (e.g., online collaboration). All these applications include streaming systems for interactive use, i.e., structured to receive as much graphical content as possible as a result of user interactions, and need a RTT delay with an order of magnitude of a few hundred milliseconds, otherwise they would not be usable (Gutwin et al., 2004).

Technological Challenges

Streaming of interactive visual contents for mobile users is very challenging. Behind the latency issues, there are others technological challenges that should be taken into account when these streaming systems are designed. First of all, video streaming based applications involve the transmission of a large amount of data. On the other hand, there are technological limitations due to the wireless network used for the ubiquitous access that have an impact on the QoS (Soldatos et al., 2005).

Different accessibility options are available, ranging from *Institute of Electrical and Electronic Engineers* (*IEEE*) 802.11 *Wireless Fidelity* (*WiFi*), IEEE 802.16e *Worldwide Interoperability for Microwave Access* (*WiMAX*), *International Mobile Telecommunications-2000* (*IMT-2000*) *Third Generation* (*3G*) networks such as *High Speed Downlink Packet Access* (*HSDPA*) or *Universal Mobile Telecommunication Service* (*UMTS*) to 2 or 2.5G networks such as *Global System for Mobile Communications* (*GSM*), *General Packet Radio Service* (*GPRS*) or *Enhanced Data rates for GSM Evolution* (*EDGE-GPRS*).

The bandwidth of wireless networks is limited, and the available bandwidth is different for different access technologies. The greater the image resolution on the encoder side, the greater the amount of data to be transmitted. By doubling the source image resolution, the occupied bandwidth might quadruplicate. Ad-hoc solutions can be used to transmit images to handheld devices, e.g., reducing the source image resolution to fit the display of the mobile terminal could save bandwidth. However, today's mobile devices support ever growing resolutions. For example, the Apple's iPad supports a 1024x768 pixels resolution at 132 pixels per inch. Therefore, technological advances in hardware tend to go against saving bandwidth.

Moreover, video communication over a dynamic environment such as mobile and wireless networks is much more difficult than over a static channel because wireless channels are unreliable, and the channel bandwidth can change over time due to congestion related issues, which involve network links or handover reasons (e.g., poor signal strength in fringe coverage areas). Bandwidth fluctuations have an impact on performance of remote visualization sessions for the delivery of interactive contents. In particular, they can dramatically impact a user's perceived QoS, strictly related to the *Quality of Experience* (*QoE*), by introducing a jerky visualization on the mobile terminal.

Furthermore, there are various types of handheld devices with different receiving capabilities and computational resources other than display sizes and supported resolutions. The use of mobile devices, in particular mobile phones, has gone from normal voice calls and sending messages to mobile TV and other streaming based functionalities. It is worth observing that the battery life of current mobile phones is not able to provide the long life needed to support these services and to implement computationally complex algorithms.

Receiving capabilities and computational resources (*Central Processor Unit* (*CPU*) speed) of a mobile device are of particular importance because if the user terminal receives image frames faster than its processing capability, frame skipping problems might occur.

All the above problems may result in a significant degradation of received video quality. Therefore, the challenge is how to efficiently stream media content over various connections with different and changing transmission conditions. Considering the above issues, adaptive transmission mechanisms are required for guaranteeing a trade-off between visual quality and smoothness of visualization.

ADAPTIVE MEDIA STREAMING SOLUTIONS

Given the numerous problems that may arise in the design of a live streaming system for interactive use (e.g., network congestions, unknown time-varying bandwidth, hand-over management, transmission errors, delay, delay fluctuations, packet losses), an in-depth analysis of the available solutions considered in the literature to tackle these issues is necessary. In particular, a media streaming system must cope with the bandwidth management, take into account the current available capacity of the network and accordingly react by changing encoding parameters. Different policies can be followed to adapt streaming parameters. However, all the solutions rely on bandwidth estimation to maximize its efficiency.

Management of Unstable Mobile Network Conditions (Bandwidth Fluctuations)

Channel bandwidth variations can strongly impact the performance of remote visualization sessions. For instance, unstable mobile network conditions and common buffering techniques can be detrimental to a user watching an on-demand video stream. Indeed, in practical viewing of on-demand videos, most users tend to frequently skip content after the start of the viewing. This kind of viewing style over wireless mobile networks may become unsatisfactory because of introducing large delays. To solve this problem, Uchihara et al. (2010) simply proposed to pre-fetch and temporarily store discontinuous partial video portions located in the future while receiving the streaming media, allowing the user to jump to any desired scene quickly. However, this solution cannot be useful for live video streaming scenarios.

Different solutions have been proposed to mitigate the effects of unstable network conditions in live video streaming scenarios. Without a network connection able to ensure a negotiable

QoS, visualization performance issues have to be tackled at the application level. The bandwidth variation must be considered to control the transmission rate.

Concerning applications for the delivery of interactive contents, an important aspect for management of bandwidth fluctuations is the scalability of encoding parameters; it is particularly valuable in communications over wireless networks because neither constant link quality nor minimum bandwidth can be easily guaranteed due the intrinsic unreliability of the medium.

In general, when a variation of the available bandwidth is detected, three scalability dimensions can be used to adapt the delivery of the visual content to the changed network state: temporal, spatial, and picture quality dimensions. The temporal dimension concern the smoothness of the visualization, i.e., the frame rate; the more *frames per second* (*fps*) are displayed, the smoother playback users can experience. The spatial dimension concerns the resolution of the delivered images, i.e., the number of pixels in a video frame. The quality dimension concern the quantization introduced by the encoding process; a higher quantization parameter decreases the amount of data transmitted at the cost of degradation in picture quality.

Conventional streaming systems perform rate control by arranging and fixing both temporal and spatial resolution before the encoding process of a video; these parameters do not change during transmission and playback. Usually, these systems act on the quantization parameter to suit to the available bandwidth, as in Ribas-Corbera and Lei (1999). This type of rate control method is widely used in widespread video coding standards of the *Moving Picture Experts Group* (*MPEG*) such as MPEG-4 Part 2 and MPEG-4 Part 10 (commonly referred to as H.264 or *Advanced Video Coding* (*AVC*)).

Rate control is also investigated in Nam et al. Al (2009), where the authors use a bandwidth estimation tool and then determine the sending rate of video streaming according to the available bandwidth of the network. This solution is investigated for video streaming over an HSDPA network, where the available bandwidth varies depending on the adaptive modulation and coding scheme; for example, in the HSDPA network the coding rate and modulation can change from *Quadrature Phase-Shift Keying* (*QPSK*) to *Quadrature Amplitude Modulation* (*QAM*), offering dynamic throughput to the user. However, the available bandwidth estimation is performed at application level; therefore, this method is also theoretically applicable to other networks. It has been shown that the proposed method to estimate the available bandwidth is capable of correctly tracking the variation of the actual throughput of the HSDPA network. Moreover, high estimation accuracy is achieved.

The adaptive rate control for streaming services used in Cha (2005) is aimed at maintaining a constant streaming bandwidth. Media packets are delivered at varying intervals that are adaptive to the current data transfer size and the available buffer size on the handheld device. Three key parts compose the rate control mechanism. First, the original media stream is split into separate audio and video streams to reduce the de-multiplexing complexity in the handset. Then, the separated streams are packed into network packets, whose size depends on the maximum transfer unit of the network, to help the mobile device by reducing the computational resources for assembling small fragmented network packets into data packets for decoding. Finally, network packets are sent back to the network according to a relative deadline that depends on the current status of the mobile device. To this purpose, time stamps for each packet are constantly updated. The authors propose two kinds of adaptive rate control mechanisms. Moreover, they carried out some experiments to compare performance of the proposed mechanisms with no rate control. Test results show that the throughput of the proposed rate control mechanisms is capable of achieving significantly better performance than no rate control.

Classic strategies to adapt the video streaming to sudden changes in available bandwidth encompass frame rate adjustment techniques. For instance, Kim et al. (2000) propose a system that continually accommodates its stream size and manages packet loss recovery in response to network condition changes. The components of an adaptive streaming system are as follows: an available bandwidth estimation and *Transmission Control Protocol* (*TCP*)-friendly, end-to-end congestion control, frame-rate control, media-aware packetization and packet loss recovery tied with congestion control, and quality recovery tools such as motion-compensated frame interpolation. One of the key features is the variable frame-rate encoding. Similarly, Song et al. (2001) developed a rate control algorithm by adjusting the encoding frame rate at the expense of tolerable time-delay. Their aim was to achieve a good balance between spatial quality and temporal quality to enhance the overall human perceptual quality, particularly focusing on low bit-rate streaming.

The combination of the above mentioned three scalability dimensions (temporal, spatial, and picture quality) identifies the bandwidth needed to transmit a video segment. The occupied bandwidth can be reduced or increased by accordingly changing the encoding parameters. For instance, the new *Scalable Video Coding* (*SVC*) amendment specified in Annex G of H.264/AVC allows streaming of videos with different quality versions as specified by the leading international standard organizations *International Telecommunication Union* (*ITU*), *International Organization for Standardization* (*ISO*) and *International Electrotechnical Commission* (*IEC*) (ITU-T & ISO/IEC Standard, 2007). A low-bandwidth bitstream, called the base layer, represents a low quality version of the video (e.g., low temporal or spatial resolution). One or more sub-streams, namely the enhancement layers, integrate with the base layer, allowing the reconstruction of the full quality video. This type of scalability allows different client devices with different processing capabilities to receive the same encoded stream with different quality settings. Moreover, the same transmission can adapt to different access networks. Indeed, enhancement layers can be dropped in case of network congestion or if the client is unable to decode the full video quality, for instance, due to the lack of enough processing capabilities. However, as stated by Lee et al. (2009), the standard does not specify what layers should be taken to achieve the best results for a specific bit-rate. Lee et al. (2009) realized a system that jointly considers picture quality, bit-rate and energy consumption to extend the adaptation of an SVC stream to lower bit-rates, while maintaining temporal stability. Their solution uses empirical results to adjust analytical optimization models able to adaptively control the spatial resolution dimension.

A different approach to tackle uncertainties in network bandwidth is used by Wang et al. (2008). The main goal of their control-theoretic approach is to meet user specified deadlines when delivering images by dynamically changing the quality of image tiles during transmission. At the same time, the system tries to maximize image quality. Each transmitted image is split into many tiles, each of which is compressed by a different quality factor. A feedback control loop dynamically adjusts the *Joint Photographic Experts Group* (*JPEG*) tile quality basing on the performance feedback.

Adaptive controls can also be employed to achieve a tradeoff between low latency and good video fluidity. As explained in the previous section this tradeoff is necessary because some interactive applications do need very low latencies, but this cannot happen at the expense of video quality. Indeed, customizing a video streaming system, e.g., a system that relies on H.264/AVC or MPEG 1/2/4 video coding, to keep the buffer as empty as possible would result in uneven trends of the playback frame rate due to buffer underflow. This issue is addressed in Gualdi et al. (2008), where the authors designed a playback frame rate control to keep the buffer occupancy between two values, with the aim to achieve the best tradeoff between low latency and good fluidity. Similarly, Hong et al. (2007) propose a buffer control scheme to tune the latency. It is based on the classification of buffer occupancy into three states: underflow, stable, and overflow. Based on this information, the inter-packet delay at the server can be adjusted to control the sending rate. These solutions do not require modification of network protocols as they act at the application layer. However, the average latency experienced by these systems remains too great and, therefore, these solutions are not suitable for high-performance streaming systems that need to prioritize delay with respect to all other streaming parameters.

A further step is faced in Kim et al. (2008) where both channel bandwidth variation and problems with disconnections or video degradation during handovers are considered. In particular, it is minimized the error propagation that occurs in these cases by performing an intra refresh method that inserts an *intra frame* (*I-frame*) soon after the handoff is detected. In this way, the bandwidth occupation is increased at the benefit of an increased image quality.

Bandwidth Estimation Tools

Streaming systems have to deal with unstable network bandwidth, and a good system should take into account bandwidth fluctuations to accordingly adapt the flow of streaming. Almost all recent solutions available in the literature have the presence of components for estimating the available bandwidth in common. Different methods have been developed to carry out the bandwidth estimation. There exist two main methodologies to analyze the network performance, known as active or passive methods. Active methods send ad-hoc packets from a sender application to a receiver application to try to understand the network state. On the other hand, passive methods are non-intrusive measurement techniques because they exploit the traffic already present in the network to estimate the available bandwidth. Most bandwidth estimation tools focus on optimizing end-to-end transport performance and are used by network engineers. In the following, some common methodologies available in the literature are presented particularly focusing on solutions that can be used by video streaming technology developers.

Carrier to Interference Noise Ratio based bandwidth estimation. A wireless link usually offers lower capacity than a wired connection. For this reason, it is possible to consider only the wireless link capacity to assess the available bandwidth on an end-to-end path that includes both wired and wireless links. This assumption is still valid for high-capacity mobile networks such as WiMAX. Kim et al. (2008) computed the available bandwidth by means of the physical data rate R_{PHY} (Mbps) that is obtained using system parameters as follows:

$$R_{phy} = D_{rate} \cdot b_{\text{mod}} \cdot c_{rate} \tag{1}$$

where D_{rate} is the data subcarrier rate and depends on the ratio of the downlink to the uplink, b_{mod} is the modulation gain and depends on the modulation used by the network, and c_{rate} is the coding rate. The *Carrier to Interference Noise Ratio* (*CINR*) is used to estimate $b_{mod} \cdot c_{rate}$. The CINR is a measurement of the signal effectiveness. The carrier is the desired signal, and the interference can be due to noise or to co-channel interference. The signal must fall into an acceptable CINR range to allow the signal receiver be able to decode the signal. The CINR range differs with the technology used (i.e., GSM, WiMAX, etc.).

After computing the physical data rate, intended as the theoretical upper bound of the available bandwidth, the authors (Kim et al., 2008) estimated the network capacity (available bandwidth R) as follows:

$$R = \frac{\left(1 - P_{overhead}\right) \cdot \left(1 - C_{overhead}\right) \cdot W_{env}}{N_{user}} \cdot R_{phy} \quad (2)$$

where R_{phy} appears to be influenced by various network parameters such as the channel resource overhead ($C_{overhead}$), the number of users (N_{user}), the ratio of the packet overhead ($P_{overhead}$) and the weighting factor W_{env} related to the environment of the subscriber station.

The current network capacity is computed both for the streaming server (uplink) and the viewer client (downlink). The available bandwidth of the streaming system is therefore computed as the minimum between the uplink and downlink capacities. This kind of CINR based bandwidth estimation is able to detect handoffs when the physical data rate drops below a certain threshold.

Packet Pair Probing Technique (*PPPT*) is a widely used, simple method to measure the end-to-end capacity of a path without any information from network routers (Jacobson, 1988). When a pair of packets with the same size L is sent in sequence over a network, they will reach the receiver with dispersion (D) that matches the transmission delay at the narrow link. In this way, the receiver can evaluate the capacity C of the link by means of the measured dispersion using the following equation:

$$C = \frac{L}{D} \quad (3)$$

However, this method presents two main drawbacks. First, the *PPPT* shows the best performance in the estimation of the channel capacity when there is no cross traffic (Jacobson, 1988). Second, there is a mismatch between the actual available bandwidth of a wireless network and the channel capacity obtained from the *PPPT*. An application of this solution to measure the available bandwidth in HSDPA networks is investigated in Nam et al. (2009).

pathChirp (Riberio et al., 2003) is an active, publicly available tool for bandwidth estimation. It is based on the concept of self-inducing network congestion by featuring exponential traffic from probes called chirps. This tool gathers a richer set of information than standard probing schemes based on packet pairs or packet trains. *pathChirp* dynamically estimates the available bandwidth by rapidly increasing the probing rate within each chirp. For instance, Arsan (2009) triggers this algorithm to provide to a rate controller the bandwidth estimation value so as to regulate the rate of both a live streaming scheduler and an adaptive encoder.

REMOTE VISUALIZATION SYSTEMS

Remote visualization is basically necessary either when the mobile device is not able to locally carry out the required data processing or when, for security issues, data cannot be moved from the remote server. For instance, volume rendering applications may have to deal with huge amount of data that are often out of reach for most mobile devices; on the other hand, it could be advisable to provide users only with representations of information when privacy or security issues are important, as for instance in the visualization of medical data. Innovative solutions to bring the gaming scenario to mobile devices using the full power of specialized remote graphic hardware are based on the remote visualization paradigm. Streaming game technologies and collaborative *Computer Aided Design* (*CAD*) based design tools on mobile devices mostly care about interactivity, i.e., RTT delays.

A remote visualization paradigm can be implemented using two different approaches. First, the whole desktop could be sent to the client device; however, in this case, small display devices are not efficiently supported (Richardson et al. 1998). Alternatively, ad-hoc solutions could be developed in order to allow the user both to remotely control

Figure 1. A remote rendering system

the application running on the server machine and to receive the visual content of the working area (Stegmaier et al., 2003; Lamberti & Sanna, 2007; Sanna & Lamberti, 2008). Figure 1 shows an example of remote visualization of a heavy weight 3D scene on a recent mobile phone.

Classic strategies to design remote visualization systems envisage the use of specialized, high-performance graphics hardware on the server side and a lightweight application on the client side (Bao & Gourlay, 2006). The server (or a cluster) renders complex (3D) scenes, creates updated frames and transmits the frame buffer to the client as a stream of images that can be either uncompressed or compressed. The first solution can be used over a local area network (where more bandwidth is generally available), whereas the latter approach is targeted to standard wide area Internet connections. In the latter case, the user can choose among different types of preconfigured parameter settings for the image encoding. A remote rendering system is composed by a rendering module that generates the 3D content, an encoder module that compresses visual data to be delivered to the network and the client application that decompresses and renders received images. To effectively reach best performance, all components of a streaming based remote rendering system have to be designed and optimized in concert by performing channel-adaptive streaming.

Generic video streaming solutions use buffering in order to avoid playback break, introducing unacceptable delays in the visualization pipeline of high-performance, interactive applications that mostly care about updated contents. At the start of a session, the buffer needs to be filled up prior the visualization of media content. Buffer-controlled adaptive video schemes such as Gualdi et al. (2008) and Hong et al. (2007) try to guarantee a tradeoff between buffer occupancy and visualization delay, but buffer underflow problems may prevent the client from keeping the buffer short enough to enhance the interactive user experience. In this case, it would be better to find custom solutions able to guarantee the lowest latency at the expense of more bandwidth. Indeed, the trend in modern mobile access networks is to allow the employment of more bandwidth by providing high data rates. In the next section, an evaluation study of an MPEG 4 live video stream system is presented to assess the latency.

Girod et al. (2002) states that "*while for voice transmission over the Internet latencies below 100 ms are achievable, video streaming typically exhibits much higher latencies, even if advanced techniques like adaptive media play-out (...) are used*" (pag. I-10). The key concept of a low delay streaming system is to send an image frame immediately after it was rendered by a server and, on the client side, to display the image frame immediately after receiving it. This concept has been also outlined by Endoh et al. (2008), proving the effectiveness of a flow of compressed still images, such as *Motion JPEG* (*MJPEG*), to keep very low latencies. In this case, the proposed streaming solution divides each frame into blocks, and compresses each block using JPEG. The system can save bandwidth usage by avoiding the dispatch of blocks that are not significantly changed with respect to the previous frame. A transmission priority for each block is computed, and then the blocks are sent in descending order of priority. The basic idea behind the priority assignment is to promote the transmission of blocks whose visual change is larger. An important advantage of this kind of solutions is given by the fact that all blocks can be independently decoded; therefore it is possible to partially update the image on the display even if all the frames are not received by a scheduled time because of the delay fluctuations. In this way, it is possible to keep the visualization delay constant.

The same approach is followed by Delgado et al. (2009) and uses a *Real-time Transport Protocol* (*RTP*) payload format for *JPEG-Compressed Video* (*JCV*). The header of the JCV format contains information about the size of the image and the data offset from images. This information is needed because JPEG images are usually much larger than the *Maximum Transfer Unit* (*MTU*) of *User Datagram Protocol* (*UDP*), and they must be fragmented into several packets before being delivered. The solution proposed by Delgado et. al (2009) gives the user the possibility to manually set three parameters; in particular, they are: resolution; frame rate; image quality.

The resolution parameter allows the user to rescale video to fit the size of the mobile device. The frame rate specifies the maximum frequency of the video frames in order to avoid saturation at the receiver. Finally, image quality reduces or enlarges the original quality of the image and impacts the size of the transmitted image and the processing time. However, manually tuning these parameters would require a skilled user, and, above all, this solution does not readapt the receiving flow based on network conditions. The work in Paravati et al. (2008) has laid the groundwork for linking these parameters with the bandwidth occupancy according to the following equation:

$$B_o = \frac{f \cdot w \cdot h \cdot C_d}{C_r} \quad (4)$$

where B_o is the theoretical bandwidth occupancy, f is the transmitted frame rate, w and h are the resolution parameters (width and height), C_d is the color depth and C_r is the compression ratio. By linking this mathematical relationship with a bandwidth estimation tool and a set of rules to prioritize some parameters over others, it is possible to adjust encoding parameters in real-time.

The challenge for a streaming system, besides supporting high-performance remote visualization sessions, is to provide users with the best trade-off of the above parameters in order to use the underlying (wireless) network connection as efficiently as possible. Moreover, the streaming system should deliver a customizable information flow to satisfy different user requirements related to subjective preferences, kinds of applications, used devices, etc.; for instance, some users of collaborative *Virtual Reality Environments* (*VREs*) (Pang & Wittenbrink 1997; Pettifer 2000) and *Multiplayer Online Games* (*MPOGs*) (Bernier, 2001; Dionne et. al 2000) may be in favor of high frame rate visualizations with respect to relatively low image qualities; on the other hand, some other users may prefer to sacrifice the interactivity in order to experience high quality and high resolution data visualizations.

Prior to analyzing two recent techniques that are able to adjust the encoding parameters of a live streaming system, results concerning a feasibility study highlighting the drawbacks related to the usage of video coding streaming techniques are presented; in particular, video coding is compared with still image compression in the specific framework of high-performance interactive visualization on mobile devices. Subsequently, an optimization based and a control-theoretic approach are analyzed.

VIDEO CODING DRAWBACKS IN INTERACTIVE VISUALIZATION SYSTEMS

Interactive 3D visualization systems based on video streaming techniques running on mobile devices have been realized in the past. In particular, Nguyen et al. (2007) developed a graphics adaptation framework with a mechanism of video streaming to overcome the shortcomings of real-time representation and to contribute interactive experiences of 3D graphics applications on handheld devices. The client viewer gives a user the possibility of navigating a 3D scene and interacting with 3D objects. The server renders a sequence of images and then encodes them into a live video stream based on the MPEG-4/H.264 standard. The media content is provided adaptively by the server in the sense that user preferences and conditions of the wireless network are taken into account. Indeed, on the client side, the user can set three encoding options: channel rate, minimizing distortion and quantizer. The first option, channel rate, is aimed at maintaining a constant frame rate that is specified by the user. If the user selects the minimizing distortion option, an alternative to the channel rate option, the system will provide a minimal video distortion for the video segment in the streaming buffer. The quantizer option allows the user to manually specify the quality of the video compression that will remain fixed during the streaming. The value of the quantizer can vary in the range from 1 to 31 in steps of 1. Instead, by selecting one of the first two options, the system requires a controller to find an optimal quantizer.

Unfortunately, researchers do not show performance in terms of RTT to evaluate the proposed video streaming solution from the point of view of high-performance interactivity. According to the authors, the system can rapidly represent a complex 3D scene on mobile devices because they directly refer to works where the whole 3D scene is downloaded to the client device, and this is a time delay process.

To overcome the limitation due to the lack of experiments aimed at assessing the degree of interactivity, Paravati et al. (2010) implemented a remote rendering scenario and carried out some tests using the MPEG-4 Part 2 video coding scheme and different network accesses (IEEE 802.11g and HSDPA). The results are compared with streaming solutions guidelines provided both by Endoh (2008) and by Delgado (2009) where a MJPEG coding scheme is used for transmission. In the following, the system overview and the main findings are outlined.

The system is composed by a rendering server with streaming functionalities and a commercially available handheld device with Windows Mobile as operating system. The rendering server has been written in C++ and is based on the OpenGL libraries; a 3D rendering application has been chosen due to its intrinsic, highly interactive constraints, but it is worth observing that the test bed is valid for any kind of high-performance, live streaming scenario. The client application is able to receive and display the MJPEG coding scheme, and it has been written in Java Micro Edition, whereas the MPEG-4 live streaming video is displayed on the mobile phone using a built-in video player.

Because the test scenario is an interactive 3D application, a crucial parameter is the RTT, evaluated as the elapsed time between sending a roto-translation command from the mobile device to the rendering server and shown on the display of the mobile terminal, of the first frame corresponding to the effect of the user command.

Timing results have been collected by connecting the mobile client to the rendering server through both a IEEE 802.11g wireless network access point and a 3G wireless network of an Italian telecom provider. Table 1 shows the average RTT delay comparison between the two implementations of the live streaming module. It is observed that the most responsive option is made by the MJPEG coding scheme over an IEEE 802.11g wireless channel, where an unnoticeable delay is introduced by the system and the user is given the impression that the 3D rendering phase is locally computed on the mobile device. Nevertheless, for both the MJPEG and the MPEG-4 streaming scenarios, the RTT delay quadruples when the traffic is conveyed over an HSDPA (3.6 Mbps) connection; however, the system remains quite interactive by using the MJPEG coding scheme, given that the delay experienced is only double the rule of thumb used by telephone industry, i.e., the RTT delay over a telephone call should be less than 100 ms. On the other hand, the RTT achieved by using the MPEG-4 coding scheme become on average an order of magnitude greater than those reached with the MJPEG scheme. This result highlights that a traditional video coding solution goes against a high-performance streaming system. For the interested reader, similar remarks can be found in Endoh (2008).

ADAPTIVE SOLUTIONS FOR INTERACTIVE VISUALIZATION

Interactive applications should provide users with the capability to adaptively control the visualization. As pointed out in Section 5, video coding streaming techniques may introduce an unacceptable delay, mainly due to buffering time. Moreover, users could be interested in customizing the visualization in terms of displayed frame rate, resolution, and quality to enhance the quality of their experience. To sustain continuous and user-centric video display on the mobile device, continuous monitoring of the network status is required. To address this concern, a real-time streaming system needs rate adaptation mechanisms able to optimize encoding parameters on the fly.

This section analyzes two novel methodologies for dealing with the above issues at the application-level, namely an optimization based technique and a control based approach.

Optimization Based Technique

The first methodology (Paravati et al., 2010b) is aimed at developing a manager for the encoding parameters at the application level able to meet a trade-off among the underlying network status, the processing capabilities of the mobile device and the user preferences.

A flow of compressed still images is continuously delivered to the client device. The manager is distributed between the client and the server; it is in charge of deciding the optimal encoding parameters by monitoring the throughput of the mobile device while receiving the stream. The throughput measured on the client side is fundamental to periodically notify the live-streaming server about both the client device's capabilities and the status of the network. If the mobile terminal cannot process a throughput larger than the

Table 1. Round-Trip Time Comparison For Different Video Coding Schemes

Access network	MPEG-4 live streaming; Average RTT	Motion JPEG live streaming; Average RTT
IEEE 802.11g	583 ms	59 ms
HSDPA 3.6 Mbps	1966 ms	214 ms

available bandwidth in downlink, the lightweight client has saturated its computational capabilities. Analogously, the throughput can be limited by the available bandwidth. In both cases, the manager should respond by changing encoding parameters according to user preferences.

The optimization based technique introduces a so-called QoE graph that relies on a formal representation of remote visualization parameters (i.e., displayed frame rate, resolution and quality). Each node of the graph represents one of the possible configurations that can be assumed by a flow of JPEG images transmitted from the remote server to the client device. Each particular configuration is expressed as a pair of values, namely *resolution* (*R*) and *quality* (*Q*) (where the term quality refers to the JPEG encoding). The edges of the graph represent possible transitions between nodes, and they are determined according to a metric that is capable of taking into account bandwidth occupation and user requirements. In particular, user requirements are characterized in terms of desired displayed frame rate, and priorities are assigned to each controllable parameter. Indeed, a user can specify the importance of the above parameters from a personal point of view by sorting them in order of relevance.

The rules for the selection of encoding parameters are incorporated in graph construction. In order to avoid sudden changes in resolution and to guarantee a smooth handover between adjacent nodes, the graph is built so as to connect a starting node to another node if the resolution of the destination node is equal to or immediately higher or lower than the resolution of the starting node; in this way, the transition is carried out without noticeable changes in the visualization. Then, the graph is periodically rebuilt to take into account the current available bandwidth. The graph is then traversed in order to reach the node (i.e., the configuration) that provides the best bandwidth occupation and matches user requirements. The metric for the determination of weights is a score function defined as:

$$w_{j,i} = \sum_{k=1}^{3} c_k \cdot \lambda_k(j) \tag{5}$$

where $w_{j,i}$ is the scoring function used to update the weights for edges directed to the *j-th* node, c_k is coefficient that takes into account the priority of the parameters (resolution, image quality, frame rate) specified by the user, and λ_k (with k=1, 2, and 3) is the contribution of each parameter (λ_1 is related to the resolution, λ_2 is related to the image quality, and λ_3 is related to the frame rate).

The contribution of the resolution is computed as:

$$\lambda_1(j) = \frac{\left(\frac{W(l)}{W(n_R - 1)} + \frac{H(l)}{H(n_R - 1)}\right)}{2} \tag{6}$$

where $l = \left\lfloor \frac{j}{n_q} \right\rfloor$, with n_q equal to the number of available quality values, *W(i)* and *H(i)* are the width and height of the *i-th* resolution and n_R is the number of available resolutions.

The contribution of the image quality is computed as follows:

$$\lambda_2(j) = \frac{Q(m)}{Q(n_Q - 1)} \tag{7}$$

where $m = j \bmod n_Q$, and *Q(i)* is the *i-th* quality value.

The contribution of the frame rate is computed as in the following:

$$\lambda_3(j) = 1 - \left|\frac{f(j) - f_d}{f_d}\right| \tag{8}$$

where f_d is frame rate specified by the user, and *f(j)* is the achievable frame rate for a particular *j-th* node. The achievable displayed frame rate depends on both the encodings parameters (resolution and image quality) and the minimum between the current available bandwidth and the experienced throughput of the device. The achievable displayed frame rate for each node is computed as follows:

$$f(j) = \frac{B}{W(l) \cdot H(l) \cdot C_d \cdot \frac{1}{C_r(R(l), Q(m))}} \quad (9)$$

where $W(l) \times H(l)$ is the resolution at the *j-th* node, $l = \left\lfloor \frac{j}{n_q} \right\rfloor$, C_d is the color depth in terms of bits per pixel and $Cr(R(l),Q(m))$ is the compression ratio of images with resolution $W(l) \times H(l)$ and quality equal to $Q(m)$ with respect to the same uncompressed image.

Control Based Solution

The control based approach (Paravati et al., 2010c) aims to take advantage of automatic control techniques to concurrently tune all the parameters involved in a remote visualization scenario. In particular, MJPEG encoding parameters (resolution, image quality, and displayed frame rate) are dynamically controlled to react to bandwidth fluctuations.

During the live streaming, it is possible to estimate in advance the theoretical bandwidth occupancy by knowing all the current encoding parameters of the streaming flow. The relationship among encoding parameters and theoretical bandwidth occupancy can be modeled as in eq. (4).

One of the main features of the controller is its sensitivity to feedback measures. Indeed, the throughput at the client side is constantly measured, and it is periodically transmitted to the server. If the measured throughput is lower than the bandwidth occupancy of the transmitted video flow, the client device is not able to receive and process all the transmitted data. In this way, the measured throughput on the client device represents the maximum throughput of the device. The maximum throughput of the device can be defined as the maximum amount of data that can be processed by the visualization device. Therefore, the throughput measure describes the amount of visualization flow that can be decoded by the mobile device in the unit of time. However, it is clear that the maximum throughput does not take into account the unused available bandwidth. Instead, the available bandwidth can be evaluated by using the techniques presented in Section 3.2.

A controller has two inputs: the reference of the controller and the measured output of the system. The reference of the controller is the theoretical bandwidth occupancy B, whereas the measured output of the system is the throughput T evaluated at the client side. The difference between these two values corresponds to a bandwidth error ΔB:

$$\Delta B = T - B \quad (10)$$

The bandwidth error symbolizes the gap between the live streaming flow created by the server and the actual visualization flow the mobile device is able to receive, decode and display in a unit of time. The feedback measure sent by the handset allows the controller to dynamically compensate both for disturbances to the system due to bandwidth fluctuations and insufficient device capabilities. Therefore, other features that the controller can exhibit are independence to the quality of the underlying network and optimal usage of available resources.

The logical operation of the controller is to continuously adjust the streaming parameters so as to keep the bandwidth error to a minimum. The controller also exhibits robustness to non-linearity of the system. This is because the considered parameters have reciprocal interactions; therefore, a change of each parameter has an indirect influence on the bandwidth error. As a matter of fact, the parameters cannot change at the same time. For this reason, the control phase has been split in cascade stages, where each of them controls the tuning of a different variable.

The client application allows the user to specify a target frame rate that is defined as the number of fps the system will try to reach (display). This information can be either preconfigured or provided by the user. Therefore, two additional stages are needed to complete the selection of the encoding parameters. During the first stage, the variation of resolution is computed as follows:

$$\Delta R = k_R \cdot \Delta f \qquad (11)$$

where ΔR is the amount of increment or decrement in resolution, Δf is the difference between the target frame rate selected by the user and the actual displayed frame rate, and k_R is computed as follows:

$$k_R = \frac{k_p}{B_R^{up} - B_R} \qquad (12)$$

where k_p is a proportional constant of a PID based system, B_R and B_R^{up} are the bandwidth occupancies experienced with the current resolution and the immediately available, higher resolution, respectively.

The last stage consists of the computation of the variation of the image quality parameter as in the following equation:

$$\Delta Q = k_Q \cdot \Delta f \qquad (13)$$

where k_Q is a proportionality coefficient, and Δf is the difference between the target frame rate selected by the user and the actual displayed frame rate.

The outputs of the controller (resolution, image quality and transmitted frame rate) are then exploited by the server to generate, compress, and stream an image based representation of the server-side working area over the network to the mobile client device.

Comparison Between the Optimization Based and the Control Based Approaches

The optimization based technique and the control-theoretic approach have been implemented and tested to analyze their behaviors. This section first analyzes the behavior of the optimization based technique under various simulated network conditions to show how MJPEG streaming parameters can change during a remote visualization session on a mobile device. Under the same assumptions, results of the tests are also collected for the control based approach to compare their performances.

The test-bed is composed by a rendering server encompassing a live streaming system that exploits an adaptive MJPEG encoding scheme to enable 3D visualization on mobile devices. The rendering server has been written in C++ and is based on the OpenGL libraries. It is composed of a Dual-Core AMD Opteron CPU 2.60 GHz, 4 GB of RAM and an NVIDIA Quadro FX 3500 graphics card. The live streaming media player has been written in Java and is compliant with the Java Micro Edition *Connected Limited Device Configuration* (*CLDC*), so as to run on most of the current mobile phones. The mobile device was connected to the rendering server through an IEEE 802.11g wireless access point during the tests. The 3D scenes from the server side can be encoded at different resolutions and different image qualities.

The experiments are designed to test the ability of the proposed solutions to adapt the live streaming parameters in different system conditions. In particular, the monitored parameters are displayed frame rate, resolution, image quality and measured throughput on the mobile device. The system conditions subject to testing are available bandwidth conditions and user preferences. The ability to adapt live streaming parameters in different network conditions is important because, in real world applications, bandwidth can significantly change during transmission. On the other hand, different user preferences in terms of target frame rate can lead to the modification of the other live streaming parameters in order to best fit the currently available bandwidth or device throughput.

The described and tested adaptive techniques do not introduce any appreciable latency to the MJPEG coding scheme presented in Section 5.

In Figure 2 we show the behavior of the optimization based technique under varying bandwidth conditions. The data have been collected on the mobile phone during visualization of the streaming flow. Four parameters were kept under observation: resolution, image quality, displayed frame rate, and measured throughput. This test aimed at showing the ability of this technique to take into account changes in network conditions. In particular, at some point of the test, the available network bandwidth of the system was forced to remain below a certain threshold. The bandwidth limitation was artificially introduced by means of a network limiter application running at the server side. The network limit was used to simulate a degradation of the underlying network performance. Therefore, the overall simulation can be divided into two time lapses.

The system initially was set to reach the best performance: no bandwidth limitations are imposed, and the only requirement for the client application was the target frame rate set equal to *15* fps at the instant *t0*.

As shown in Figure 2, from the beginning to the instant *t1*, the mobile client was able to reach the target frame rate value, which was on average maintained during this lapse of time. At the same time, the throughput is a parameter that reflects

Figure 2. Live streaming parameters monitored during changing network conditions by using an optimization based technique

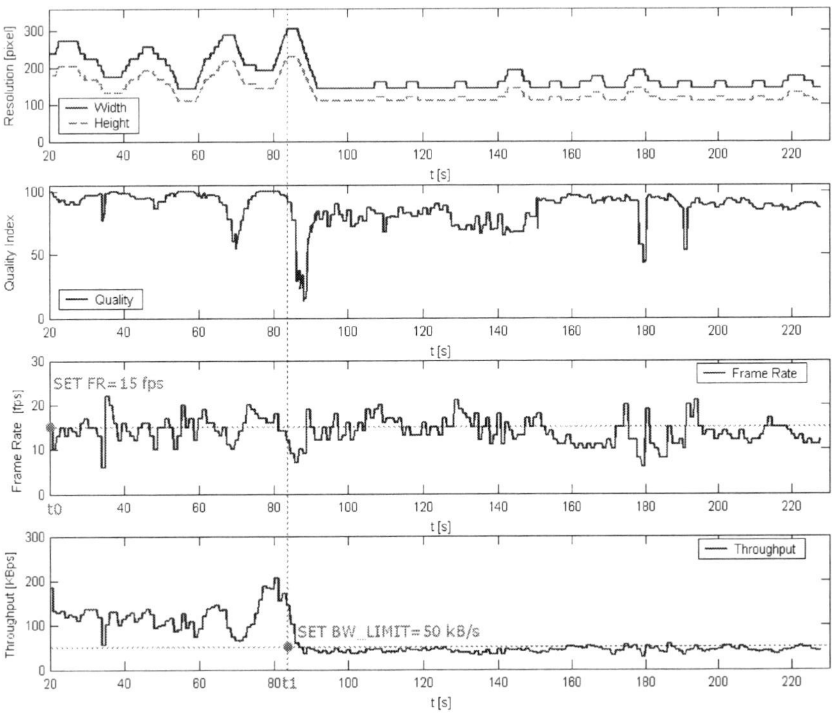

the variable bit-rate nature of this kind of coding scheme. On average, the throughput was greater than *100 KBps*, and, therefore, the image quality and resolution were able to achieve high values on average. However, this solution suffers from the drawback of being extremely sensitive to changes yielding to oscillations that are particularly visible in the trend of the resolution parameter during this first lapse.

At the instant *t1* a network limitation has been artificially introduced at the server side. As a consequence, the available network bandwidth suddenly drops down and this fact is confirmed by the throughput measurements. A network limiter application reduced the available bandwidth to remain below the threshold equal to *50 KBps* for the remainder of the simulation, as shown in the throughput graph after the instant *t1*. The reduced bandwidth level causes the frame rate at the client device to drop, and the system immediately reacts by reducing the resolution and the image quality levels. As a result, the frame rate increases by reaching a steady state where the target frame rate is again achieved, on average.

Figure 3 shows the ability of the optimization based technique to react to changes in user preferences, and, in particular, to changes in the target frame rate specified by the user at run-time.

The system initially has been set to reach a target frame rate equal to *12* fps (at the instant *t0*). During the first lapse of time (between the instant *t0* and the instant *t1*), the mobile viewer experienced the visualization task by performing a measured displayed frame rate that guaranteed, on average, the target frame rate. Relatively high values of resolution and image quality have been reached during this first part of the test.

At the instant *t1* the target frame rate has been changed to *22* fps by the user through the graphical user interface of the mobile phone. As a result, the live streaming encoding parameters gradually reduced to allow the displayed frame

Figure 3. Live streaming parameters monitored during the adaptation of changing target frame rate requirements by using an optimization based technique

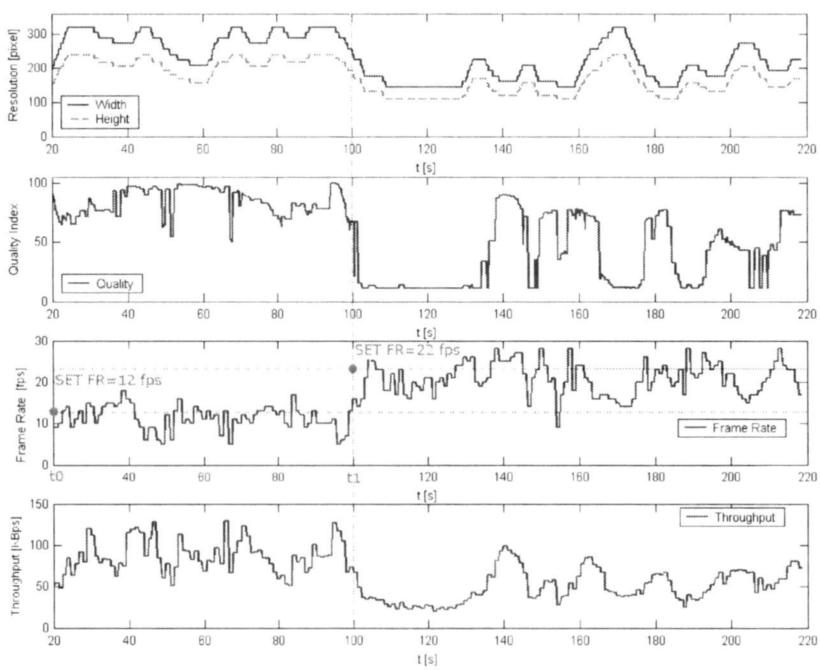

rate to increase. In this way, during the second part of the test (after the instant *t1*) the frame rate reached the new target value specified by the user, on average. Nevertheless, oscillations while managing the encoding parameters are clearly visible in the graphs. The resolution increases when the quality decreases and vice versa. In spite of the correct behavior while changing the target frame rate, the parameters continue to oscillate, leading to instability issues in the system. This behavior has a negative influence on the visualization from a qualitative point of view. However, this situation can be avoided by using a closed-loop controller that provides finer controls.

In Figure 4 and Figure 5 we show the behavior of the proposed, control-theoretic approach. The same conditions and the same parameters of the tests carried out for the optimization based solution, shown in Figure 2 and Figure 3, were used to perform the tests on the controller.

Figure 4 shows the ability of the controller to take into account changes in network conditions. In particular, at some point of the test, the available network bandwidth of the system has been forced to remain below a certain threshold, as was done for the optimization based scenario. At the beginning of the test (at the instant *t0*), the system was instructed to maintain a constant displayed frame rate equal to *15* fps.

As shown in Figure 4, from the instant *t0* to the instant *t1*, the client viewer on the mobile device was able to reach a target frame rate value of 15 fps, which was on average maintained during this lapse of time. The image quality remained stable around a value of 50, and the resolution was able to achieve high values, on average. It is worth observing that, unlike the previous case, this solution is less prone to fluctuations.

At the instant *t1*, the network bandwidth drops suddenly. Again, a network limitation has

Figure 4. Live streaming parameters monitored during changing network conditions by using a control-theoretic solution

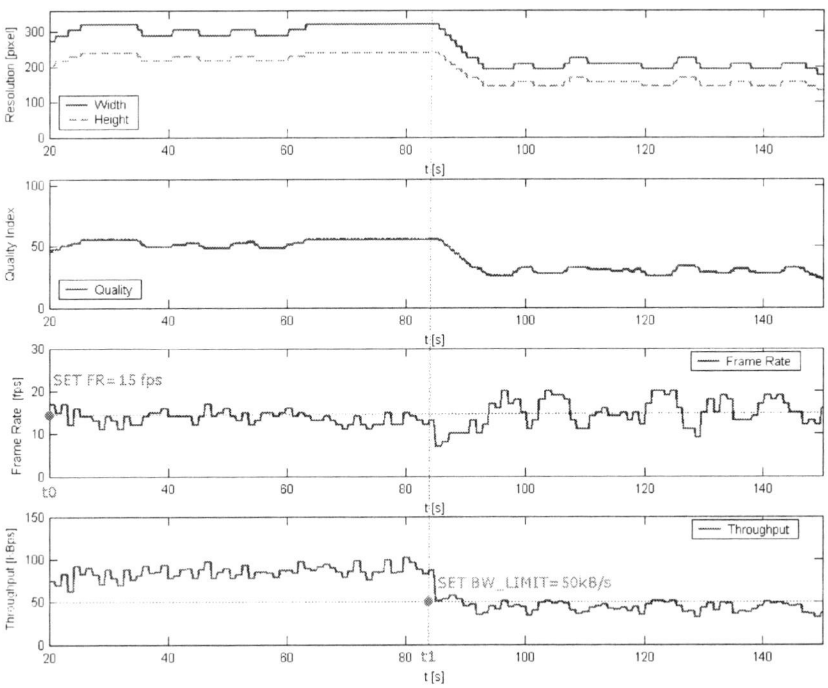

Figure 5. Live streaming parameters monitored during the adaptation to changing target frame rate requirements by using a control –theoretic approach

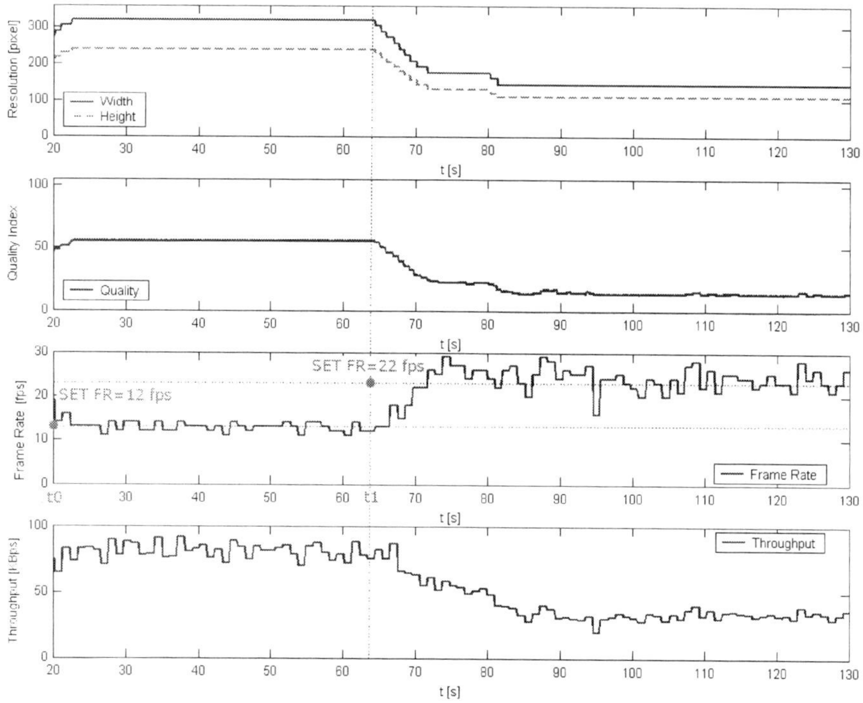

been artificially introduced on the server side. A network limiter application reduced the available bandwidth to remain below *50 KBps* for the remainder of the simulation, as shown by the throughput graph after the instant *t1*. Also, in this case, the reduced bandwidth level causes the frame rate on the client device to drop below 10 fps, and the system reacted by reducing the resolution and the image quality levels. As a result, the displayed frame rate increases by reaching again a steady state where the selected target frame rate is guaranteed.

Figure 5 shows the behavior of the proposed control-theoretic approach when user preferences are subject to change. In particular, this figure highlights the ability of the controller to follow a specified target frame rate. The results of this test are dual with respect to the test for the optimization based technique, shown in Figure 3.

From the instant *t0* to the instant *t1*, the system maintained a target frame rate of *12* fps. During this lapse of time, the system was steady around relatively high values of resolution and middle level of image quality.

The target frame rate has been manually changed to *22* fps at the instant *t1* by the user. As a result, the live streaming encoding parameters gradually reduced to allow the displayed frame rate of increasing. In this way, during the second part of the test (after the instant *t1* till the end of the test) the frame rate averaged the new target value specified by the user. Again, the controller based result outperformed the optimization based technique in terms of achieving a steady state for the whole duration of the live streaming.

FUTURE RESEARCH DIRECTIONS

Due to the lack of video coding schemes able to guarantee very low latencies, more research efforts are required in this field to develop near-real-time video encoding solutions that meet the aggressive requirements of remote visualization architectures. Control-theoretic approaches can be applied to video coding schemes (e.g. MPEG) to continuously adjust video streaming parameters (e.g. the quantization scale, the *Group Of Picture* (*GOP*) size, the occurrence of I-frames, and so forth) with the aim of controlling the end-to-end latency and buffering time. The impact of extremely reducing the buffer in video streaming techniques to keep very low latencies needs deep investigation. On the other hand, changing on the fly encoding parameters requires tests and simulations to correlate objective video quality metrics with subjective video quality.

Low latency video streaming is particularly useful in interactive remote rendering applications (e.g. navigation of three-dimensional virtual worlds, inspection of complex 3D objects and scenes), where the hard job is moved from *thin* client devices (e.g. smart phones, PDAs) to server components. The techniques investigated in this chapter mainly provide an analysis of the issues encountered by connecting a single mobile client to a single streaming/rendering server. Other types of applications require the simultaneous fruition of real-time content by multiple users at the same time. Indeed, shared visualization environments can represent an effective means to enhance collaborative work in engineering and design tasks. In this context, further developments may concern the investigation of issues related to the management of this type of interactive collaborative applications. For example, shared visualization in engineering and scientific tasks involves the use of different mobile and desktop based devices to support collaborative work, e.g. in a 3D visualization session for mobile devices each streaming server can be used to support more clients. In these situations, the number of simultaneous users, as well as the number of concurrent visualization sessions supported by the servers, could represent two additional parameters that may have to be considered by the illustrated techniques. Thus, future works will investigate the impact of these (as well as other) factors on the discussed metrics, so that the proposed framework can provide a collaborative experience ever closer to user needs.

CONCLUSION

This chapter is focused on high-performance, live streaming systems to mobile devices. In particular, interactive applications are considered because they introduce hard constraints to be met when low-delay systems have to be designed. Remote visualization systems, where 3D scenes are streamed to the clients, represent an example of highly interactive applications that can run on handheld devices.

The interactive nature of this kind of application and the unstable network conditions make the design of live streaming system challenging. Ad-hoc solutions are considered to reduce the RTT delay experienced during the visualization and to improve the quality perceived by the user when this service is supplied. In particular, the drawbacks related to the use of video streaming techniques with buffering functionalities within the context of interactive applications are outlined. Related works in the literature addressing interactive streaming functionalities by using MJPEG coding schemes are introduced. To sustain continuous and user-centric video display on the mobile device, continuous monitoring of the network status is required. Typically, if a mobile device is used for receiving and displaying data from a streaming server, the latter will act as a fast sender while the latter acts as a slow receiver, respectively, due to substantial differences in their computational capabilities. To address this concern, real-time streaming systems integrating rate adaptation mechanisms able to optimize encoding parameters on the fly have been presented. This chapter also presents some metric that can be used to estimate the available bandwidth on the considered wireless channel. Furthermore, two up-to-date techniques, namely an optimization based technique and a control based approach, that are able to react to bandwidth fluctuations are presented here and analyzed in terms of speed and smoothness in providing users an optimal tuning of the delivered visual content. The two methodologies for dealing with the above issues work at the application-level and are able to dynamically set the encoding parameters at the server side; the considered parameters are frame resolution, visualization frame rate and image quality. The proposed solutions are able to tune the above parameters according to network performance and device capabilities by means of feedback measures coming back from the client device. The feedback measure is the device throughput. Experiments carried out on the designed streaming architecture showed the efficiency and flexibility of the proposed techniques, in terms of reaction to bandwidth changes in the network.

ACKNOWLEDGMENT

This article is part of a work developed within the frame of the project LITES "Led based intelligent street lighting for energy saving" (G.A. no. 238916) co-funded by the EC under the Competitiveness and Innovation Framework Programme - ICT Policy Support Programme.

REFERENCES

Arsan, T. (2009, November). Model based predictive rate controller for video streaming. *IEEE International Conference on Electrical and Electronics Engineering*, II-347-II-351.

Bao, P., & Gourlay, D. (2006). A framework for remote rendering of 3-D scenes on limited mobile devices. *IEEE Transactions on Multimedia*, *8*(2), 382–389. doi:10.1109/TMM.2005.864337

Bernier, Y. W. (2001). Latency compensating methods in client/server in-game protocol design and optimization. In *Proceedings of the 15th Games Developers Conference*.

Cha, H., Lee, J., Nang, J., Park, S.-Y., Jeong, J.-H., Yoo, C., & Choi, J.-Y. (2005). A video streaming system for mobile phones: Practice and experience. [The Netherlands: Springer.]. *Wireless Networks*, *11*(3), 265–274. doi:10.1007/s11276-005-6610-y

Delgado, M., Medina, J., Ruiz-Lozano, D., & Vila, A. (2009). Integrating video streaming from heterogeneous sources in mobile devices. *Proceedings of the 4th IEEE International Conference on Ubiquitous Information Technologies & Applications* (pp. 1-6).

Dionne, C., Lavoie, M., & Trueman, K. (2000). NetZ - Multiplayer architecture for online games. In *Proceedings of the Game Developers Conference*.

Eisert, P., & Fechteler, P. (2007, July). Remote rendering of computer games. In *Proceedings of the IEEE International Conference on Signal Processing and Multimedia Applications*.

Elkin, N. (2010). Seven key trends in mobile usage. *eMarketer Report*, September 2010.

Endoh, K., Yoshida, K., & Yakoh, T. (2008). Low delay live video streaming system for interactive use. *6th IEEE International Conference on Industrial Informatics* (pp.1481-1486).

Fan, X., Xie, X., Zhou, H.-Q., & Ma, W.-Y. (2003). Looking into video frames on small displays. *Proceedings of the 11th ACM International Conference on Multimedia* (pp. 247-250).

Girod, B., Kalman, M., Liang, N. J., & Zhang, R. (2002). Advances in channel-adaptive video streaming. *Proceedings of International Conference on Image Processing, 1*, I-9–I-12.

Gualdi, G., Prati, A., & Cucchiara, R. (2008). Video streaming for mobile video surveillance. *IEEE Transactions on Multimedia, 10*(6), 1142–1154. doi:10.1109/TMM.2008.2001378

Gutwin, C., Benford, S., Dyck, J., Fraser, M., Vaghi, I., & Greenhalgh, C. (2004). Revealing delay in collaborative environments. In *Proceedings of ACM Computer Human Interaction* (pp. 503–510).

Hamasaki, S., & Yakoh, T. (2008, July). Implementation and evaluation of decorators for delayed live streaming video on remote control system. In *6th IEEE International Conference on Industrial Informatics* (pp. 1220–1225).

Hong, Y.-S., Kim, J.-H., Kim, Y.-H., & Hong, J.-S. (2007, November). A buffer-controlled adaptive video streaming for mobile devices. *IEEE International Conference on Convergence Information Technology* (pp. 30-35).

Hua, K. A., Tantaoui, M. A., & Tavanapong, W. (2004). Video delivery technologies for large-scale deployment of multimedia applications. *Proceedings of the IEEE, 92*(9), 1439–1451. doi:10.1109/JPROC.2004.832954

ITU-T & ISO/IEC Standard (2007). Advanced video coding for generic audiovisual services. *Annex G, scalable video coding H.264 & 14496-10.*

Jacobson, V. (1988). Congestion avoidance and control. In *Proceedings of ACM SIGCOMM* (pp. 314-329).

Kim, H.-S., Nam, H.-M., Jeong, J.-Y., Kim, S.-H., & Ko, S.-J. (2008). Measurement based channel-adaptive video streaming for mobile devices over mobile WiMAX. *IEEE Transactions on Consumer Electronics, 54*(1), 171–178. doi:10.1109/TCE.2008.4470040

Kim, J., Kim, Y.-G., Song, H., Kuo, T.-Y., Chung, Y.-J., & Kuo, C.-C. J. (2000). TCP-friendly Internet video streaming employing variable frame-rate encoding and interpolation. *IEEE Transactions on Circuits and Systems for Video Technology, 10*(7), 1164–1177. doi:10.1109/76.875520

Kim, J.-W., Choi, B.-D., Park, S.-H., Kim, K.-K., & Ko, S.-J. (2002, June). Remote control system using real-time MPEG-4 streaming technology for mobile robot. *Digest of Technical Papers of International Conference on Consumer Electronics* (pp. 200–201).

Kim, J.-W., Kwon, G.-R., Lee, J.-S., Kim, N.-H., & Ko, S.-J. (2006). Spatio-temporal video transcoder for streaming over mobile communications networks. *IEICE Transactions on Communications, 89-B*(10), 2678–2686. doi:10.1093/ietcom/e89-b.10.2678

Lamberti, F., & Sanna, A. (2007). A streaming based solution for remote visualization of 3D graphics on mobile devices. *IEEE Transactions on Visualization and Computer Graphics, 13*(2), 247–260. doi:10.1109/TVCG.2007.29

Lee, H., Lee, Y., Lee, J., Lee, D., & Shin, H. (2009). Design of a mobile video streaming system using adaptive spatial resolution control. *IEEE Transactions on Consumer Electronics, 55*(3), 1682–1689. doi:10.1109/TCE.2009.5278043

Nam, H.-M., Shin, K.-S., Jeong, J.-Y., Kim, H.-S., & Ko, S.-J. (2009). Bandwidth estimation based video streaming for mobile devices over HSDPA network. *Digest of Technical Papers of International Conference on Consumer Electronics*, 1-2.

Natarajan, S., & Ganz, A. (2008, August). Efficient force feedback transmission system for tele surgery. *30th Annual International Conference of the IEEE Engineering in Medicine and Biology Society* (pp. 3245 – 3248).

Nave, I., David, H., Shani, A., Tzruya, Y., Laikari, A., Eisert, P., & Fechteler, P. (2008, April). Games@large graphics streaming architecture. *IEEE International Symposium on Consumer Electronics* (pp. 1–4).

Nguyen, C., Le, M.-T., Yoon, D.-I., & Kim, H.-K. (2007). A graphics adaptation framework and video streaming technique for 3D scene representation and interaction on mobile devices (LNCS). In *Proceedings of the 4th international conference on Universal access in human computer interaction: coping with diversity* (pp. 481-490). Berlin-Heidelberg, Germany: Springer-Verlag.

Pang, A., & Wittenbrink, C. (1997). Collaborative 3D visualization with CSpray. *IEEE Computer Graphics and Applications*, *17*(2), 32–41. doi:10.1109/38.574676

Paravati, G., Celozzi, C., Sanna, A., & Lamberti, F. (2010). A feedback based control technique for interactive live streaming systems to mobile devices. *IEEE Transactions on Consumer Electronics*, *56*(1), 190–197. doi:10.1109/TCE.2010.5439144

Paravati, G., Sanna, A., Lamberti, F., & Ciminiera, L. (2008, April). A novel approach to support quality of experience in remote visualization on mobile devices. In *Eurographics 2008 short paper Proceedings* (pp. 223–226).

Paravati, G., Sanna, A., Lamberti, F., & Ciminiera, L. (2010a). On quality of experience in remote visualization on mobile devices. *International Journal of Mobile Human Computer Interaction*, *2*(1), 1–20. doi:10.4018/jmhci.2010100601

Paravati, G., Sanna, A., Lamberti, F., & Ciminiera, L. (2010b). *An adaptive control system to deliver interactive virtual environment content to handheld devices. Mobile Networks and Applications*. The Netherlands: Springer.

Pettifer, S., Cook, J., Marsh, J., & West, A. (2000). Deva3: Architecture for a large-scale virtual reality system. In *Proceedings of ACM Symposium in Virtual Reality Software and Technology* (pp. 33-40). New York, NY: ACM Press.

Ribas-Corbera, J., & Lei, S. (1999). Rate control in DCT video coding for low-delay communications. *IEEE Transactions on Circuits and Systems for Video Technology*, *9*(1), 172–185. doi:10.1109/76.744284

Riberio, V., Riedi, R., Baraniuk, R. G., Navratil, J., & Cottrell, L. (2003). PathChirp: Efficient available bandwidth estimation for network paths. *Passive Active Measurement Workshop*.

Richardson, T., Stafford-Fraser, Q., Wood, K. R., & Hopper, A. (1998). Virtual network computing. *IEEE Internet Computing*, *2*(1), 33–38. doi:10.1109/4236.656066

Sanna, A., & Lamberti, F. (2008). *3D visualization on mobile devices. Handbook of research on user interface design and evaluation for mobile technology* (pp. 556–573). Hershey, PA: Information Science Reference.

Soldatos, J., Vayas, E., & Kormentzas, G. (2005). On the building blocks of quality of services in heterogeneous IP networks. *IEEE Communications Surveys & Tutorials*, *7*(1), 69–88. doi:10.1109/COMST.2005.1423335

Song, H., & Kuo, C.-C. J. (2001). Rate control for low-bit-rate video via variable-encoding frame rates. *IEEE Transactions on Circuits and Systems for Video Technology*, *11*(4), 512–521. doi:10.1109/76.915357

Stegmaier, S., Diepstraten, J., Weiler, M., & Ertl, T. (2003). Widening the remote visualization bottleneck. In *Proceedings of 6th International Symposium on Image and Signal Processing and Analysis*.

Uchihara, N., Kasai, H., Suzuki, Y., & Nishigori, Y. (2010). Asynchronous prefetching streaming for quick-scene access in mobile video delivery. *IEEE Transactions on Consumer Electronics*, *56*(2), 633–641. doi:10.1109/TCE.2010.5505981

Wang, X., Chen, M., Huang, H.-M., Subramonian, V., Lu, C., & Gill, C. D. (2008). Control based adaptive middleware for real-time image transmission over bandwidth-constrained networks. *IEEE Transactions on Parallel and Distributed Systems*, *19*(6), 779–793. doi:10.1109/TPDS.2008.41

Zanaty, P., Brscic, D., & Frei, Z. (2008, May). 3D visualization for intelligent space: Time-delay compensation in a remote controlled environment. *Conference on Human System Interactions* (pp. 802–807).

ADDITIONAL READING

Berc, L., Fenner, W., Frederick, R., & McCanne, S. (1996). *RFC2035: RTP Payload Format for JPEG-compressed Video*. RFC from Internet Society.

Carter, R. L., & Crovella, M. E. (1996). Measuring bottleneck link speed in packet-switched networks. *Performance Evaluation*, *27-28*, 297–318.

Cheshire, S. (1996). *Latency and the quest for interactivity*. White paper commissioned by Volpe Welty Asset Management, L.L.C., for the synchronous person-to-person interactive computing environments meeting, San Francisco, CA.

Ding, J.-W., Lin, C.-T., & Huang, K.-S. (2006). An adaptive reception scheme for handheld devices supporting mobile video streaming services. *Digest of Technical Papers of International Conference on Consumer Electronics* (pp. 141-142).

Hamasaki, S., & Yakoh, T. (2008). Implementation and evaluation of decorators for delayed live streaming video on remote control system. *6th IEEE International Conference on Industrial Informatics* (pp. 1220-1225).

Hu, N., & Steenkiste, P. (2002). *Estimating available bandwidth using packet pair probing (Tech. Rep.)*. Pittsburgh, PA: Carnegie-Mellon University, Pitt School of Computer Science.

Iqbal, R., & Shirmohammadi, S. (2009). A cooperative video adaptation and streaming scheme for mobile and heterogeneous devices in a community network. *IEEE International Conference on Multimedia and Expo* (pp. 1768 – 1771).

Kim, G., & Yoo, G. (2009). Streaming session mobility across multiple devices in mobile IPTV environments. *Proceedings of the 4th International Conference on Ubiquitous Information Technologies & Applications* (pp. 1-6).

Prasad, R. S., Murray, M., Dovrolis, C., & Claffy, K. (2003). Bandwidth estimation: Metrics, measurement techniques and tools. *IEEE Network*, *17*(6), 27–35. doi:10.1109/MNET.2003.1248658

Rezaei, M., Wenger, S., & Gabbouj, M. (2005). Video rate control for streaming and local recording optimized for mobile devices. *IEEE 16th International Symposium on Personal, Indoor and Mobile Radio Communications* (pp. 2284-2288).

Schwarz, H., Marpe, D., & Wiegand, T. (2007). Overview of the scalable video coding extension of the H.264/AVC standard. *IEEE Transactions on Circuits and Systems for Video Technology*, *17*(9), 1103–1120. doi:10.1109/TCSVT.2007.905532

Vassiliou, V., Antoniou, P., Giannakou, I., & Pitsillides, A. (2006, September). Requirements for the transmission of streaming video in mobile wireless networks. *International Conference on Artificial Neural Networks* (pp. 528-537).

Wu, S.-Y., & He, C. E. (2009). (in press). QoS-aware dynamic adaptation for cooperative media streaming in mobile environments. *IEEE Transactions on Parallel and Distributed Systems*.

Yashiro, D., Yakoh, T., & Ohnishi, K. (2009, June). End-to-end flow control using PI controller for force control system over TCP/IP network. *7th IEEE International Conference on Industrial Informatics* (pp. 194-199).

Zheng, B., & Atiquzzaman, M. (2003). A novel scheme for streaming multimedia to personal wireless handheld devices. *IEEE Transactions on Consumer Electronics*, *49*(1), 32–40. doi:10.1109/TCE.2003.1205453

Chapter 9
Using MapReduce Framework for Mobile Applications

Adam Dou
University of California, USA

Vana Kalogeraki
Athens University of Economics and Business, Greece

Dimitrios Gunopulos
University of Athens, Greece

Taneli Mielikainen
Nokia Research Center, USA

Ville H. Tuulos
Nokia Research Center, USA

ABSTRACT

Most of today's smart-phones are geared towards a single user experience, whether it is reading a book, watching a movie, playing a game or listening to music. However, there has been a shift towards providing a more complex and social experience: applications are being developed and deployed to help users connect and share information with each other. These applications allow people to keep track of their friends' statuses in real time, or to help them navigate around traffic congestion. While exciting, most such applications are currently being developed in an ad-hoc nature, reinventing and duplicating a lot of work to support their distributed operations. In this work, we present our framework, Misco. A platform for developing distributed applications for mobile smart-phones. We also explore some existing solutions, applications and related systems. We then discuss some of the many future research paths and show that solutions like ours are just the beginning.

DOI: 10.4018/978-1-61350-144-3.ch009

INTRODUCTION

Smart-phones are everywhere. There are currently over four billion active subscribers and that number of subscribers continues to grow. Smart-phones are also becoming more powerful. The latest generation of smart-phones boasts 1 *gigahertz* (*GHz*) Snapdragon processors, 512 *megabytes* (*MB*) of *Random Access Memory* (*RAM*) and 32 *gigabytes* (*GB*) of persistent storage (HTC, 2010). Network connectivity is also expanding and getting faster with *International Electrical Electronic Engineers* (*IEEE*) 802.11n, *Wireless Fidelity* (*WiFi*) and *Fourth Generation* (*4G*) cell networks. In addition to their increasingly powerful resources, smart-phones are being equipped with motion sensors, proximity sensors, *Frequency Modulation* (*FM*) receivers, *Global Positioning System* (GPS), digital compass, cameras and microphones.

All these factors combine to make smart-phones a very lucrative platform for developing distributed applications. Phone sensors provide environmental data such as sound, connectivity, movement, images and social information from user input, user interaction, contact lists and locations. Even more complex sensors are being developed and embedded into phones, such as sensors that detect deceases (Breslauer, Maamari, Switz, Lam & Fletcher, 2009) and air pollution (Honicky, 2010).

To take advantage of these phones, development and deployment of distributed applications must be simplified. There are many factors leading to the complicated nature of distributed applications: concurrency, resource allocation, software distribution, and device and network failures. These issues are further compounded by the hurdles of developing software on any of the current major smart-phones. In order to leverage the power of these devices, we must provide programming constructs that simplify the development and deployment of the applications.

We propose to use MapReduce as distributed computation framework for mobile devices. Since its introduction, MapReduce (Dean & Ghemawat, 2004) has grown immensely in popularity, supporting a wide array of applications spanning machine learning, simulations and media processing (Chen & Schlosser, 2008). MapReduce is being heavily used by prominent companies such as Google, IBM, Yahoo and Facebook to tractably process their large amounts of information. Although MapReduce started as a framework geared to run on systems in large data centers, it has been successfully implemented for other environments such as *Graphics Processing Units* (*GPUs*)(He, Fang, Luo, Govindaraju, & Wang, 2008), shared memory systems (Ranger, Raghuraman, Penmetsa, Bradski, & Kozyrakis, 2007) and even JavaScript clients on browsers (Grigorik, 2009). There have been some recent explorations with implementations of MapReduce on smart-phones (Dou, Kalogeraki, Gunopulos, Mielikainen & Tuulos, 2010; Marinelli, 2009; Elespuru, Shakya & Mishra, 2009). We feel that MapReduce's support for the weak connectivity model of computations across open networks makes it very suitable as a framework for smart-phones and mobile devices.

In this chapter, we first explore the current state and recent trends of smart-phones and a few recent projects focused on using mobile devices. Then, we explain the basics of MapReduce and discuss some of the recent advancements and several MapReduce implementations. We present Misco, a MapReduce framework we have developed for smart-phones and finally look at some of the issues and future work in this area.

BASICS OF MAPREDUCE FRAMEWORK

Smart-phones have been becoming increasingly powerful and are the fastest growing segment in the mobile devices market (Gartner, 2010). A simple look at the evolution of successive generations of iPhones show that their memory and storage has been doubling every year and their processing power has doubled in two years. In terms of these resources, the progression in smart-phones appears to follow Moore's law. Network speeds have also been increasing with recent migrations from *Third Generation* (*3G*) networks which provide 0.2 ~ 14 Mbps to 4G networks which provide 0.1 ~ 1 Gbps (Moray Rumney, 2008).

Recently, there has been a lot of attention placed on using the phones as a low-cost alternative for more specialized machinery to measure things like pollution, medical and decease detection. The N-SMARTS (Honicky, 2010) project aims a building a distributed system of phones for monitoring the environment. One of their research areas is the development of *Micro electro mechanical Systems* (MEMS) based particulate mass sensors which can be embedded in phones(Breslauer, Maamari, Switz, Lam, & Fletcher, 2009)have built a mobile phone mounted light microscope which can be used for light microscopy, a simple, cost-effective and vital method for diagnosing and screening hematologic and infectious diseases. These efforts in creating more powerful and low cost sensors for smart phones are only expanding the phone's capabilities as a mobile sensing platform.

Mobile Applications and Projects

Currently, developing applications for smart-phones is not an easy ordeal and there are major obstacles to develop for any of the big three smart-phone platforms: Symbian, iPhone and Android. First, each vendor provides platform specific tools that must be used to develop for and deploy on their devices. Each phone also has its own unique program life cycle and flows. This means that the way memory is managed, how a program starts and stops and how interrupts are handled are all different for each phone and vastly different from developing on more traditional desktop and server environments. In order to write efficient code for Android devices, object instantiation and internal getters and setters must be avoided and "virtual" is preferred over "interface". Android also had a slightly different implementation of the *Java Virtual Machine* (*JVM*) and shell tools which are incompatible with desktop or server JVMs, making it difficult to port existing Java applications. For Symbian and iPhone applications, expensive developer certificates and accounts are required in order to install application. Being a closed system, iPhone provides application installation and log viewing within the XCode *Integrated Development Environment* (*IDE*) but does not support execution of arbitrary shell commands or the creation of files. The SymbianOS is currently undergoing an open sourcing effort but as for yet has not made too much progress.

Even after a developer has gone through the effort of learning the new programming paradigm for smart-phones and acquired the tools and software certificates, writing distributed applications is still not an easy undertaking. There are many things to design for when creating a distributed application: concurrency issues, communications and device and network failures are among a few of the many issues.

One of our major motivations for developing a MapReduce framework that can be used on mobile devices is to address this difficulty in developing distributed applications on smart-phones. The MapReduce framework provides a flexible platform for developing applications without having to worry about the underlying distributed nature of the application.

We are not looking to reproduce the typical massive data backed applications which are currently running on MapReduce in data centers. Current phones do not have the processing power, network stability and most importantly, battery power to make much impact on these types of applications. Instead, we are more interested in the applications which take advantages of the strength of smart-phones and look at several current projects which leverage these strengths.

The first types of applications we look at are social application. These applications are used to connect people and allow them to share information. Dartmouth has several such projects: SoundSence (Lu, Pan, Lane, Choudhury & Campbell, 2009), CenceMe (Miluzzo, et al., 2008) and BikeNet (Eisenman, Miluzzo, Lane, Peterson, Ahn & Campbell, 2009) are just a few. SoundSence is a project that uses a combination of supervised and unsupervised learning techniques to classify both general sounds like voice and music and also to discover novel sound events specific to users. SoundSence is designed for scalability and to run on the limited resources of phones. Likewise, CenceMe is also developed to run on cellphones, but its purpose is to allow members of a social network to share their sensing presence and status with friends in a secure and automatic manner. CenceMe captures the user's status by detecting her activity (e.g. sitting, walking, meeting friends), disposition, habits and surroundings automatically through classification algorithms. BikeNet allows cyclists to share information about themselves and the path they transverse in real-time using bike-to-bike sharing or through third party checkpoints. Location based social applications are also gaining popularity. Foursquare (foursquare, 2010), an application which allows users to check into venues and gain badges, has gained over 500,000 users in its first year. These types of applications take advantage of the intimate connection between a smart-phone and their users.

These systems can leverage the MapReduce technology to radically improve the quality of the data collected over time and from multiple users. The MapReduce framework allows the user to collect and process data locally (through programming functions implemented and executed on the phones) and then data collected from multiple phones can be processed together in order to produce results gathered over time and from multiple users. For projects like SoundSence, CenseMe and BikeNet, the local tasks can perform basic classifications (or just feature extraction) of the user's actions or surroundings based on the sensed data, while data gathered from multiple phones can be processed together at the servers to produce a more accurate view of the data. Another very important feature is that the data collected from multiple users can then be re-distributed back to the phones for more accurate processing.

The second types of applications we are interested in are environmental and monitoring applications. These are more geared towards detecting or monitoring conditions in a region. There are many traffic monitoring projects underway which use smart-phones. Two such projects are the Berkeley Mobile Millennium Project (Hoh, et al., 2008) and MIT CarTel project (Hull, et al., 2006). The Berkeley project is a traffic monitoring system which uses the GPS in smart-phone to gather traffic information, processes it and then distributes it back in real-time in an effort to mitigate traffic delays. CarTel is also uses smart-phones and automobiles to collect and process data to be delivered to an internet server. In addition to monitor and mitigate traffic congestion, CarTel also use data collected to monitor and classify road surface conditions (Eriksson, Girod, Hull, Balakrishnan & Madden, 2007). Other projects such as PEIR (Agapie, et al., 2008) use location data samples from phones to calculate personalized estimates of environmental impact and exposure. These applications which use sensed data to produce an area mapping of values can be easily modeled to use MapReduce to process data.

MapReduce Programming Model

MapReduce(Dean & Ghemawat, 2004) was popularized by Google as a programming model for performing transformation on very large data sets. The model itself was motivated by the map and reduce functions of functional programming language and is shown in Figure 1. To use the model, an application implements a map and a reduce function. The map function takes a piece of the input data and outputs intermediate key/

Figure 1. A typical MapReduce model, the user is only responsible for providing the input data, the Map and Reduce functions. The Map and Reduce functions then operate on the input data in parallel

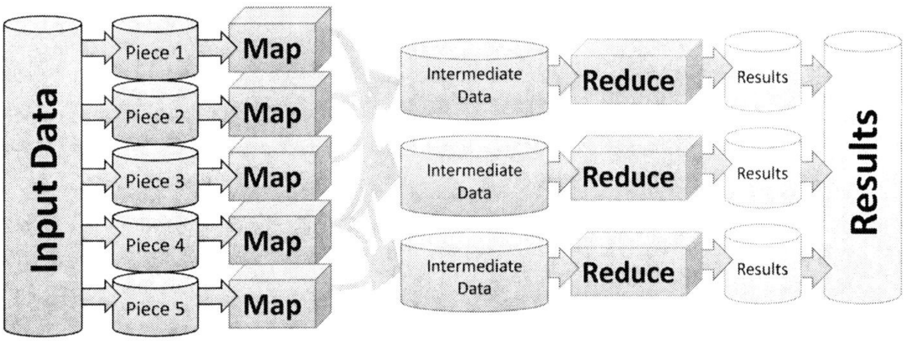

value pairs. These intermediate key/value pairs are then grouped by key using a partitioning function and sorted if necessary. The reduce function then processes a group of key/value pairs and generates the desired output data. The framework automatically parallelizes the execution of the functions, the distribution of the function code and data and handles the failures of faulty nodes.

MapReduce Systems

Currently, Hadoop(Hadoop, 2009)is the most popular open source implementation of MapReduce. It is developed for applications that work with thousands of nodes and petabytes of data. Hadoop is inspired by and follows the design of the Google MapReduce (Dean & Ghemawat, 2004) and Google File System (Ghemawat, Gobioff, & Leung, 2003) papers. The system is aimed at data center environments, with many rack-aware optimizations that take into account geographical clustering of servers. Hadoop also provides its own *Hadoop Distributed File System* (HDFS) (HDFS, 2009)which it uses to store input and output data, although it has been recently extended to support other file systems such as *Hypertext Transfer Protocol* (HTTP).Scheduling in Hadoop is performed by a Job Tracker which pushes tasks to idle Task Tracker nodes.Each Task Tracker has a number of available slots for tasks and each task takes up one of these slots. If the tracker node fails or times out, the task is reassigned to another Task Tracker.

Disco(Tuulos, 2010) is an open source Erlang implementation of MapReduce for clustered server environments. Applications for Disco are typically written in Python, which allows for complex code behavior with very compact code. Disco comes with a built-in distributed storage system (Disco Distributed Filesystem) and a distributed indexing system (Discodex) for querying of large data sets.

Due to its popularity, there are many efforts to extend and optimize Hadoop. YAHOO has been a major contributor to the Hadoop codebase and is actively researching methods to allow for real-time MapReduce. This advancement would be very useful for applications such as personalization, user feedback, malicious traffic detection and real-time search which all require very fast response and scalability. In search, if the output data sets are made available to the system before the user executes her next search, adaptations can be made to the search models based on the user's intent. One approach used to provide this type of rapid feedback possible is MapReduce Online (Condie, Conway, Alvaro, Hellerstein, Elmeleegy & Sears, 2009), which modifies the MapReduce architecture so that data can be pipelined between operators. This change extends MapReduce beyond batch processing and allows for early partial results, continuous queries and stream processing.

The MapReduce framework has also been successfully implemented in other environments, such as on multi-core and multiprocessor systems (Phoenix (Ranger, Raghuraman, Penmetsa, Bradski & Kozyrakis, 2007)) and on GPUS (Mars (He, Fang, Luo, Govindaraju & Wang, 2008)). Phoenix uses the MapReduce model to automatically provide parallelization on shared-memory systems. Unlike traditional server cluster based approaches, Phoenix is less concerned with network access latencies and is more focused on working with low level memory buffers to provide increased data throughput. Particular attention is paid to the way data is stored in the shared memory buffers to make sure that there is no interference between different tasks and to minimize unnecessary copying or movement of data. Similarly, Mars is also concerned by low level optimizations and workarounds specific to GPUs such as managing thread parallelism, coalesced data access, and file manipulation. Mars must deal with GPU particularities like determining the number of threads and thread groups to use in order to maintain high occupancy and to provide custom library implementations to handle variable sized data. Both these systems provide high performance MapReduce implementations to their respective target architectures by lowering the operating overhead of the framework through low level and very specific techniques. Although low level optimizations are also important in our setting to reduce operating overheads, our environment is more influenced by network latencies, which we have chosen to focus on for our initial implementation.

MapReduce for Mobile Phones

There are currently a few existing MapReduce implementation targeted at smart-phones: Hyrax (Marinelli, 2009), an implementation for a master's project, a research prototype implementation (Elespuru, Shakya & Mishra, 2009) and our own Misco implementation (Dou, Kalogeraki, Gunopulos, Mielikainen & Tuulos, 2010). All three systems were developed independently around the same time frame between 2008 and 2010.

Hyrax is a mobile MapReduce platform derived from Hadoop and supports cloud computing on Android smart-phones. The developers for Hyrax attempted to port directly from Hadoop, but had to make several modifications to the code to allow it to run on mobile phones. The first issues they overcame was the differences between Android's JVM and the JVMs found on desktops and servers. Secondly, Hadoop was not coded with efficiency in mind, as evidenced by its heavy use of interfaces and inheritance. While interfaces and inheritance make code more maintainable, they incur heavy computational overheads and are not suitable for Android devices. Further, Hadoop makes liberal use of *Extensible Markup Language* (XML) for communications, which is flexible but requires very high processing overheads to parse. Due to these problems, Hyrax is largely a proof of concept system and due to its minimal changes to the Hadoopcodebase, suffers from poor computation performance results. For example, Android only allocates 16 MB of memory for an application, but with Hadoop and Hyrax's monitoring overhead, only 1 MB remains to be used for buffering MapReduce output. In comparison, our system was designed from the beginning for use in systems with limited resources and our framework system uses very little overhead. However, because our system is relatively new, it doesn't support many of the more advanced features of Hadoop, such as automatic data replicas, which may need to be significantly re-worked for mobile systems.

Elespuru, Shakya and Mishra (2009) deployed a research prototype used to study the feasibility of using smart-phones as a part of a MapReduce system. The prototype is developed for iPhones and uses Objective-C for the client and a server implemented in both PHP and Ruby. Their system uses a task requesting architecture similar to ours where the workers actively requests tasks from the server instead of the server assigning tasks to the workers. However, they only use the phones for performing map tasks and leaves the reduce tasks to be performed on the server. The researchers, using data throughput as a benchmark, showed that iPhones are able to consistently perform at only one order of magnitude lower than traditional desktop clients. This contribution to total processing power of smart-phones is considerable and a large portion of computation can be offloaded if there are enough devices.

Recently, there has been some interest in using JavaScript (Grigorik, 2009)to drive MapReduce clients. This would allow any device with a JavaScript enabled browser to participate in a MapReduce task. Due to JavaScript's popularity and pre-installation in all major web browsers (from desktops to netbooks to smart-phones), this idea can potentially add millions of extra computing resources to any MapReduce project. Also, efforts into making JavaScript engines more efficient (Eich, 2008) will greatly benefit any such JavaScript driven systems.

THE MISCO MOBILE MAPREDUCE FRAMEWORK

Traditional MapReduce implementations have been primarily focused on processing huge data sets in large data centers. These are not the types of applications we are aiming for with Misco. While the processing and network resource of smart-phones are growing at an incredible rate, they are currently not capable of replacing traditional MapReduce tasks. Instead, we present Misco as an abstraction framework for enabling developers to concentrate on the functionality of their applications rather than wrestle with the complexities of phone development environments and distributed applications. We believe the types of applications developed on Misco will take advantage of the new types of capabilities which are unique to mobile devices and smart-phones. These new mobile MapReduce nodes can become the producers of data rather than just processors of pre-collected data.

These are the types of applications we are currently targeting: Monitoring and social applications. However, like the introduction of the MapReduce framework itself, we are only limited to our own imaginations and there may be many more application domains which we have not thought about, or even considered possible.

Monitoring represents a large class of applications for which smart-phones are particularly well suited. The ubiquity and sensing capabilities of phones allow them to provide very good coverage over large areas, particularly where there is high population density. For these types of applications, the phones can be thought of as taking the place of traditional sensor network nodes.

Traffic Monitoring applications can be used to detect areas of high density and congestion. These areas can then be used to help guide user and shorten travel times. Phones with GPS can be used to detect traffic conditions, and then used to avoid further congestion. Some works such as the Berkeley traffic project already perform such monitoring using specialized software developed for cell phones.

Environmental Monitoring applications can be used to plot different levels of noise, temperature, seismic activity or pollution on a map. The resulting maps of these measurements can then be used to raise alerts for abnormal patterns or to study conditions throughout a city.

Social applications are another popular class of applications that people are interested in and are particularly suited for smart-phones. Many users are very intimately connected to their phones. They carry them around, store contact lists and keep track of upcoming events in calendars. The closeness of users and their phones provides even more value by using the user's context to customize content.

Similarity Based Sharing applications can be used to detect and automatically share similar multimedia items between users: music, pictures and videos. These items can be from similar in style, time and location or contents.

Event Summarization applications can be used to produce a fuller experience from sporting events, conferences or political gathers. Data from multiple people can be gathered together to generate a more complete view.

Location and Status Based applications such as the aforementioned CenceMe project can be used to notify friends of your current status. Other location based services can provide recommendations for venues in a given area or notify you when your friends are nearby.

There are many unique characteristics of phones and their mobile environment which must be taken into account when designing a distributed system for them. The first and largest concern is the limited available battery power. Although we are very concerned with energy, this is not a focus of our research. There are many other research groups focused on energy efficient computing and their efforts are orthogonal to ours and their techniques can be applied to our system independently. However, there are some basic things we keep in mind when designing our systems. It is almost always better to perform local computation and storage than to transfer data over the network and Wi-Fi connections are generally more energy efficient than cellular networks. We also take into consideration that network transfers are not free and service providers frequently charge usage fees.

Mobility and failures are also things we need to consider. Frequent failures occur due to device failures, software malfunctions and bad network coverage. Additionally, mobility also acts as a failure mode. A phone is considered failed if it is disconnected from the network for an extended period of time.

Unlike server machines and even general desktop computers, phones have a more personal connection with their users. Users generally own the phones they use and carry their phones with them at all times.This connection provides many advantages and disadvantages for using phones for distributed applications. Because phones are personal property, it is generally in the owner's best interests to keep their devices functioning properly. This eliminates the need for costly maintenance and power consumption of these devices. Also, the individual ownership of the devices encourages the development of portable and open standards without proprietary lock-in. A further advantage comes from the connection between owner and device. People are still the best at determining what events and things are interesting whether it's scenes to take pictures of or locations to congregate at. This user captured information is implicitly exciting and can be used for applications many people are interested in.

However, there are quite a few disadvantages to using privately owned devices. The most significant would be the reluctance to provide personal computation resources for use in part of a distributed system. This can be addressed through providing applications which users are interested in participating in such as traffic or social applications, or by providing incentives for contribution. Another drawback of using mobile phones as a basis for applications is coverage. Since phones are with their users, there are likely areas with little or no coverage. This can be addressed by use hybrid systems composed of additional devices such as a wireless sensor network.

A Concrete Example

As part of our project, we developed an application which uses Misco as its processing core: Dowser. Dowser is used by the user to locate nearby areas of good internet connectivity. Figure 2 shows some sample data and calculated centroids on the UC Riverside campus. Dowser is composed of three main components:

Data collection component runs in the background of our smart-phones and periodically gathers and stores local connection strength information along with a GPS location and the current timestamp.

Figure 2. A map of the UC Riverside campus and some connectivity data points we collected on 5 separate phones. The black dots represent the centroids calculated by our distributed k-means clustering algorithm

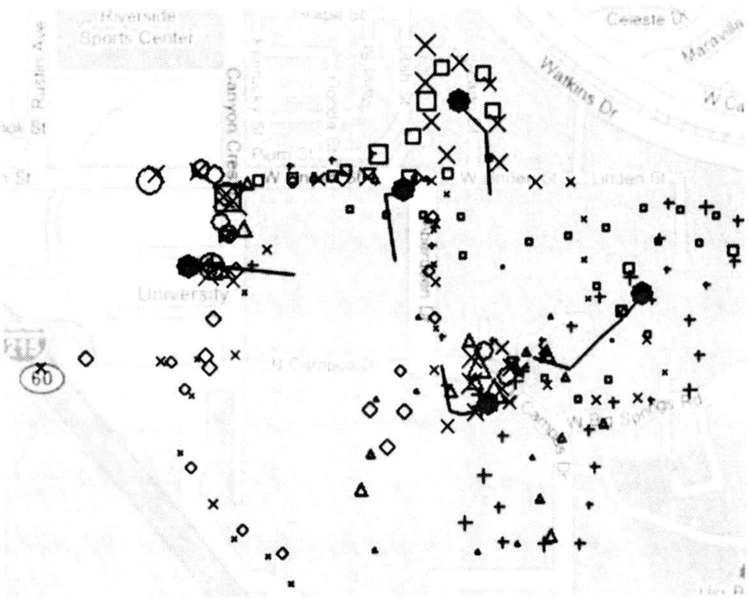

Data processing component which is a distributed k-means clustering algorithm developed to run on top of MapReduce and use Misco.

User Interface (UI) component which visualizes the user's current location and directs the user to nearby locations where areas with many good connections are located.

Focusing on the data processing component, we wrote a simple MapReduce application which operates on the collected data on the phones in such a way that we can calculate the k-means clustering without actually transferring any data off of the originating phones. The only data exchanged were locally calculated centroid locations information on each individual phone. This application demonstrates that we can perform computations where users do not need to send their connected data and save on network usage and also provides a degree of privacy for their location information. For the last stage of the algorithm, a last round of MapReduce is run to disseminate the final results to the phones and they are then able to use these results in the UI.

Misco System Model

The Misco system is composed of a set of N applications $A = A_1 ... A_N$ and a set of M workers $W = W_1..W_M$. Each application A_j consists of input data which must be processed via a number of map task and a number of reduce tasks. The number of map tasks correspond to the number of pieces the input data is split into $[T_{map}]$ and the number of reduce tasks correspond to the number of partitions $[T_{reduce}]$. An application defines the number of pieces the input file should be split into and also defines the number of reduce partitions.

The tasks operate independently of each other and are executed in parallel on multiple workers. Multiple workers can reside on a single machine and operate independently of each other. The input for the reduce tasks depend on the output of the map tasks, so every map task in an application must be completed before any reduce tasks from that application can start. A worker operates on one task at a time until the worker fails or the task is completed. Once a task is complete, the worker can request another task. Since all tasks are independent from each other, the new task may be from the same or a different process.

An application can be started by users adding them into the system or can be triggered automatically or periodically by an external process. The application is considered to have been completed once each map and reduce task has been successfully completed once. Each application may also define a deadline before which the application should complete by. This deadline may be used by some schedulers to determine how to assign tasks to workers.

Communication Design

One of the biggest challenges faced when designing Misco for mobile devices is the communication method used between phones and the server to transfer data and instructions. This communication problem, while important for traditional systems, is not as problematic as traditional systems are typically located in well connected data centers with high bandwidth, pairwise connectivity between each node. Many systems also provide a shared file system as part of their implementation, similar to HDFS, where Hadoop stores input and output data. For our mobile, smart-phone system, network usage is very costly in terms of both time (low bandwidth) and energy usage. For our system, we decided that our phones do not communicate with each other, but instead, only communicate with a centralized server for both instructions and data. This is a realistic model of typical smart-phone usage as they generally connect to the Internet and an Internet server through either Wi-Fi or a service provider's data plan (e.g. 3G).

All communications in our system happens using HTTP. We chose this route due to the ubiquitous nature of the protocol. It is very well supported by many devices, especially the devices with enough computation power to support processing. Another advantage of HTTP is it's built in end-to-end error correction and message ordering. For basic instruction passing between the server and workers, plain text HTTP requests and response can be used. We can use a multi-part POST request for file uploads from the user or phone to the server.

In addition to the communication technology, there are two main approaches to communicating between two parties: polling based and interrupt based. For our design, we chose to use a polling based approach despite the advantages interrupt based communication provides. The biggest attraction of interrupt based communication (and biggest drawback of polling) is the communication overhead. Less communication is required for interrupt based communication because data is sent only when it is required.

However, to support an interrupt based communications model, we would need to implement a server to run on each phone to constantly listen for incoming requests or messages. This is not practical or desirable for many reasons: it adds complexity to the software required on the phone and uses up limited memory resources. Also, to implement such a system, each phone has to be reachable from any other device in the network; this is unrealistic given the complexities imposed by firewalls and network mappings. For example, the IP and port number mapping to the phones are not consistent or guaranteed by the different service providers. We have attempted to use *Short Message Service* (SMS) messages as a means to provide an interrupt based method to send instructions to the workers, but this method is not reliable and not all phones are equipped with such messaging plans or contain *Subscriber Identity Module* (*SIM*) cards.

In our design, the worker will simply poll the server for work whenever they are idle, the server will respond with work if available or "None" if there are no tasks. The phones will idle for a period of time and then retry requesting for a task. The implementation for such a polling method is extremely simple, and the idle polling frequency can be adjusted to try and match the task arrival frequency, but if it is not calibrated correctly, there can be inefficiencies resulting in energy waste and periods where idle capacity is not being used.

Server Design and Implementation

One of the fundamental design goals of our system is scalability. Smart-phones are about one magnitude less powerful than desktop computers, but there are a lot more phones than computers. Current MapReduce clusters consist of thousands to tens of thousands of server nodes. In the case of phones, we would want to reach numbers in the hundreds of thousands or millions. One of the bottlenecks for current systems is the server which coordinates the tasks. We want to make our server as lightweight as possible. To this end, instead of monitoring all the workers and checking heartbeat messages for failures, we do not care about the status of the workers. We are only concerned with the status of the MapReduce tasks. In this way, we do not even need to know which workers or how many workers are in our system.

The Misco master server, shown in Figure 3, is composed of several components responsible for storing application data, assigning map and reduce tasks to workers and interacting with the user for creating application and providing results. The server sits at a known IP address and listens to a pre-specified port. The server is implemented in Python and uses the *BaseHTTPServer* and *BaseUDPServer* combined with the *ThreadingMixIn* to provide multi-threaded request handling. Misco uses custom request handlers to deal with user and worker requests. To download data and Python module files from the server, the worker simply navigates to the proper URL containing the file name, similar to a standard HTTP server.

The process of creating an application in the server is initialized by making a HTTP request to the server. This request can be user generated through a form submission by a user on a browser, or by an external application which supplies the required fields. The server expects an application name, the input data file, the python module file, how to divide the input file and the number of partitions for the reduce phase. The user can also provide optional parameters for their map and reduce functions to use and provide timing details such as deadlines and estimated execution times. Additional parameters are described in more detail later in the discussion of our example application and timing details are described when we elaborate on scheduling.

Once this request has been submitted to the server, it will create the data structures to store the application's execution information. First, the input data is split into a number of pieces with each piece numbered correspondingly to a map task (each piece of data is the input to a map task). Different schedulers are free to use different data structures to keep track of tasks and their completion, but in our most basic scheduler, we use an array to keep track of which tasks have been completed and the number of workers working on any particular task. This information is then used when we decide on the next task to assign to a worker. A similar data structure is created to keep track of both map and reduce tasks. The number of map tasks is determined by the number of files the input is split into and the number of reduce tasks is determined as the number of partitions the user specified. These two structures are then passed into the scheduler component which places it into the pool of available applications.

For our design, the scheduling module is very flexible and can be replaced very easily. In our initial implementation, we use a very simple *First In, First Out* (FIFO) scheduling technique, which is very similar to what many established MapReduce implementations are using. When a worker requests work from the server, an application, which has not yet been completed, will be picked from the pool of applications (FIFO order, in our case). Then, a task which has not been completed will be picked from that application and its details will be returned to the worker. This worker will be noted in the data structure. Note

Figure 3. TheMisco Server architecture. The Server is composed of two important internal components: the scheduler and the data storage. The scheduler keeps track of the progress of applications and is responsible for deciding which tasks to assign to workers. The data storage is responsible for holding the initial input, intermediary map results, final reduce results and Python module code for each application. Finally, the HTTP allows users and workers to interact with the server

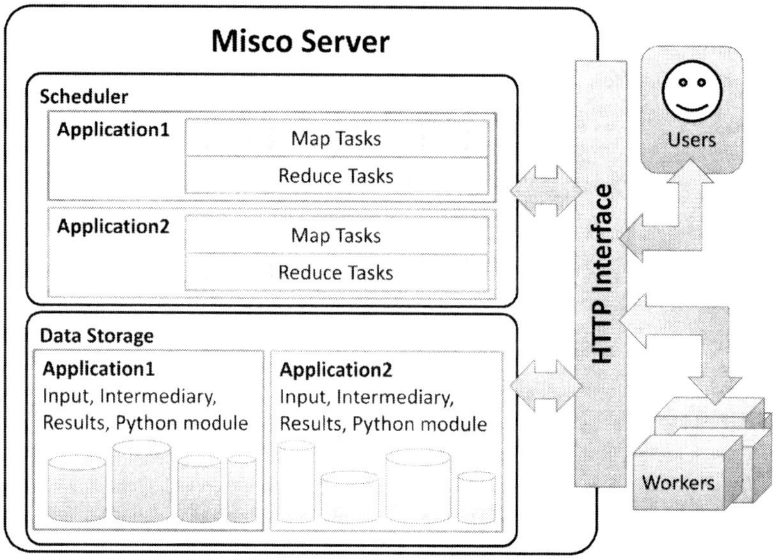

that all the map tasks from an application must be completed before any of it is reduce tasks can start. This restriction is a part of the traditional MapReduce model, but there are recent research efforts to bypass this (Condie, Conway, Alvaro, Hellerstein, Elmeleegy & Sears, 2009).

For more complicated schedulers, more complex algorithms are used for choosing which application and task to assign to workers. In addition to user provided timings, the scheduler can choose to track worker statistics such as processing power, failure rates and resource availability and use these factors to determine a scheduling technique to optimize its target parameters. Some schedulers will be concerned with meeting user specified application deadlines while others may only care about data throughput. The statistics and operations of more complicated schedulers are out of our scope and further discussions on scheduling will only consider our basic FIFO scheduler.

When a worker has completed a task, it will upload the results to the server. The server will pass this information to the scheduler, which can then mark off the completed task. The uploaded data from a map task is intermediary data and this data will be used as input for reduce tasks. The results from reduce tasks are considered to be the final results of the application. Once every task in an application is completed, the application is considered done and the results are considered complete.O ur server has the ability to notify an external process that the application has been completed by sending a *User Datagram Protocol* (*UDP*) message to an external process. This feature is very useful for automatic applications which want to perform further actions once the results are ready. Dowser is one such application which we discuss later.

Data storage on the server is straightforward, for each application, we create a folder to hold its Python module file. We then create a separate folder for each of: the input data (split into separate files), the intermediary data returned from map tasks and the final result data returned from reduce tasks. To access any of these files, the worker simply requests it at an HTTP *Uniform Resource Locator* (*URL*).

Worker Design and Implementation

Compared to the server, the worker is relatively simple as it is not concerned with scheduling. Instead, it is only concerned with processing a single task at a time. Figure 4 shows an architectural diagram of the worker. The worker is also implemented in Python and can be run on any device that supports Python and Python's networking Library.

Once the worker script is executed, no further user intervention should be necessary unless the script or device fails and must be restarted. The worker knows how to contact the server a priori and will immediately start requesting tasks. Whenever the worker is not processing a task, it is available for work and will contact the server with a request for work, passing the server the worker ID (for our phones, we use the *International Mobile Equipment Identity* (IMEI) number as an ID). The server will respond to the request with either "None" when there are no tasks for the worker to process or with a Python dictionary data structure containing information about the input data file to use, the function (map or reduce) to perform and any additional parameters the user specified when the application started. The worker will *eval()* this dictionary and download the required files from the server.

Once the worker has downloaded the files it needs, it will perform a dynamic import on the Python module file containing the map, reduce and accessory functions. The proper function will be executed with the input file passed in as a parameter. For additional flexibility, the user can supply an *InputReader* function for map tasks so that custom input data files can be used instead of the default behavior of reading the input file one line at a time. The user can also supply a

Figure 4. The architecture for the Misco Worker. The worker requests tasks from the worker and only keeps track of the data for the current task it is working on. The processor uses the application's Python module files to process the input data for the task and saves the results to files which are then uploaded to the server

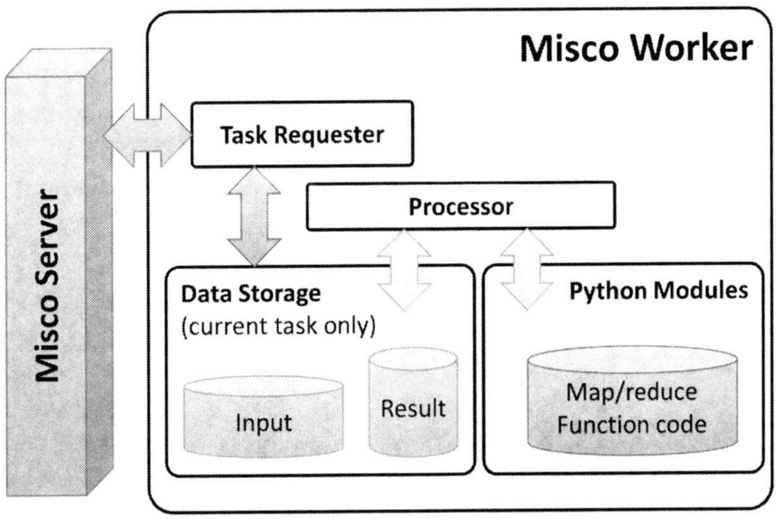

custom combiner function which can be used to compress or pre-aggregate output data from the map function before sending the results to the server. An example of a useful pre-aggregate is demonstrated by the word count example, where we want to count the occurrences for each word in the document. The map function for *WordCount* simply takes each word and emits it as <*word*, 1> and then the numbers are added together in the reduce function. However, for common words, there may be a lot of duplicates (e.g. <"the", 1>). It would be much more efficient to compress the intermediary data right after mapping by adding duplicates together, so if there are five instances of "the" in the input data for a map task, it would only output <"the", 5>.

The results from map tasks are stored in separate files with each file corresponding with a reduce partition. The results of a reduce task is a single results file. When the task is complete, the worker will upload the results to the server and request a new task.

Data storage on the worker is even simpler than the server. The worker creates a single folder for each task. Downloaded input data is placed into this folder and the results generated are also placed into this folder. Once the results from the task are submitted to the server, this folder and its content are deleted.

There are several local optimizations we perform on the workers. First, we allow for caching of the Python module files because a worker may be performing the same computation on different input data sets for the same application or even for different applications. If a worker determines that it already has a local copy of the module, it will not re-download it from the server. We have also found that if we store too many files on the phones, they have trouble with file locking and will often times not allow files to be deleted later. Thus, whenever we finish a task, we delete all generated data files before starting the next task.

Experimental Results

We performed a variety of experiments to determine the performance and resource usage of our MapReduce design and implementation. Our experimental test bed consists of 30 Nokia N95 8GB smart-phones (Nokia, 2010), a Linksys WRT54G2 router and a commodity desktop machine as the server. Figure 5 shows a portion of our testbed. The phones are equipped with ARM 11 dual *Central Processing Units* (*CPUs*) at 332 MHz, they support wireless 802.11b/g networks, Bluetooth, cellular 3g and has 90 MB of available main memory and 8GB of built flash memory for persistent storage. The server is a Windows XP desktop machine with a Pentium-4 2Ghz CPU and 640 MB of main memory, this is more than enough to resources to act as a Misco server and does not introduce any bottlenecks. The server has a 100Mbps wired connection to the router and each phone is connected to the router using IEEE 802.11g.

The application we chose to exercise our system is the very well-known distributed Grep application where each worker search through different sections of a large text file for a matching word or expression. Such an application can be used to analyze user web logs to find places of interest within a city based on browsing history. This application is representative of a general MapReduce application and fully exercises all aspects of the system, demonstrating its performance and resource usage. More complex applications differ mainly in the processing of their map and reduce tasks but do not affect the system's underlying execution.

For our tests, we spawn 10 instances of our application. The input size and number of tasks were varied for the different instances with input sizes between 100KB to 1MB and tasks between 16 and 73. Additionally, we set different deadlines for the applications, varying between 100s and 550s to simulate both tight and loose timing constraints. These deadlines are used when scheduling tasks

Figure 5. A part of our actual test bed: ten Nokia N95 smart-phones and a Linksys WRT54G2 router

in our *Earliest Deadline First* (EDF) scheduler. The failure of the worker nodes were set to follow a Poisson distribution with an adjustable failure arrival rate. Each experimental data point was arrived at by running the experiment 3 times and then averaging the results.

In our first set of experiments, we wanted to determine the resource overhead used by our system and the feasibility of running our system on smart-phones. We used the Nokia Energy Profiler (Nokia, 2009) to obtain memory, power and CPU measurements.

To measure Misco's memory overhead, we measured the amount of available main memory before and after loading Misco on our phones. This memory usage includes the data structures implemented by our system and also any additional Python libraries which are loaded. Misco only requires 800KB of memory, this is much less than the MBs of memory Hyrax required and is less than 1% of the 90MB of total available memory. We then looked at the memory requirements when running our application, shown in Figure 6. There is increased memory usage at 125, 165, 205 and 235 from the data structures used by the map tasks. The memory usage at 240, 280 and 315 are due to data structures built during the reduce task. This extra memory usage is dependent on what types of data the application needs to store and process.

Figure 7 shows the power usage of our application. Processing tasks with full CPU utilization uses 0.7 watts while access the network draws 1.6 watts, more than twice the power. The power spikes in Figure 7 correspond to the beginning and end of tasks when the worker is downloading or uploading data to the server.

When processing tasks, Misco tries to use all the resources that it can. Depending on other applications running on the phone or processor throttling policies, this can mean that Misco can use very little or 100% of CPU resources.

We implemented several experiments for Local performance. In this set of experiments, we wanted to see the performance of our system when processing a single task. Table 1 shows several measurements we took locally on a phone while it was processing different sized map tasks (similar results were obtained for reduce tasks). The *download* time reflects the time it takes to download the input data from the server to the phone and *upload* time is the time required to upload the results from the phone to the server. *Cleanup* time is for the time spent deleting the downloaded input file, the results and any temporary files that were

Figure 6. The memory usage on a single phone during the run of a distributed grep application. This phone processed four map tasks, followed by three reduce tasks. The timing of the tasks can be seen from the increased memory usage

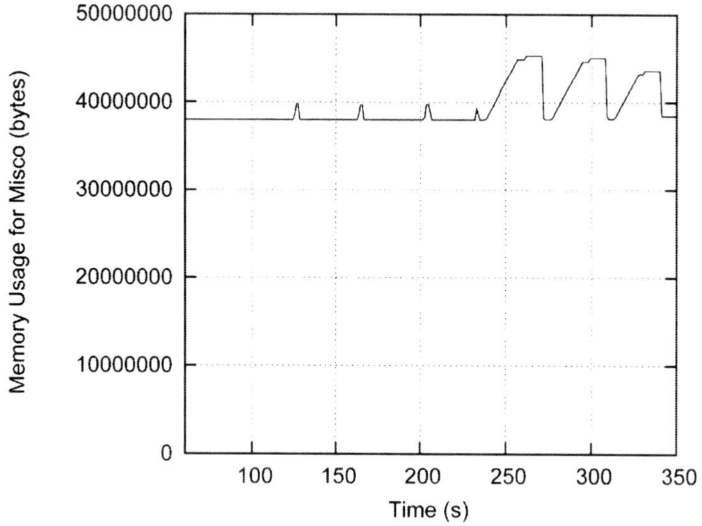

Figure 7. The energy usage of our system on a single phone during the run of an application. This phone processed four map tasks, followed by three reduce tasks. The starts of the tasks can be seen from the spikes in power usage when the phone accesses the network

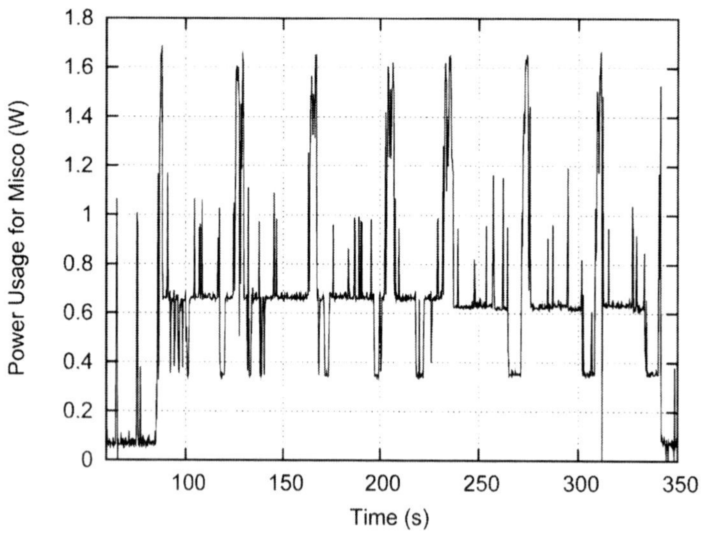

generated. Clearly, the measurements show that the majority of the time spent on a task is performing computations and only a small fraction of the time is required for downloading, uploading and cleanup. This demonstrates that our system is quite efficient for the phones we are considering.

For end-to-end execution experiments, we wanted to see how our implementation performs

Table 1. The performance of a single task locally on a phone with different sized inputs

Input (KB)	Down (s)	Up (s)	Clean (s)	Compute (s)
Map Tasks				
62	0.5	1.2	0.5	8
94	0.6	1.2	0.6	10
125	0.7	1.4	0.7	11
Reduce Tasks				
96	2	0.42	0.05	3.8
150	3	0.5	0.04	5
192	4	0.7	0.04	7

on a cluster of smart phones. In addition to just processing the distributed Grep application, we also wanted to measure the effects of phone failures and see what effects different scheduling algorithms have on the performance. To simulate phone failure, we varied the probability of any given task failing. If a task is simulated to fail during its execution, it will erase all its results, and wait for a preset recovery period before requesting more work from the server. This failure rate was the same for all phones. We have also implemented two basic scheduling methods for picking a task to assign to a worker from an application. The random approach is to just pick any task which has not yet been completed. The sequential approach is to order the tasks and assign tasks which have not been completed in their predetermined order.

Figure 8 shows the resulting end-to-end times for the two different schedulers as the number of failures is increased. The end-to-end times is the time required for all the applications in the system to complete. When failure rates are low, the performance is quite good with an end-to-end time of 250 s. As the failure rates increase, the performance of the system degrades exponentially as there are fewer phones available to do meaningful work. This behavior is as expected because each addition to the failure rate affects the total computation power of the system more severely as the failure rate increases. Our results also show that a sequential scheduler performs 40% better than a naive random scheduler at lower failure rates and 20% better at higher failure rates. This is because the sequential scheduler assigning less duplicate work to multiple workers. However, this effect is less pronounced as the failure rate increases as more of the workers are likely to fail any given task.

FUTURE RESEARCH DIRECTIONS

The idea of using smart-phones as part of distributed applications is fairly new and only recently been feasible due to technological advances in portable electronics. As a relatively new area, there are many areas that need to be explored such as energy, data locality, heterogeneity, real-time applications and security.

One of the major concerns for developing systems for mobile devices is power usage. The biggest culprit of energy usage is network access. This fact makes for a very convincing case to fully apply the principles of data locality. Instead of transferring data to different machines for processing, try to perform the processing where the data is already located. This concept is not new and has already been investigated for data centers and there are methods to reduce data usage. Our smart-phone based model is different. The majority of feasible application for our setting is not just using the phones as processing units for raw data. Instead, the phones themselves are often the generator for the data, making it even more important for us to take advantage of data locality.

Conserving energy is not just constrained to our mobile space. It is also an area of study with a broad range of applications from electric vehicles to large data centers. Orthogonal research is being conducted on new, more efficient ways to generate and store energy. Renewable energy sources such as solar power and human gener-

Figure 8. The end-to-end performance of our test applications when the failure rates of individual tasks are varied across the phones. The end-to-end time is the total time required to finish processing all applications. We also test the difference in completion times when we use a different method for assigning tasks from applications to workers

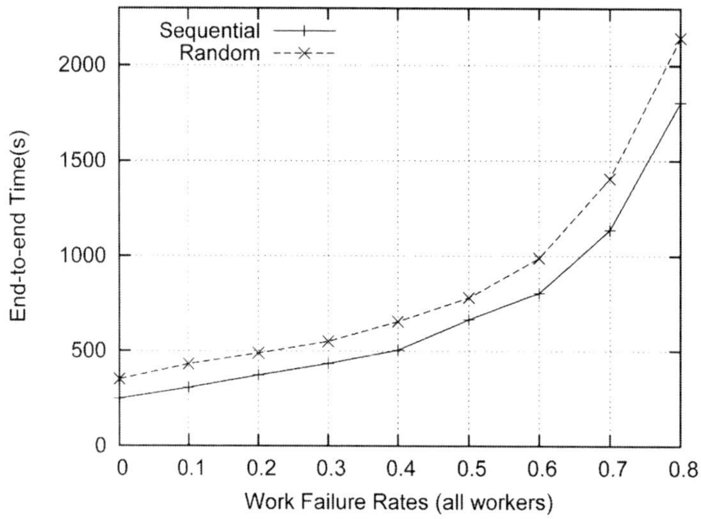

ated kinematic energy are being developed to help power mobile electronics. Other interesting ideas such as wireless power supplies are also helping to solve this problem of providing energy. Research in providing more efficient energy will not just benefit our system, but has a large impact on many other areas.

Since MapReduce is built on top of a collection of very loosely coupled components, it is reasonable to try increasing the performance and processing power of the system by allowing more devices to join the system. It would be relatively easy to extend the system to run on a variety of different devices with vastly different capabilities and processing powers. However, it is not clear how the work would be distributed, how capabilities of the system will be discovered and how to best allow the system to work together. For example, it would be impractical to process very large amounts of data on cell phones. We would ideally like to have a large server farm do this processing. Conversely, it would be very difficult to use servers to gather and monitor noise over a city. Studying how to build harmonious and efficient heterogeneous systems is key to allowing for more portable and useful distributed systems.

There has been some research in speeding up processing and providing incomplete or imprecise results. There are other approaches that speed up execution by optimizing the scheduling of applications and tasks. Research to providing faster results is very important for many of the monitoring applications, which must detect realtime problems, and for user facing applications, which must return results to the users within a reasonable time frame.

CONCLUSION

Mobile phones are becoming more powerful and more commonplace. Applications developed for these phones continue to evolve and advance. However, development tools for creating the types of distributed social and monitoring applications that people are demanding are difficult to use. In this chapter, we looked at several current mobile applications and introduced the MapReduce framework as a solution for creating distributed applications for smart-phones and other devices.

We have shown through our experimental test-bed of Nokia N95 8GB smart-phones that we are able to do useful work. We have also demonstrated through our prototype application Dowser, that applications can be easily developed around our framework. As new developments for mobile devices progresses rapidly, we expect our system to be able to fully use any addition resources given to it and for its performance to scale linearly.

There are many areas where it is not practical to use our mobile system, such as for processing very large quantities of data. The design of any system is based on a set of tradeoffs, on one end, specialized systems, such as databases, are highly optimized to perform specific computations. On the other end, completely open platforms allow for very flexible and complex applications, but only very basic optimization is possible. Using MapReduce as a basis for a distributed application framework was our choice on this spectrum of flexibility versus optimizations.

MapReduce has been proven to be very useful in large scale server environments and has spawned many more applications than was predicted. We hope that providing such a platform for mobile devices would also lead to similar benefits.

ACKNOWLEDGMENT

This research has been supported by the European Union through the Marie-Curie RTD (IRG-231038) project, the SemsorGrid4Env project and the MODAP project, and by AUEB through a PEVE2 project.

REFERENCES

Agapie, E., Chen, G., Houston, D., Howard, E., Kim, J., & Mun, M. ...Rahimi, M. (2008). Seeing our signals: Combining location traces and Web-based models for personal discovery. *HotMobile*. Krakow, Poland. New York, NY: ACM. doi>10.1145/1411759.1411762

Breslauer, D., Maamari, R., Switz, N., Lam, W., & Fletcher, D. (2009). Mobile phone based clinical microscopy for global health applications. *PLoS ONE*, *4*(7), e6320. doi:10.1371/journal.pone.0006320

Chen, S., & Schlosser, S. W. (2008). *MapReduce meets wider varieties of applications*. Pittsburgh, PA: Intel Labs Pittsburgh Tech Report.

Condie, T., Conway, N., Alvaro, P., Hellerstein, J. M., Elmeleegy, K., & Sears, R. (2009). *MapReduce online*. USA: University of California at Berkeley.

Dean, J., & Ghemawat, S. (2004). *MapReduce: Simplified data processing on large clusters* (pp. 137–150). San Francisco, California: In OSDI.

Dou, A., Kalogeraki, V., Gunopulos, D., Mielikainen, T., & Tuulos, V. H. (2010). *Misco: A MapReduce framework for mobile systems*. Samos, Greece: PETRA.

Eich, B. (2008, August 23). *TraceMonkey: JavaScript lightspeed*. Retrieved on August 24, 2010, from http://weblogs. mozillazine.org/ roadmap/ archives/ 2008/ 08/ tracemonkey_ javascript_ lightsp.html

Eisenman, S., Miluzzo, E., Lane, N., Peterson, R., Ahn, G. S., & Campbell, A. A. (2009). *BikeNet: A mobile sensing system for cyclist experience mapping*. TOSN.

Elespuru, P. R., Shakya, S., & Mishra, S. (2009). *MapReduce system over heterogeneous mobile devices*. *SEUS* (pp. 168–179). Newport Beach, California: Springer.

Eriksson, J., Girod, L., Hull, B., Balakrishnan, H., & Madden, S. (2007, July 22). *Pothole Patrol (P2)*. Retrieved on August 24, 2010, from MIT Cartel: http:// cartel. csail. mit. edu/ doku. php?id=p2_ pothole_ patrol

Foursquare (2010). *foursquare*. Retrieved 08 24, 2010, from http://foursquare.com/

Gartner (2010, Aug 12). *Gartner says worldwide mobile device sales grew 13.8 percent in 2nd quarter of 2010, but competition drove prices down*. Retrieved on August 24, 2010, from Gartner: http:// www.gartner.com/ it/ page.jsp?id=1421013

Ghemawat, S., Gobioff, H., & Leung, S.-T. (2003). *The Google file system*. Bolton Landing, New York: SOSP.

Grigorik, I. (2009, March 03). *Collaborative MapReduce in the browser*. Retrieved August 24, 2010, from igvita.com: http:// www.igvita. com/ 2009/ 03/ 03/ collaborative- map- reduce- in-the- browser/

Hadoop (2009, July 18). *Hadoop MapReduce*. Retrieved August 24, 2010, from Hadoop: http:// hadoop. apache.org/ mapreduce/

HDFS. (2009, July 18). *Hadoop distributed file system*. Retrieved August 24, 2010, from Hadoop: http://hadoop. apache. org/hdfs/

He, B., Fang, W., Luo, Q., Govindaraju, N. K., & Wang, T. (2008). *Mars: A MapReduce framework on graphics processors*. Toronto, Canada: PACT.

Hoh, B., Gruteser, M., Herring, R., Ban, J., Work, D., & Herrera, J. ...Jacobson, Q. (2008). Virtual trip lines for distributed privacy preserving traffic monitoring. *In MobiSys '08: Proceeding of the 6th International Conference on Mobile Systems, Applications, and Services* (pp. 15-28). Breckenridge, Colorado. *doi:10.1145/1378600.1378604*

Honicky, R. (2010, September 24). *N-SMARTS: Advanced sensing*. Retrieved September 24, 2010, from N-SMARTS: http:// www.cs.berkeley.edu/ ~honicky /nsmarts/ sensing. shtml

HTC. (2010). *Droid Incredible Technical Specifications*. Retrieved 08 24, 2010, from http:// www. htc.com/ us/ products/ droid-incredible-verizon/ #tech-specs

Hull, B., Bychkovsky, V., Chen, K., Goraczko, M., Miu, A., & Shih, E. ...Madden, S. (2006). CarTel: A Distributed Mobile Sensor Computing System. *SenSys*. Boulder, Colorado.

Lu, H., Pan, W., Lane, N. D., Choudhury, T., & Campbell, A. T. (2009). *SoundSense: Scalable sound sensing for people-centric sensing applications on mobile phones*. Kraków, Poland: MobiSys.

Marinelli, E. E. (2009). *Hyrax: Cloud Computing on Mobile Devices using MapReduce*. Unpublished Masters Thesis, Pittsburgh, PA.

Miluzzo, E., Lane, N. D., Fodor, K., Peterson, R. A., Eiserman, S., & Lu, H. ...Campbell, A. (2008). Sensing meets mobile social networks: The design, implementation and evaluation of the CenceMe application. *SenSys*. Raleigh, North Carolina.

Moray Rumney. (2008). *IMT-advanced: 4G wireless takes shape in an olympic year*. Agilent Technologies.

Nokia (2009, April 1). *Nokia energy profiler*. Retrieved on August 24, 2010, from http:// www.forum.nokia.com/ info/ sw.nokia.com/ id/ 324866e9-0460-4fa4-ac53-01f0c392d40f/ Nokia_ Energy_ Profiler.html

Nokia (2010). *Nokia N95 8GB*. Retrieved on August 24, 2010, from http:// www.nokiausa. com/ find-products/ phones/ nokia-n95-8gb/ specifications

Ranger, C., Raghuraman, R., Penmetsa, A., Bradski, G., & Kozyrakis, C. (2007). *Evaluating MapReduce for multi-core and multiprocessor system*. Phoenix, Arizona: HPCA.

Tuulos, V. H. (2010, May 27). *Disco: Massive data - Minimal code*. Retrieved on August 24, 2010, from http://discoproject.org/

ADDITIONAL READING

Abdelzaher, T., Anokwa, Y., Boda, P., Burke, J., Estrin, D., & Guibas, L. (2007). *Mobiscopes for human spaces*. Pervasive Computing.

Andersen, D. G., Franklin, J., Phanishayee, A., Tan, L., & Vasudevan, V. (2009). *FAWN: A fast array of wimpy nodes. SOSP*. Big Sky.

Armbrust, M., Fox, A., Griffith, R., Joseph, A. D., Katz, R. H., & Konwinski, A. (2009). *Above the clouds: A Berkeley view of cloud computing*. Berkeley, CA: Berkeley Press.

Beach, A., Gartrell, M., Xing, X., Han, R., Lv, Q., & Seada, K. (2009). *SocialFusion: Context-aware inference and recommendation by fusing mobile, sensor, and social data*. Boulder, CO: University of Colorado.

Buyya, R., Yeo, C. S., Venugopal, S., Broberg, J., & Br, I. (2009). Cloud computing and emerging IT platforms: Vision, hype, and reality for delivering computing as the 5th utility. *Future Gener. Comput. Syst.*

Chu, C.-t., Kim, S. K., Lin, Y.-a., Yu, Y., Bradski, G., & Ng, A. Y. (2007). *MapReduce for machine learning on multicore*. Vancouver, Canada: NIPS.

Deng, L., & Cox, L. P. (2008). *Nericell: Rich monitoring of road and traffic conditions*. Raleigh, NC: SenSys.

Eagle, N., & Pentland, A. (2006). Reality mining: Sensing complex social systems. *Personal and ubiquitous computing*.

Gaonkar, S., Li, J., Choudhury, R. R., Cox, L., & Schmidt, A. (2008). *Micro-blog: Sharing and querying content through mobile phones and social participation. Mobisys*. Breckenridge.

Hamilton, J. (2009). *Cooperative expendable micro-slice servers (CEMS): Low cost, low power servers for Internet-scale services. CIDR*. Asilomar.

Malek, S., Mikic-Rakic, M., & Medvidovic, N. (2005). A style-aware architectural middleware for resource-constrained, distributed systems. *Software engineering*.

Mtibaa, A., Chaintreau, A., Massoulie, L., & Diot, C. (2007). The diameter of opportunistic mobile networks. *CoNEXT' 07*. New York, NY.

Patel, S. N., Kientz, J. A., Hayes, G. R., Bhat, S., & Abowd, G. D. (2006). *Farther than you may think: An empirical investigation of the proximity of users to their mobile phones. UbiComp*. Orange County.

Reddy, S., Mun, M., Burke, J., Estrin, D., Hansen, M., & Srivastava, M. (2010). *Using mobile phones to determine transportation modes*. TOSN.

Shankar, P., Nath, B., Iftode, L., Ananthanarayanan, V., & Han, L. (2010). *SBone: Personal device sharing using social networks*. USA: Rutgers University.

Sousa, J. P., & Garlan, D. (2002). *Aura: An architectural framework for user mobility in ubiquitous computing environments*. Montréal, Canada: WICSA.

Zaharia, M. Konwinski, y., Joseph, A. D., Katz, Y., & Stoica, I. (2008). Improving MapReduce performance in heterogeneous environments. *OSDI*. San Diego.

Chapter 10
Context-Aware Multimedia Services for Mobile Lecture Streaming

Ulrike Lucke
University of Potsdam, Germany

ABSTRACT

Applications and infrastructures of the future will have to be intelligent to some extent, which means that they should automatically adapt to the current situation and intention of the user. This regards basic interoperability issues in heterogeneous environments as well as advanced behavior of software. Two technological trends contribute to solve this problem: service-orientation and context-awareness. We present a system architecture and realization that makes use of these trends. Its scope is education in general, or the interconnection of face-to-face and online learning in particular. The focus of this article is to explain the technical architecture behind these applications. Its validity is demonstrated with the help of selected use cases.

INTRODUCTION

Education is a yielding field for research, not only since most researchers at Universities are also educators and may organize or even design their educational arrangements, themselves. This is true for several disciplines, including Computer science. Current research in this area covers, for instance, aspects of mobility, adaptivity, or integration of educational infrastructures. Today, educational applications are characterized by a large number of multimedia in various types and formats. Besides descriptive text, there may be schema drawings, explanatory sound files, photographs, or videos – combined in a web page, slide show, or printable script. Plenty of research

DOI: 10.4018/978-1-61350-144-3.ch010

tried to tease out flexible, adaptive, and easily reusable teaching and learning content from existing document formats. This was leading to the *eXtensible Markup Language* (*XML*) technology as the preferred mechanism, and to enormous efforts for creation of such material (Wallace & Levinson, 2006). This brought up several form of so-called rapid e-learning content production, among this the capture, archiving, and playback of traditional face-to-face lectures, also referred to as *e-lectures* (Chandra, 2007). While production of e-lectures is easy, current research in this field is mainly focused on issues of semantic analysis, cross-system integration, and mobile use.

Along with the increasing number of *mobile devices* owned by students and staff, there is a strong demand for mobile services at today's universities. This includes not only education, but also access to research facilities and administrative systems (like campus management). Delivery of light-weight Web sites and other simple client interfaces can be sufficiently handled on mobile devices. However, complex platforms and rich media content require special attention due to limitations in network connection and interaction features. For instance, accessing an e-lecture stored in a traditional learning management system implies navigation to the respective course, opening its file area, searching for the desired recording, downloading the file, and opening it in the associated player. Accordingly, it takes some effort for the teacher to upload his e-lectures. Thus, depending on the designs of the website and the client device, publication and retrieval of an e-lecture can be a hard job.

Technological progress can improve this situation, twofold: Considering small, repeatable units of the system functionality as *services* helps to mash-up distinct parts of the related platforms in order to automate some of the users' activities (Ogrinz, 2009). Furthermore, analysis of *context* conditions helps to estimate what the users might want or need in the given situation (Loke, 2006). This article presents some results of our developments in this field. We implemented an infrastructure for mobile and context-aware access to several multimedia services, including streaming of e-lectures. We found that both teachers and students are a quite critical audience, and their continuous stimuli for releasing improved versions of educational systems create a productive field for experiments and developments.

Background

Several universities offer *e-lectures* of complete curricula. These are used by students to make up for lectures they could not attend, and to prepare themselves for exams. Recently, there are some projects to deploy complete studies to remote participants in foreign countries with the help of e-lectures (Donick & Tavangarian, 2010). Furthermore, with our "Media-based Junior Studies" we are delivering regular lectures from university studies in several disciplines (like chemistry, history, medicine, computer science) to high-performing scholars in grades 10 to 12 at selected regional secondary schools. The intention is to bridge the gap between different teaching and learning cultures and thus to prepare the scholars for university courses in order to decrease the number of drop-outs (Thomanek et al., 2009). Several other universities have similar programs, but limited to regular face-to-face lectures. This is not feasible in a region of large territory but lower population density, where it takes scholars two or more hours to reach the next university. Thus, we decided to produce e-lectures for the most suitable courses of targeted studies, and scholars (so-called junior students) can work through these in schools or at home, supervised both by their local teachers as well as remote tutors at university. Face-to-face lectures are reduced to few events. The IT infrastructure of this system consists of a recording, archiving, and streaming platform at our University, which is accessed from various stationary and mobile clients. We use the tool *Lecturnity* for production of e-lectures, because it can fully import traditional *Powerpoint* presentations (in contrast to pixel-based screen grabbers), offers a full-text search on all content, and exports several file- and stream-based formats. For streaming, we use a *Darwin Server Unlimited*, which supports several data formats and a large number of simultaneous client connections.

Beside this, we are involved in a major PhD program "*Multimodal Smart Appliance Ensembles for Mobile Applications*" (*MuSAMA*). It is located in the fields of *mobile, smart multimedia* and *pervasive computing*. The aim of the program is to design and create an infrastructure for intelligent, context-aware support of mobile users. This includes aspects like intuitive interaction, intention recognition, device interoperability, or energy awareness (Heider & Kirste, 2006). Several PhD students from different research groups across computer science and electrical engineering are working together in order to assemble their solutions to an innovative framework. Here, we are creating an infrastructure for seamless interconnection of different applications, networks, and devices to a homogenous ensemble. The targeted groups of users are students and staff at institutions like a university, in order to support their daily activities by a pro-active, intelligent IT infrastructure. Typical scenarios are lectures, seminars, or meetings.

Furthermore, we carried out several activities in the field of *Service-Oriented Architecture* (*SOA*), especially its application to the infrastructure of a university. SOA has proven to be a mature mechanism to bridge between different programming languages, operating systems, and hardware platforms. Services communicate with the help of standardized protocols. Service consumers do not need a-priori knowledge on service providers, but receive contact information on the currently most suitable service from a central broker. We found that complex infrastructures like in a university can gain huge benefit from SOA (Tavangarian & Lucke, 2009). Many existing systems for education, research, administration, and for private use (or at least parts of them) can be connected using services. Dynamic assignment of appropriate service providers opens up new possibilities for the user that could not be realized with static system interconnections, e.g. for automated integration of external content sources into the document collection of a course. The highest potential lies in inter-institutional scenarios beyond the borders of single administrative domains, were traditional integration mechanisms are exceeding their limits.

The combination of all these projects brought up the possibility of mobile, context-aware use of lecture streaming (and other services) across the campus and beyond (Lehsten et al., 2010). Our solution consists of a central service infrastructure – the so called university service bus –, a variety of service providers (including lecture recording, audio and video streaming, communication, and context management), and several components for use of services (as a part of existing platforms or as new applications). As an exemplary result, we are able to provide the late student a live-stream of the ongoing lecture he should be attending on his mobile device (along with some information on how to get to the lecture hall) without the need to heavily interact with the learning platform for this purpose, and to turn off the stream right when he enters the lecture hall. Of course, repeated playback of archived lecture recordings is possible, too. This was a challenging work since there were many types of network and service technologies we had to consider, like Bluetooth on mobile phones and Ethernet or *Wireless Local Area Network* (*WLAN*) in the university network.

Research Challenges

Two key directives for our work were: To integrate the currently existing infrastructure at the university (in terms of learning platform, media archive, multimedia equipment, client devices, and so on). To make the most use of modern mobile appliances (like smart phones and *Personal*

Digital Assistants (*PDAs*)) without forcing the user to install new software on it.

These conditions set some limits regarding the obtainable functionality of the system as well as the intended simplicity of the resulting architecture, but at the same time they ensure a broader use of the developed services. In this way, we had to face three major challenges:

1. How to bridge between different network technologies?

Standard networked environments – not only in a university – are based on Ethernet, the *Internet Protocol* (*IP*), and the *Transmission Control Protocol* (*TCP*) or the *User Datagram Protocol* (*UDP*). This includes wired as well as wireless networks. Backbone technologies like *Asynchronous Transfer Mode* (*ATM*) are outside the focus of this article. But, in the consumer area there is a number of additional wireless technologies like Bluetooth, which is supported by most mobile phones. Though some devices also support WLAN (and thus, IP) users do often switch it off because of high energy consumption, or device features are best used over Bluetooth. So, we needed a mechanism to fully harness Bluetooth devices and their capabilities in traditional IP-based networks, and vice-versa.

2. How to bridge between different service technologies?

The SOA hype during the last decade brought up a number of incompatible service technologies. This is not only a matter of IP-based or Bluetooth-based services. Moreover, in the IP sector there is a number of competing technologies like Web Services and Jini – each with special pros and cons, with characteristic application domains. There is no common denominator of their features or language constructs. That's why a mechanism for interoperability of different service technologies is needed, with a focus on openness for future developments in this still developing field.

3. How to model, gather, manage, and interpret context in a service-based system?

Intelligent system behavior requires not only static knowledge on the user and his tasks, but also dynamic information on his current situation and the surrounding environment. Context (as the amount of all this information) can be gathered from a variety of sources. But, systematical processing of context requires a basic model of what data is available and how it relates to each other. Besides smartness of an application itself, there are many special points to consider context in network and service layer interoperability. We needed a flexible and extendable mechanism for this.

These questions lead to the structure of the remainder of the article. We are starting with an overview of related work in the next section. Afterwards, we provide an architectural overview of the system. This includes our networking model and the resulting communication platform, as well as implemented mechanisms to bridge heterogeneity on network and service level. The next section is dedicated to some use cases that are of interest in the context of this book, namely different services for media delivery. We identify requirements that come along with the use of mobile devices and present our respective approaches. To sum up our experience in developing and deploying mobile, context-aware multimedia services, we present a simple categorization of how to use context in service-based systems. We claim that this model may help to design new services, regardless of the area of application. Finally, we present ongoing work as well as some further ideas, and give a short summary.

Related Works

There is plenty of research in the area of mobile, context-aware multimedia. Since the focus of this chapter is on architectural issues of context-awareness and interoperability in mobile environments, we would like to point out selected work with a contribution to these questions. Related work includes various stand-alone tools as well as complex infrastructures. Some well-known and sophisticated examples are:

The AMBIENCE project (Teixeira & Faihe, 2002) created context-aware multimedia environments. It was focused on intuitive interaction with a pervasive infrastructure in the home entertainment market. Though several approaches for dealing with multimedia content and transmissions were designed and analyzed, a systematical mechanism for integrating an open pool of multimedia services was missing.

The Mediascapes project (Stenton et al., 2007) allows triggering multimedia content based on physical locations. Users can upload content to a server and tag it with GPS information. This is used when another client enters the place in order to present the respective image, video, or sound. Mediascapes is a powerful tool for a variety of applications, but it is designed for handheld devices only and requires special software on a host PC and the mobile device. A similar idea lies behind M-Studio (Pan et al., 2002) which also provides location-based video sequences to mobile devices. It comes along with a comfortable authoring kit to link content with spaces. Moreover, it is not limited to GPS for gaining positioning information and thus is also suitable for indoor experiences. QoSDREAMS (Naguib et al., 2001) also provides similar mechanisms. Beyond the recent two examples, it unifies types of location data and events in a systematical approach with an efficient architectural model. Also, it adds quality of service as a parameter for multimedia transmission. However, heterogeneous networking and the variety of context conditions besides location are not considered.

There are a lot of developments in the area of mobile lectures. Lecture recordings and lecture streaming, respectively; have proven to be a mature mechanism for large-scale educational settings (Ulrich et al., 2010). Encoding techniques and quality of service parameters can be adapted to given conditions even for use on mobile devices (Liu & Choudary, 2004). Moreover, there are some sophisticated interaction techniques that focus on temporal or spatial conditions (Hürst & Meier, 2008; Huber et al., 2010), audio & video conferencing (Adorni et al., 2007). To the best of our knowledge, these developments do not offer a systematical approach for targeting the diversity of devices and tools that can be found in educational settings, and thus do not address interoperability issues.

Another approach is focused in MPEG-21, associated with the DANAE project (Timmerer & Hellwagner, 2005). It offers several features for adaptivity of multimedia content in terms of coding techniques and device capabilities. This is an important point for interoperability in a heterogeneous environment, but in turn there are no mechanisms for context-awareness regarding the user's situation and personal needs.

GAIA (Román et al., 2002) is a middleware for active spaces. It offers sophisticated support for managing distributed resources, devices, and physical objects in relation to the tasks and the current context of the user. GAIA is a very powerful example for pervasive computing infrastructures with intuitive interaction in a smart environment, but it strictly requires applications to adhere to its own design and communication principles. This limits its scope.

We see that many mechanisms for adjusting the transmission and presentation of multimedia content to the current context of the user already exist. To the best of our knowledge, the uniqueness of our system lies in the key directives stated above: integrating existing IT infrastructures and keeping hands off client devices to a maximum extent. Other characteristics of our system are the broad re-usability of our (multimedia and other) services in different platforms, along with the simplicity of providing or invoking a service.

Besides the above mentioned results of research, there are several other fields which are of certain importance to our work. To name just a few: context modeling and management (Bettini et al., 2010), service-orientation on embedded devices (Driscoll & Mensch, 2009), multimedia encoding and packaging strategies (Pagani, 2008; Grosky et al., 2009), mobile ad-hoc networks (Tonguz & Ferrari, 2006), and mobile learning in general (Byrne et al., 2007). We did not explicitly focus on these issues, but rather utilized the results of the current state of the art in these fields.

THE ARCHITECTURAL APPROACH

This section describes our general thoughts and developed mechanisms to tackle the challenges identified above. The first subsection explains a model for systematically structuring complex networks. Afterwards, our implementation of this model in terms of a dedicated hardware component is presented. Basic issues of interoperability on network and services layer are discussed in the third and fourth subsection.

Structuring Heterogeneous Environments

An abstract model of the network organization helps to clearly identify single layers and to assign dedicated features and components to them. This is crucial for mobility, since mobile network access is not self-evident just because of broad availability of wireless networks. Rather, heterogeneity often prevents from interoperability of systems components. That's why we will tackle this problem by defining and providing an infrastructure to bridge the gaps between single technologies. Considering typical networks one can identify areas of homogenous technologies that are put together to heterogeneous structures, and finally larger systems organized across such local networks. Thus, we defined a virtual overlay structure for heterogeneous environments consisting of the three layers, as visualized in Figure 1 (Dressler et al., 2007).

A *cell* is the smallest unit in the concept. It is built from devices connected by a single type of network, like Ethernet, WLAN, Bluetooth, or ZigBee. On the one hand, cells are comparatively easy to manage due to their homogeneity. On the other hand, for wireless technologies the

Figure 1. Homogeneous network cells are hierarchically structured to a meshwork of heterogeneous ensembles (local) and communities (remote)

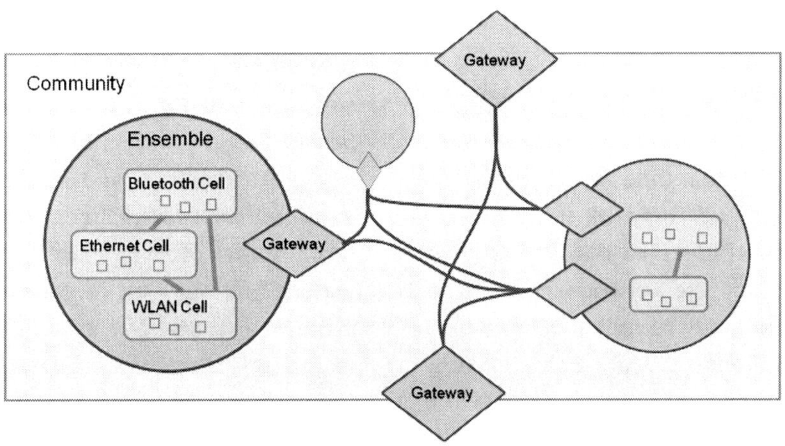

set of devices within reach can dynamically change, so that members may leave or enter the cell. That's why mechanisms of self organization are of special interest on this layer.

An *ensemble* is the second layer in the concept. It consists of several local cells that are combined to a heterogeneous structure. This can take place either by devices that are member of more than one cell, or by a dedicated network component (gateway). Mechanisms to bridge between network technologies are required here, which enables transparent communication of devices. Moreover, due to existence of a stable gateway service-orientation can be established in this layer, so that single devices can offer their resources and functions as services to other devices.

A *community* is the third and highest layer of the concept. It represents a remote interconnection of several ensembles sharing the same groups of users and applications. A community is formed by ensemble gateways that share information with each other, for instance in a scalable and fault-tolerant peer-to-peer network. This allows devices to deploy and invoke services for/from remote ensembles, and the autonomy of foreign administrative domains remains untouched.

In its entirety, such a system enables straightforward realization of pervasive computing scenarios, because of the transparent integration of different network and device types, sensor nodes, and various services for context management, smart applications, and so on also in a broader environment.

The General Purpose Access Point

As stated above, mechanisms are necessary to bridge between different network technologies. This requires a more detailed consideration of available elements in the network, in order to define how bridging may be organized. Imagine a typical scenario in a lab, meeting room, or work place. There are traditional Ethernet devices like *Personal Computer* (*PC*), projector, and printer. Moreover, there can be a WLAN with notebook and PDA. The notebook also has a Bluetooth connection to a mobile phone and some circuitry like sensor nodes. A visualization of this scenario is given in Figure 2. The Ethernet, WLAN, and Bluetooth cells can be understood as *horizontal* network structures, while devices with more than one network interface (which are the notebook and the access point) also span a *vertical* structure. These can be used for spontaneous, step-by-step interconnection of cells, as depicted in the left hand part. Another possibility is to introduce a dedicated network component with a vertical network structure covering all existing technologies, as shown in the right hand part of Figure 2.

Spontaneous interconnection using available devices with more than one network interface is the most flexible solution. But, it is hard to implement and requires special software on every such node. Routing of messages through a larger network of this form is very complex, since no static routes are given, not even the same route may work for request and response, and the high number of passed technology bridges may cause performance problems. Also, due to its dynamic nature it is hard to maintain, error-prone, and not reliable.

Static interconnection using a central instance can be compared with the access point in a WLAN cell, but with network interfaces for different technologies. Such an infrastructure is suitable for all scenarios with pre-defined locations (in-house, on-campus), and it is even feasible for ad-hoc settings in the field (e. g. for events). This solution allows the clients to fully rely on its native network technology, and thus it is transparent also to all higher layers. Direct bridging between the technologies of two communication partners as well as routing of messages can take place in an efficient manner. The system can be opened to additional technologies just by a central modification without any effects on other clients.

Figure 2. Horizontal and vertical network structures can be facilitated to bridge between cells. Alternatively, a dedicated network component can be used for bridging

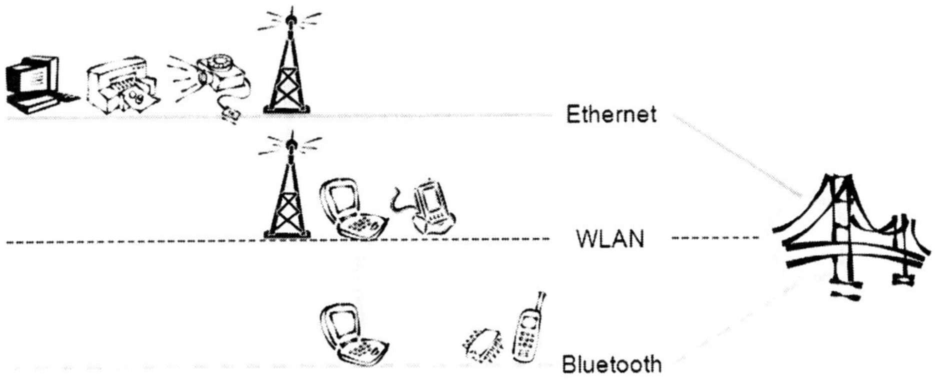

We developed the so called *General Purpose Access Point* (*GPAP*) for bridging between technologies (Zender et al., 2009a). The GPAP is depicted in Figure 3. The basic system is an embedded Linux box with slots used for network interfaces. It acts as a translator on the level of ensembles, and as a gateway between the ensembles of a community. Also, it is the stable point in the network structure and can thus act as the broker in the established service-oriented architecture.

Network Level Interoperability

The challenge on network layer is to unify existing mechanisms of different network technologies for addressing a node and for sending/receiving messages. Therefore, the GPAP consists of a basic infrastructure for exchanging messages in distributed environments, called *Heterogeneous Context-Based Routing* (*HCBR*) (Dressler et al., 2008). The core can be extended with plug-ins, namely for adding a specific network technology and for adding enhanced functionality to the system. This is depicted in Figure 4.

Network plug-ins provide functions for acquiring and releasing resources associated with the given interface, for sending and receiving messages on this channel, and for general information about this interface. Our prototype includes network plug-ins for Ethernet, WLAN, Bluetooth, and ZigBee. Several point-to-point bridges are available from previous research (for instance,

Figure 3. The General Purpose Access Point subsumes network interfaces of various technologies and thus enables communication of different client devices

Figure 4. The network layer of the GPAP includes a basic mechanism for message exchange as well as plug-ins for specific network technologies and additional functionality

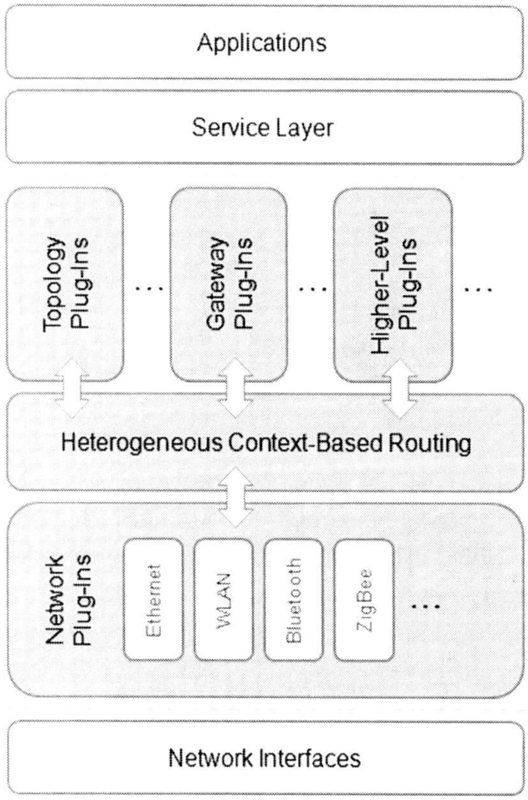

Choi & Cho, 2005) that could be followed in our implementation. Moreover, virtual network interfaces are used for simulating and evaluating network traffic in order to evaluate the system.

Among additional HCBR plug-ins, discovery and management of the *topology* plays an important role. This is realized using frequent Hello messages in order to find a) other GPAPs in the community, and b) individual devices in local network cells. The results of detected nodes are managed in tables containing the IDs, network names, and addresses of these nodes. This information is spread through all neighbored GPAPs in order to keep their tables up to date.

Moreover, there are plug-ins for realization of *gateways* on the level of cells. Basically, this requires the GPAP to map between virtual device addresses in the ensemble and network-specific addresses inside a cell, and to act as a proxy on behalf of foreign devices. For instance, Ethernet devices appear as Bluetooth devices in a Bluetooth cell, and vice versa. Frequent data alignment with topology plug-ins guarantees successful communication even in case of mobile clients, realizing handover between different cells.

Finally, several functions associated with the management of context are subsumed in the group of *higher level plug-ins*. This includes, for instance, measurement of certain context information in the network (like quality of service parameters). The GPAP service layer is based on these functions. In the reverse direction, information on user behavior or intentions (in relation to the topology of ensembles and cells) can be passed from applications down to the network layer in order to optimize routing of messages. But, this is not yet part of the existing prototype.

Service Level Interoperability

On service layer, an abstraction mechanism called *Service Technology-independent Architecture* (*STiA*) has been developed, which subsumes the features and mechanisms of various SOA realizations and allows us to translate data and requests/responses between them. The advantage of this approach is the low complexity of translations (Zender et al., 2010a). In contrast to this, consumer-side (or provider-side) abstraction requires all clients (or servers) to be aware of all possible SOA technologies, and for n technologies used in the m systems one would need $n \cdot m$ translation components (Uribarren et al., 2006; Pantazoglou et al., 2006). The concept of central unification requires only n translators in the central infrastructure, as depicted in Figure 5. All consumers and providers may fully confine themselves to their native technologies.

This requires our architecture to fully implement the so called SOA triangle of provider, broker, and consumer including a sophisticated service bus.

All elements that offer any functionality of certain benefit to the system act as a *provider*. This includes, for instance, the media equipment in a lab, sensor nodes in the environment, or parts of application software. These components are enhanced with a service wrapper, encapsulating their functionality as a service and offering it in the network.

All elements that make use of such functionality act as a *consumer*, regardless which platform or device they are based on. Besides dedicated service interfaces, existing applications can invoke a service by plug-ins or internal code modifications for mapping conventional function calls to the process of dynamically selecting and calling a service.

The core of our SOA is the so called *university service bus*. All services available in the system are announced and invoked using this infrastructural component. It is implemented on

Figure 5. The concept of central SOA unification involves fewer translators than consumer-side or provider-side abstraction mechanisms, but makes a central service bus necessary

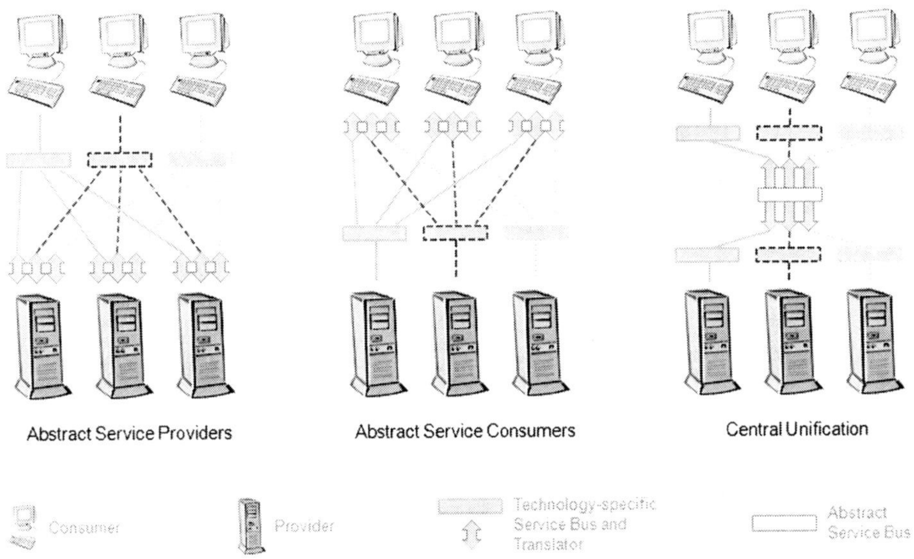

the GPAP. Besides a broker as central mediator between consumers and providers, it includes mechanisms for service announcement and discovery. This is done using an abstract, XML-based vocabulary for service description. Moreover, data interchange with other GPAP service layers is realized. Technology-specific plug-ins are responsible for bridging from this abstract layer to service technologies.

Thanks to the clearly distinguished architectural layers, and to the abstraction in terms of services, new services, service types, providers, and consumers can be added to the system easily and without any structural modifications. Our overall vision is the realization of a fully service-oriented university, as depicted in Figure 6.

Our implementation is focused on Web Services, internally realized with Java. Nevertheless, STiA supports several other service technologies, particularly Jini, Bluetooth, and Bonjour. Given the limited extent of our current prototype, we decided to use the simple mechanisms of the *Web Services Inspection Language* (*WSIL*) as a broker. In the course of further extensions a switch to a more powerful, but rather complex *Universal Description, Discovery, and Integration* (*UDDI*) directory is possible without any difficulty. We consider the server of the central learning platform or the university's web server to be the best choice for the broker as central point of access, since their name and/or address is well-known to all clients.

SELECTED USE CASES

We use the developed architecture to decide which stream to play, when to play a stream, and to realize accompanying message transfer in a heterogeneous environment.

Synchronous and Asynchronous Lecture Streaming Service

One of our key services in the field of mobile multimedia is the streaming of e-lectures (Zender et al., 2009b). This includes the recording, archiving, and simultaneous deployment of a face-to-face lecture. Our system does not only allow for synchronous integration of remote par-

Figure 6. The university service bus integrates different consumers and providers of services

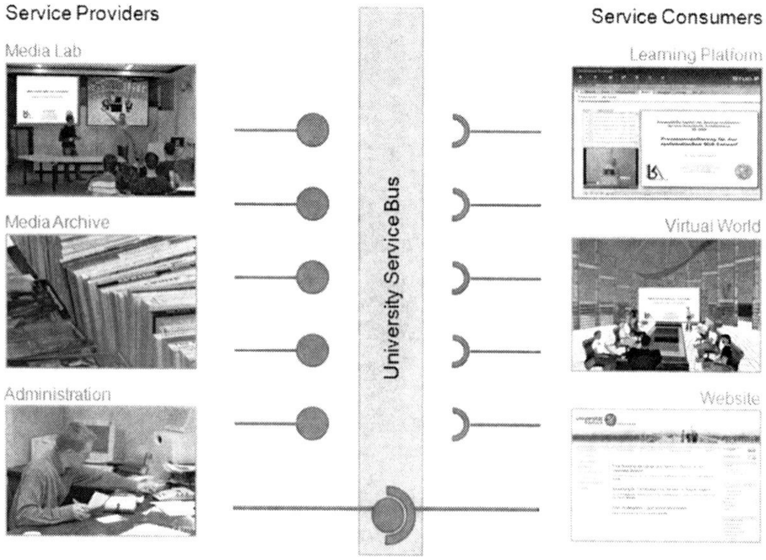

ticipants both listening (passively) and discussing (interactively). This affects also some remote control of media equipment, like selection from different audio/video streams and from different camera angles and zooms. Moreover, archived e-lectures can be deployed on-demand, but due to the missing counterpart on-site only in form of asynchronous listening. Though accessible from mobile devices, this use case does not provide special features for mobile applications like location awareness or adaptive GUIs. Nevertheless, it is the basis for the context-aware scenario in the next subsection and shall thus be explained here for better understanding.

The media control center in the lab on-site acts as a *provider* of the media services. It contains an audio and video crossbar switch, which assigns all available inputs/outputs. This is controlled using a wireless panel via the *Internet Control System Protocol* (*ICSP*). Instead of this panel, our service wrapper sends such ICSP commands to the switch and thus is able to control the media streams even from remote. Other services have been developed for audio/video streaming. This includes both live streams of ongoing lectures as well as on-demand streams of archived e-lectures from a central media server. Though the internal realization is completely different, the use is fully transparent for the consumer. The user may only recognize additional interaction possibilities for live streams. We use a Helix Server Unlimited as streaming server, since it offers various data formats and simultaneous client connections, in contrast to other solutions.

As exemplary *consumers* of these services we realized interfaces for the learning platform *Stud. IP* and the virtual 3D world *Second Life*. *Stud.IP* offers web-based support for face-to-face lecturing with a sophisticated management of institutions, courses, and content. It is open-source and hosted directly by the university, which allowed us to easily extent the PHP-based system. When opening the file area of a course, our SOA extension automatically sends a request to the broker and searches for available streams of this course. The live and on-demand streams that were found are offered to the user in different formats. After selecting one of these streams an embedded player is opened directly in the learning platform. This allows the student to follow the lecture both synchronously and asynchronously without any media breaks, and the list of streams is always update without manual upload by the teacher or an administrator. We used *Asynchronous JavaScript and XML* (*AJAX*) and Stud.IP's internal stylesheets as a technological base. In this way, our plug-in guarantees continuous actuality of the displayed stream list and a consistent layout also for future software versions. Figure 7 shows some screenshots of the modified user interface: the ongoing search for suitable services, the list of available streams, and the embedded e-lecture player (please note that all screenshots of the developed/modified systems are in German language. Nevertheless, the aim of the figures should be recognizable). The only activity that we require from the teacher is to tag the recording with the name of the lecture and the course.

Though this is a dedicated solution for Stud. IP, the developed plug-in is light-weight thanks to the used SOA concept and thus can be straightforward transferred to other PHP- or HTML-based platforms. Additional services (like for the communication between participants of a lecture) can be easily added. A further reduction of the effort to integrate other systems could be achieved by mechanisms for platform-independent GUI design, like *Web Services for Remote Portlets* (*WSRP*) or *User Interface Markup Language* (*UIML*).

A different approach was necessary for our integration of *Second Life* (*SL*). This is a complex world model that is completely hosted, simulated, and rendered on servers of the vendor Linden Lab. Also, there is no possibility for own extensions in the dedicated SL client. That's why we introduced a surrogate as a transmitter between available services and Web Service protocols

Figure 7. The integration of the e-lecture streaming service into the learning platform Stud.IP consists of an automated detection of suitable streams as well as an embedded player

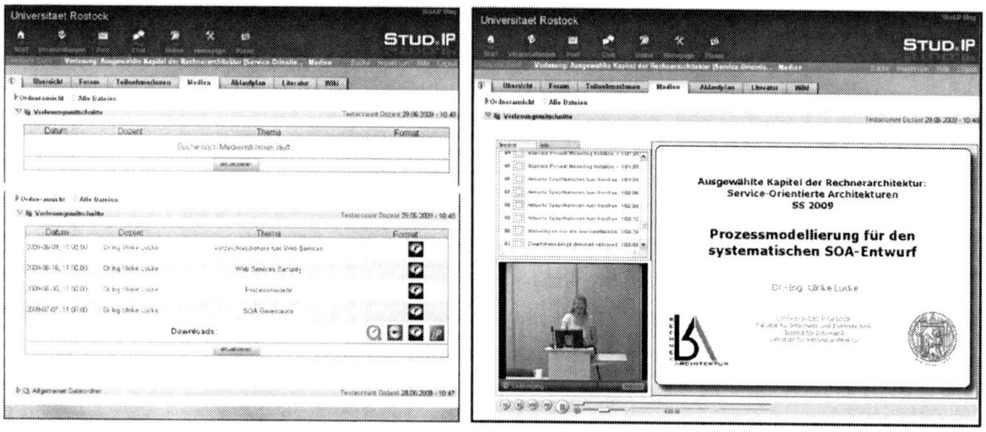

on the one hand and SL's native communication mechanisms on the other hand. We modeled the historic light-house and so called "tea pot" of Rostock-Warnemünde as landmark which is memorable and easily recognizable. With the help of the *Linden Scripting Language* (*LSL*) these objects have been extended by interactive functions for a *HyperText Transfer Protocol* (*HTTP*) connection to the surrogate and thus to the SOA. Events can be triggered by mouse clicks, timers, or messages on the SL-internal communication channels. An HTTP request includes the name of the acting avatar, the invoked service, and optional parameters. The returned response is translated by an LSL handler into textual messages to the user, or changes in appearance or behavior of virtual 3D objects. Figure 8 shows two views of our developed SL representation.

The real-life lecture (no matter if live or from archive) is mapped to the screen and speakers in SL, and the teacher is present as an avatar. Interactive control of media equipment takes place using a virtual keyboard, which opens a so called head-up display for selection and assignment of media streams.

The added value of the virtual world for education is not only the re-modelling of face-to-face scenarios, or streaming of lectures. A big advantage is the almost unlimited changeability and immediate presence of the environment, i. e. the high degree of immersion (Zender et al., 2009). This permits to model scenarios that are

Figure 8. The integration of the e-lecture streaming service into the virtual 3D world Second Life required designing new interaction techniques and communication architectures

inaccessible in real-life (like dangerous areas, micro- or macroscopical levels, historical scenes), and theoretical facts can be clearly demonstrated. In contrast to a learning platform there arise media breaks since the platform is externally hosted and cannot be freely controlled. As an example, for access control in Stud.IP the same mechanisms as in the lecture hall (like a campus-wide *Lightweight Directory Access Protocol* (*LDAP*)) can be used. In SL, these data is not available, which requires a mapping between user accounts. Our solution makes use of the SL group concept: registered users of our group are labeled as trustworthy. Thus, they are allowed to fully control the virtual and real media equipment, and to permit rights for a lecture to other avatars. Unknown avatars are grouped as guests and can only use lecture streaming and communication services.

The contribution of this use case for mobile learning is that independent from where a user is going to access the system (one of the presented portals above, a traditional streaming client, or any future development) this endpoint can be seamlessly integrated without any additional efforts. The service broker brings together existing streams and requesting clients. Thus, the benefit of our service-based solution has an impact on server-side implementations and on workflows to be performed by the teacher that do not need to consider special characteristics of mobile use.

Mobile, Context-Aware Lecture Streaming Service

A mobile user of these services might benefit from the consideration of context information. By capturing and combining different context sources (like the user's position, his schedule, his personal profile, the current time, the status of the public transport system, geographical data, room status, and so on) we are able to offer the user an intelligent support that helps to achieve a maximum of functionality and comfort in accessing the system services while reducing his interaction with the platform to a minimum. Unfortunately, many traditional applications like learning platforms are not suitable for mobile use. This is with regards to the *Graphical User Interface* (*GUI*) design (not fitting to small screens), to the interaction patterns (too many clicks, too tiny click-items), and to context awareness (static application logic).

Our prototype of a context-based mobile client for the learning platform Stud.IP shows that context can be used to adjust traditional applications to modern mobile devices (Lehsten et al., 2010). We had to extend the system on server- and on client-side. However, it works to a limited extent without modifications at the client. We implemented a small browser plug-in that detects the user's physical position, the current time, and the identity of the user. This context information is sent to the server, along with the regular HTTP request for accessing the learning platform. Here, we inserted a GUI proxy. It is responsible for two tasks: First, it generates a re-designed web site of the learning platform adjusted to the capabilities of the user's current device. Second, it estimates the user's intention from the given context and automatically opens the respective area in the learning platform. Figure 9 shows the resulting screenshots.

After login the user receives a reduced Stud. IP interface fitting to the dimensions of his mobile device. This is realized by defining typical classes of mobile devices (along with mean values for screen size and computing/communication capacity) and associated web interface variants and stylesheets. Context is considered as follows. An internal context transformation service is used to derive the user's logical position (e. g. building and room number) from the physical coordinates. Together with the current time and the user's timetable the system can determine if the user is inside an ongoing lecture, on the way to a lecture, or in any other situation. This is done using a rule-based knowledge system implemented with *Drools* (Browne, 2009) A dedicated context server was introduced to the system that is re-

Figure 9. The context-based mobile Stud.IP client realizes a selection of interface elements and a redesign to adjust a traditional learning platform to mobile computing scenarios

sponsible for gathering context information from the learning platform and the mobile devices, and for transforming these facts to applicable knowledge by using defined rules. In contrast to the service-based front-end, for internal communication with the context engine a message-oriented middleware is used for better performance. Accordingly, the website of the lecture, the public transport connections from the current location to the lecture hall or the live stream of the lecture are provided. Any change of the context has an effect on the interface; e. g. when the user enters the lecture hall the transmission of the stream is stopped.

The mobile browser plug-in as well as the context processing and reasoning component is implemented in Java, while Stud.IP is based on the *HyperText Markup Language* (*HTML*) and the *PHP Hypertext Preprocessor*. For reasoning we use the business rule management system Drools including a self defined set of rules. This is applied, for instance, to assign streams to courses, to decide on the user's position or his current activity. If the user did not install the browser plug-in and context information is not available, the system only adjusts the GUI to the device capabilities.

Cross-Platform Messaging Service

The services presented above are designed for educational purposes. There are on-site students, remote students, offline students, and the teacher. Social effects come to play in the virtual world to a much larger extent than during (isolated) use of a learning platform. That's why we decided to accompany the SL scenario by a messaging service that connects the mobile user with the on-site audience in the lecture hall. Otherwise, there is the SL client software as the only feedback channel, which requires the lecturer (or someone else in the room) to coordinate the discussion between real and virtual participants. This includes repeating of all spoken words in the SL chat, which is a hard job.

Alternatively, our system offers an innovative cross-technology communication concept (Zender et al., 2009b). The GPAP translates between the Web Services and IP-based communication over Ethernet used in our service architecture on the one hand side, and Bluetooth as well as native services on mobile phones of the users on the other hand side. We implemented a chat service running in the learning environment on-site. A messaging client on the users' devices enables them to find other users by service discovery. SL also provides a chat interface to their users. We developed a virtual cell phone as messaging object which is able to find other messaging objects. If an avatar is "wearing" this object, it connects to the SL surrogate through HTTP and requests for the list of available messaging services. Furthermore, it registers itself as messenger and the surrogate provides a corresponding Web Service. Both in SL and on the Bluetooth phones the group of available chat partners is displayed as textual list, and the user (or his avatar, respectively) is able to send messages to each chat partner no matter where he is. Figure 10 shows the interfaces on a cell phone and in SL.

To bring together the two environments we make use of the university service bus that contains abstract STiA services. Every real messaging client provides its ability to receive messages in terms of a service, and the same does the surrogate for every virtual messenger. These services are transformed into STiA services. This is performed by the GPAP (for the real-life messaging clients) and the SL surrogate (for the virtual environment).

Furthermore, all devices continuously search for new messaging services on the service bus and provide them to their users. Thus, virtual messaging services appear as Bluetooth messaging services in the real environment and Bluetooth services as virtual messaging services in the virtual world. This transformation remains transparent to the users and allows everybody, whether in the real or the virtual environment, to communicate with everybody else. Thus, we created some kind of pervasive (or even immersive) walkie-talkie. Lectures can not only be followed as a passive listener, but also as an active participant in a constructive discussion – even across platforms, no matter if on mobile or stationary devices. The pedagogical value of such activation of the learner is significant. New devices can be integrated easily by using their own SOA and network technology as well as their own user interface.

USING CONTEXT IN SERVICES

In order to stimulate the deployment of new services, especially in combination with context-awareness, we have designed a simple categorization of using context in services based on our experiences in system development. Contextual information can be used for a selection which aspect of an application should be brought forward, as well as for adjusting internal parameters of its functionality. Also, traditional services and context services can be brought together to adjust a system to the users' needs. The case studies underline that

Figure 10. The cross-platform message exchange bridges between a mobile phone (Bluetooth) and the Second Life chat functions (Ethernet, Web Services)

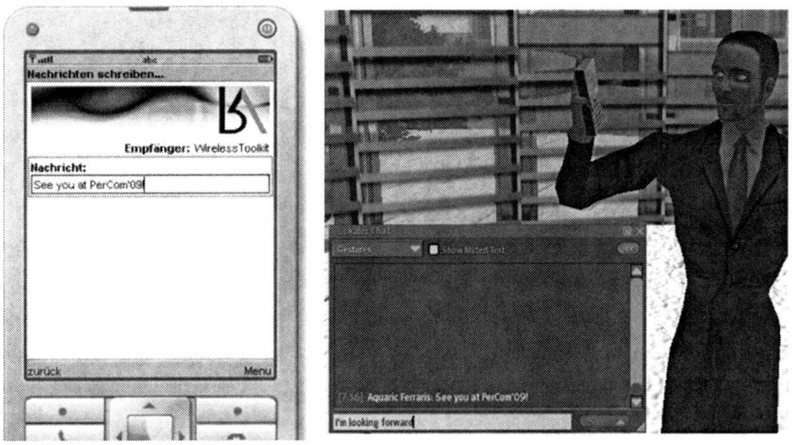

strategies for the integration of context and SOA highly depend on the area of application (Zender et al., 2010b). Nevertheless, we identified three strategic directions.

Context services are a fundamental infrastructure component which provides context to a system. In contrast to direct transmission of context information by the client when invoking a service, context services are utilized by different parts of the system itself during aggregation (i. e. orchestration) of basic services to higher-level services. Context services are the simplest way to systematically integrate context into a system.

Context-dependent service discovery leads to a subset of available services that best match the application-specific requirements. Furthermore, it avoids unnecessary dissemination of service requests. Mostly, this strategy is pursued for large amounts of similar simple or even elementary services. The contextual requirements have to be published together with the service itself. But, the provider of the contextual information doesn't have to be the same as the service provider, and the service itself doesn't have to be context-aware. Particularly, *Quality of Service* (*QoS*) parameters (e.g. service availability or network bandwidth and latency) can be acquired automatically by enhanced brokers or further network components.

Context-dependent service use requires the implementation of the service itself to be context-aware. This strategy needs an initial step of modification or even re-implementation towards context-awareness. This is hard for existing applications, especially when these are not service-oriented and thus difficult to maintain. Often, context-based service use is applied only to small parts of system functionality. Anyway, the contextual information has to be considered already during service design.

While context-based service use is domain- as well as application-specific, we are able to automate the generic enrichment of existing services with dedicated contextual information for a context-based service discovery. However, availability of basic context services is an important pre-condition for systematical enrichment of SOA with context. The GPAP unites our approaches to integrate context information both on network and service layer. It provides several points to gather context information and to interpret it, e.g. for context-aware packet routing or service brokerage.

FUTURE RESEARCH DIRECTIONS

Further work is dedicated to three issues: To extend the amount of services that is available in our system. This includes basic services like authentication, authorization, and accounting, but also complex application services in the areas of education and research. To further intensify our activities in merging SOA with context-awareness. Based on further case studies, we are developing methods for integrating service and context models in order to automatically enhance service orchestration with context information. To support the creation of process models, which are a major prerequisite for substantiated service architectures, with the help of generalized reference models that can be easily tailored to the individual needs.

We strongly believe that the combination of services and context will make an important contribution to developing intelligent infrastructures and applications for the future.

CONCLUSION

Besides entertainment, education is a popular field of application for mobile multimedia. Derived from traditional e-learning with stationary PCs, mobile devices offer huge possibilities to extend the learning process in terms of independency from time and place beyond face-to-face settings in labs or classrooms. While the creation of document-based learning material is complex and time-consuming, especially when different needs of mobile devices shall be considered, recording and archiving of common lectures enjoys increasing popularity. So called e-lectures can be easily created, managed, and deployed to the learners. However, current challenges are a close integration with existing infrastructures (i. e. hardware and software interfaces that are used by teachers and students) as well as intelligent support for the current situation and goals of the users (i. e. context-awareness) without forcing them to use special hardware and/or software. This brings up the following questions: How can be dealt with heterogeneous environments, in terms of network and software? And how can be dealt with context information? In this article, a solution to meet these challenges was presented.

Based on a multi-layer networking model, we developed an infrastructure to bridge between different network and service technologies. This allows for flexible and transparent integration of various devices and applications. We developed service wrapper and plug-ins for a number of existing systems, namely: media equipment in a lab, web-based learning platform, virtual 3D world, and mobile phone. Also, we identified several possibilities to gather context information and to make use of it during selection and invocation of services. As a result, we are able to deploy traditional services with an added value of adjusting them to the current situation and presumed intention of the user. Related to the topic of this book, we presented four examples: service-based streaming of e-lectures to different platforms, context-based extension of e-lecture streaming, cross-platform messaging service, and individualized music service. In addition, we sketched the general possibilities to make use of context in a service-oriented architecture.

ACKNOWLEDGMENT

The author wants to thank her colleagues Raphael Zender, Enrico Dressler, and Philipp Lehsten at the University of Rostock for their comprehensive and valuable contributions to the described system. Our interchange of ideas was always inspiring and productive. Without their intense participation our system could not gain the described complexity. Furthermore, several graduate students were involved in the implementation. We appreciate their dedication and creative urges. The work was partially supported by the German Research Foundation (*Deutsche Forschungsgemeinschaft, DFG*) under grant "*Research Training Group 1424*".

REFERENCES

Adorni, G., Coccoli, M., Fadda, C., & Veltri, L. (2007). Audio and video conferencing tools in learning management systems. In *Proceedings of 3rd International Conference on Automated Production of Cross Media Content for Multi-Channel Distribution* (pp. 123-129). Los Alamitos, CA, USA: IEEE.

Bettini, C., Brdiczka, O., Henricksen, K., Indulska, J., Nicklas, D., Ranganathan, A., & Riboni, D. (2010). A survey of context modelling and reasoning techniques. *Pervasive and Mobile Computing*, *6*(2), 161–180. doi:10.1016/j.pmcj.2009.06.002

Browne, P. (2009). *Jboss Drools Business Rules*. Birmingham, UK: Packt Publishing.

Byrne, P., Tangney, B., Sanchez-Villalon, P., Ortega, M., Wu, L., & Chang, K. (2007). Mobile Learning. *IEEE Distributed Systems Online*, *8*(7), 3. doi:10.1109/MDSO.2007.44

Chandra, S. (2007). Lecture video capture for the masses. In *Proceedings of the 12th Annual SIGCSE conference on Innovation and technology in computer science education* (pp. 276-280). New York, NY, USA: ACM.

Choi, H., & Cho, D. (2005). Takeover: A new vertical handover concept for next-generation heterogeneous networks. In *Proceedings of Vehicular Technology Conference* (pp. 2225-2229). Los Alamitos, CA, USA: IEEE.

Donick, M., & Tavangarian, D. (2010). Qualitatives evaluationsverfahren für interkulturelle e-learning-Szenarien am beispiel des online-MSc "Visual Computing". In *Proceedings Grundfragen Multimedialen Lehrens und Lernens: E-Kooperationen und E-Praxis* (pp. 273-285). Münster, D: Waxmann.

Dressler, E., Zender, R., Lucke, U., & Tavangarian, D. (2007). A new architecture to realise pervasive communities. In *Proceedings 7th International Workshop on Innovative Internet Community Systems* (pp. 10-17). München, Germany: University of the German Federal Armed Forces.

Dressler, E., Zender, R., Lucke, U., & Tavangarian, D. (2008). A new Architecture for heterogeneous context based routing. In *Advances in Computer Science and Engineering – Proceedings of 13th International CSI Computer Conference* (pp. 526-534). Heidelberg, Germany: Springer.

Driscoll, D., & Mensch, A. (2009). Devices profile for Web services (DPWS), Version 1.1. *OASIS standard.* Retrieved from http:// docs.oasis- open. org/ ws-dd/ dpws/ 1.1/ os/

Grosky, W., Zhang, C., & Chen, S. (2009). Intelligent and rervasive multimedia systems. *IEEE MultiMedia, 16*(1), 14–15. doi:10.1109/ MMUL.2009.10

Heider, T. & Kirste, T. (2006). Building smart environments: The ensemble challenge. *IM - Information Management & Consulting, 21*(4), 6-12.

Huber, J., Steimle, J., Olberding, S., Lissermann, R. & Mühlhäuser, M. (2010). Browsing e-lecture libraries on mobile devices: A spatial interaction concept. In *Proceedings of International Conference on Advanced Learning Technologies* (pp. 151-155). Los Alamitos, CA, USA: IEEE.

Hürst, W., & Meier, K. (2008). Interfaces for timeline-based mobile video browsing. In *Proceedings of the 16th International Conference on Multimedia* (pp. 469-478). New York, NY, USA: ACM.

Lehsten, P., Zender, R., Lucke, U., & Tavangarian, D. (2010). A service-oriented approach towards context-aware mobile learning management systems. In *Proceedings 8th IEEE Annual Int. Conference on Pervasive Computing and Communication Workshops* (pp. 268-273). Los Alamitos, CA, USA: IEEE.

Liu, T., & Choudary, C. (2004). Content-aware streaming of lecture videos over wireless networks. In *Proceedings of 6th International Symposium on Multimedia Software Engineering* (pp. 458-465). Los Alamitos, CA, USA: IEEE.

Loke, S. (2006). *Context-aware pervasive systems: Architectures for a new breed of applications.* Boca Raton, FL, USA: Auerbach Publications. doi:10.1201/9781420013498

Naguib, H., Coulouris, G., & Mitchell, S. (2001). Middleware support for context-aware multimedia applications. In *Proceedings of the IFIP Tc6 / Wg6.1 3rd International Working Conference on New Developments in Distributed Applications and Interoperable Systems* (pp. 9-22). Deventer, The Netherlands: Kluwer.

Ogrinz, M. (2009). *Mashup patterns: Designs and examples for the modern enterprise.* Amsterdam, NL: Addison-Wesley Longman.

Pagani, M. (2008). *Multimedia technology and networking.* Hershey, PA, USA: Information Science Reference.

Pan, P., Kastner, C., Crow, D., & Davenport, G. (2002). M-studio: An authoring application for context-aware multimedia. In *Proceedings of the 10th ACM International Conference on Multimedia* (pp. 351-354). New York, NY, USA: ACM.

Pantazoglou, M., Tsalgatidou, A., Athanasopoulos, G., & Pilioura, T. (2006). A unified approach for the discovery of Web and peer-to-peer services. In *Proceedings of International Conference on Web Services* (pp. 901-902). Los Alamitos, CA, USA: IEEE.

Román, M., Hess, C., Cerqueira, R., Ranganathan, A., Campbell, R. H., & Nahrstedt, K. (2002). A middleware infrastructure for active spaces. *IEEE Pervasive Computing / IEEE Computer Society [and] IEEE Communications Society, 1*(4), 74–83. doi:10.1109/MPRV.2002.1158281

Stenton, S. P., Hull, R., Goddi, P. M., Reid, J. E., Clayton, B. J., Melamed, T. J., & Wee, S. (2007). Mediascapes: Context-aware multimedia experiences. *IEEE MultiMedia, 14*(3), 98–105. doi:10.1109/MMUL.2007.52

Tavangarian, D. & Lucke, U. (2009). Pervasive university — A technical perspective. *IT - Information Technology, 51*(1), 6-13.

Teixeira, D., & Faihe, Y. (2002). In-home access to multimedia content. In *Proceedings of the 10th ACM International Conference on Multimedia* (pp. 49-56). New York, NY, USA: ACM.

Thomanek, A., Schönfeldt, C., Donick, M., Schwelgengräber, W., & Tavangarian, D. (2009). Media-based junior studies in context of the "New Learning Culture". In *Proceedings of e-Learning Baltics 2009* (pp. 27-36). Stuttgart, Germany: Fraunhofer.

Timmerer, C., & Hellwagner, H. (2005). Interoperable adaptive multimedia communication. *IEEE MultiMedia, 12*(1), 74–79. doi:10.1109/MMUL.2005.7

Tonguz, O. K., & Ferrari, G. (2006). *Ad-hoc wireless networks*. Hoboken, NJ, USA: Wiley. doi:10.1002/0470091126

Ullrich, C., Shen, R., Tong, R., & Tan, X. (2010). A mobile live video learning system for large-scale learning—System design and evaluation. *IEEE Transactions on Learning Technologies, 3*(1), 6–17. doi:10.1109/TLT.2009.54

Uribarren, A., Parra, J., Uribe, J. P., Makibar, K., Olalde, I., & Herrasti, N. (2006). Service oriented pervasive applications based on interoperable middleware. In *Proceedings of International Workshop on Requirments and Solutions for Pervasive Software Infrastructures* at *4th International Conference on Pervasive Computing.* Dublin, Ireland.

Wallace, D. & Levinson, A. (2006). The XML e-learning revolution: Is your production model holding you back? *Learning Solutions e-Magazine*, 08/2006.

Zender, R., Dressler, E., Lucke, U., & Tavangarian, D. (2009a). Multi-level interoperability for pervasive communication networks. In *Proceedings of the 7th IEEE International Conference on Pervasive Computing and Communication* (pp. 1-3). Los Alamitos, CA, USA: IEEE.

Zender, R., Dressler, E., Lucke, U., & Tavangarian, D. (2009b). Pervasive media and messaging services for immersive learning experiences. In *Proceedings of the 7th IEEE Annual International Conference on Pervasive Computing and Communication Workshops* (pp. 1-6). Los Alamitos, CA, USA: IEEE.

Zender, R., Lucke, U., & Tavangarian, D. (2010a). SOA interoperability for large-scale pervasive environments. In *Proceedings of the 24th IEEE International Conference on Advanced Information Networking and Applications Workshops* (pp. 545-550). Los Alamitos, CA, USA: IEEE.

Zender, R., Lucke, U., & Tavangarian, D. (2010b). Integrating context awareness with SOA: Case studies for service-oriented and context-based architectures. In *Proceedings of the 23rd International Conference on Architecture of Computing Systems Workshops* (pp. 143-150), Berlin-Offenbach, Germany: VDE.

ADDITIONAL READING

Abowd, G. D. (1999). Classroom 2000: An experiment with the instrumentation of a living educational environment. *IBM Systems Journal*, *38*(4), 508–530. doi:10.1147/sj.384.0508

Amor, D. (2001). *Internet future strategies: How pervasive computing services will change the world.* Upper Saddle River, NJ, USA: Prentice Hall.

Austerberry, D. (2004). *The technology of video and audio streaming*. Burlington, MA, USA: Focal.

Bersin, J., Howard, C., O'Leonard, K., & Mallon, D. (2009). *Learning management systems 2009.* Oakland, CA, USA: Bersin & Associates.

Clark, R. C., & Mayer, R. E. (2007). *e-Learning and the Science of Instruction: Proven Guidelines for Consumers and Designers of Multimedia Learning*. Hoboken, NJ, USA: Wiley.

Erl, T. (2007). *SOA Principles of Service Design*. Upper Saddle River, NJ, USA: Prentice Hall.

Josuttis, N. M. (2007). *SOA in practice: The art of distributed system design*. Sebastopol, CA, USA: O'Reilly.

Kalasapur, S., Kumar, M., & Shirazi, B. (2006). Evaluating service oriented architectures (SOA) in pervasive computing. In *Proceedings of the 4th Annual IEEE International Conference on Pervasive Computing and Communications* (pp. 276-285). Washington, DC, USA: IEEE.

Kemp, J., Livingstone, D., & Bloomfield, P. (2009). SLOODLE: Connecting VLE tools with emergent teaching practice in Second Life. *British Journal of Educational Technology*, *40*(3), 551–555. doi:10.1111/j.1467-8535.2009.00938.x

Lampi, F., Kopf, S., Benz, M., & Effelsberg, W. (2008). A virtual camera team for lecture recording. *IEEE MultiMedia*, *15*(3), 58–61. doi:10.1109/MMUL.2008.43

Macedonia, M. (2007). Generation 3D: Living in virtual worlds. *IEEE Computer*, *40*(10), 99–101.

Mertens, R., Knaden, A., Thelen, T., & Vornberger, O. (2005). Integration of LMS and lecture recording systems: Requirements and potentials. *I-Com.*, *4*(3), 52–55. doi:10.1524/icom.2005.4.3.52

Moore, D., Thome, M., & Haigh, K. (2008). *Scripting your world: The official guide to second Life scripting*. Hoboken, NJ, USA: Wiley.

Piskurich, G. M. (2003). *The AMA handbook of e-learning: Effective design, implementation, and technology solutions*. New York, NY, USA: Mcgraw-Hill.

Repp, S., Groß, A., & Meinel, C. (2008). Browsing within lecture videos based on the chain index of speech transcription. *IEEE Transactions on Learning Technology*, *1*(3), 145–156. doi:10.1109/TLT.2008.22

Satyanarayanan, M. (2001). Pervasive computing: Vision and challenges. *IEEE Personal Communications*, *8*(4), 10–17. doi:10.1109/98.943998

Schulmeister, R. (2010). Students, Internet, e-learning and Web 2.0. In *Looking toward the future of technology-enhanced education: Ubiquitous learning and the digital native* (pp. 13–36). Hershey, PA, USA: IGI Global.

Steinmetz, R., & Nahrstedt, K. (2002). *Multimedia fundamentals*. Upper Saddle River, NJ, USA: Prentice Hall.

Steinmetz, R., & Nahrstedt, K. (2004). *Multimedia applications*. Berlin, Germany: Springer.

Steinmetz, R., & Nahrstedt, K. (2004). *Multimedia systems*. Berlin, Germany: Springer.

Weber, A., Rufer-Bach, K., & Platel, R. (2007). *Creating your world: The official guide to advanced content creation for Second Life*. Hoboken, NJ, USA: Wiley.

Weiser, M. (1991). The computer for the 21st Century. *Scientific American, 265*(3), 94–104. doi:10.1038/scientificamerican0991-94

Zhu, F., Mutka, M. W., & Ni, L. M. (2005). Service discovery in pervasive computing environments. *IEEE Pervasive Computing / IEEE Computer Society [and] IEEE Communications Society, 4*(4), 81–90. doi:10.1109/MPRV.2005.87

Chapter 11
Recent Advances and Challenges in Wireless QoE-Aware Multimedia Streaming Systems

Rosinei Oliveira
Federal University of Pará, Brazil

Ádamo L. Santana
Federal University of Pará, Brazil

João C. W. A. Costa
Federal University of Pará, Brazil

Carlos R. L. Frances
Federal University of Pará, Brazil

Elisangela Aguiar
Federal University of Pará, Brazil

Paulo Bezerra
Federal University of Pará, Brazil

Allan Costa
Federal University of Pará, Brazil

Eduardo Cerqueira
Federal University of Pará, Brazil

Antônio J. G. Abelém
Federal University of Pará, Brazil

ABSTRACT

It is expected that multimedia applications will be the most abundant application in the Internet and thousands of new wireless and mobile users will produce and share multimedia streaming content ubiquitously. In this multimedia-aware system, it is important to assure the end-to-end quality level support for video and voice applications in wireless systems. Traditional Quality of Service techniques assure the delivery of those services with packet differentiation assurance and indicate the impact of multimedia traffic only on the network performance; however, they do not reflect the user's perception. Recent advances in multimedia are exploring new Quality of Experience approaches and including metrics and control schemes in wireless networking systems in order to increase the user's satisfaction and optimize network resources. Operations based on Quality of Experience can be used as an indicator of how a networking environment meets the end-user's needs and new assessment and packet control approaches are still important challenges. This chapter presents an overview of the most recent advances and challenges in assessment and traffic conditioner procedures for wireless multimedia streaming systems.

DOI: 10.4018/978-1-61350-144-3.ch011

In addition, an intelligent packet dropper mechanism for IEEE 802.11e systems is proposed and evaluated by using the Network Simulator 2, real video sequences and Evalvid tool. The benefit and the impact of the proposed solution is evaluated by using well-know objective and subjective Quality of Experience metrics, namely, Peak Signal-to-Noise Ratio, Video Quality Metric, Structural Similarity Index and Mean Option Score.

INTRODUCTION

Recently, advances in multimedia and mobile communications have emerged to offer a novel and comfortable living style for users. In this context, the delivery of multimedia content, such as video streaming, anytime, anywhere and with an end-to-end quality level support is a key requirement. This fact explains the increase of wireless networking standards, such as the *Institute of Electrical and Electronics Engineers* (*IEEE*) 802.11 (Lidinsky, Chambers & Jeffree, 2001) and IEEE 802.16 (Marks, Stanwood, Eklund & Chang, 2004) as well as the emerging new multimedia streaming applications.

With respect wireless systems, the IEEE 802.11 standard provides communication coverage limited to an area of 100m and can also operate in mesh mode, IEEE802.11s (Akyildiz & Wang, 2005), to increase the coverage area. The IEEE 802.16 was designed to work in outdoor scenarios, with a range of up to 50km and rates of up to 75 Mbps for architectures fixed (IEEE 802.16d), and coverage of up to 4km and bandwidth up to 15 Mbps for mobile devices (IEEE 802.16e). The wireless facility can allow the ubiquitous access of multimedia content with low operational cost. It is expected that video-based services will account for 50 percent of all consumer network traffic in 2012 and 80 percent in 2020.

In order to keep and attract customers, wireless operators must also provide quality level assurance for multimedia applications in order to maximize the user's satisfaction and the usage of network resources, while increasing the profits of network providers. However, wireless and multimedia-aware *Quality of Service* (*QoS*) assessment and management schemes must be implemented to fulfill such important requirement.

To cope with QoS issues in *Wireless Local Area Network* (*WLAN*), the IEEE802.11e working group was created, where the draft version brought new *Media Access Control* (*MAC*) improvements incorporated in the IEEE 802.11 standard (IEEE 802.11e, 2010). To provide QoS assurance, eight *User Priorities* (*UP*s) were defined. Each packet is assigned to an UP and mapped to an *Access Category* (*AC*). Each AC is directly mapped to a queue, where several queues have different priorities, and applications are assigned to them according to requirements, policies, content, among other parameters.

In the case of broadband access in *Wireless Metropolitan Area Network* (*WMAN*), the *Worldwide Interoperability for Microwave Access* (*WiMAX*) system (IEEE 802.16 standardized the architecture for all-*Internet Protocol* (*IP*) networks) is the most attractive solution to last mile connectivity to Internet with quality level assurance. The WiMAX system provides differentiated levels of QoS for multimedia applications, based on the combination of a set of communication service classes, supported in both wired IP-based and wireless IEEE 802.16-based links. In the former, network elements with standard IP QoS models, such as *Differentiated Services* (*DiffServ*), can be configured to guarantee QoS support for sessions crossing wired links. In the latter, several IEEE 802.16 QoS services are defined to provide packet differentiation in the wireless interface (Andrews, Ghosh & Muhamed, 2007).

As presented above, the current wired and wireless techniques that aim to maximize the quality level of multimedia services in a networking system are focused only on QoS aspects. QoS-based schemes define a set of network level (and packet level) measurement and control operations to guarantee the distribution of multimedia content over heterogeneous networks with an acceptable quality level (Zapter & Bressan, 2007).

Existing QoS metrics, such as packet loss rate, packet delay rate and throughput, are typically used to indicate the impact on multimedia streaming quality level from the network's point of view, but do not reflect the user's experience. Consequently, these QoS parameters fail in capturing subjective aspects associated with human perception (Perkis, Munkeby & Hillestad, 2006).

In order to overcome the limitations of current QoS-aware multimedia networking schemes respect to human perception and subjective-related aspects, recent advances in multimedia-aware system *Quality of Experience* (*QoE*) approaches, have been introduced (Takahashi, Hands & Barriac, 2008). Hence, new challenges in emerging networks involve the study, the creation and the validation of QoE measurements and optimization mechanisms to improve the overall quality level of multimedia streaming content and the usage of scarce wireless network resources (Mu, Cerqueira, Boavida & Mauthe, 2009).

The QoE applicability scenarios, requirements, evaluations and assessment methodologies in multimedia systems have been investigated by several researches and working groups, such as *International Telecommunication Union – Telecommunication Standardization Sector* (ITU-T,2010), *Video Quality Experts Group* (*VQEG*) (VQEG,2010) and *European Technical Standard Institute* (*ETSI*),*Committee for Speech, Transmission, Planning and QoS* (*STQ*) (ETSISTQ, 2010).

In the emerging future multimedia networks, QoE assessment solutions and metrics are needed to measure the performance of multimedia streaming applications with human-based precision. Currently, there are subjective and objective measurement approaches to evaluate the quality level of multimedia content from the user´s point of view. Additionally, new network and application-sensitive mechanisms are required to optimize network resources and increase the end-to-end quality level of multimedia streaming. The results of QoE research can be used as an extension to the traditional QoS schemes, in the sense that QoE provides information regarding the delivered multimedia service from the user's point of view. Examples of control mechanisms that will be included in QoE support in wired and wireless systems are new routing approaches, base station selection process and traffic conditioners.

In this chapter, an overview of the most recent advances and challenges in wireless multimedia streaming systems, with focus on QoE measurements and packet control proposals will be addressed. In order to demonstrate the benefits and efficiency of QoE solutions on controlling the quality level of multimedia streaming, simulation experiments were carried out by using the *Network Simulator* 2 (ns2) (NS2, 2010), well-know real video sequences and the Video Quality Evaluation Tool-set Evalvid (Klaue, Rathke & Wolisz, 2003). Objective and subjective QoE metrics were used to analyzed the proposed mechanisms in a multimedia-aware wireless system, namely, *Peak Signal-to-Noise Ratio* (*PSNR*), *Video Quality Metric* (*VQM*), *Structural Similarity Index* (*SSIM*) and *Mean Option Score* (*MOS*).

The remainder of this chapter is organized as follows. Section 2 discusses a survey of main metrics of QoE. Section 3 presents the state of the art approaches used to evaluate the QoE in video streaming. The optimization approach for streaming video over wireless networks, and its main features are described in section 4. In section 5, the scenario presented is detailed, as well as the tools used and results. Section 6presents futures research directions in wireless multimedia systems. Finally, section 7 presents the conclusions.

QUALITY OF EXPERIENCE METRICS

Traditional techniques that aim to maximize the quality level of multimedia streaming are based only on QoS metrics. They define a set of management operations and measurement at the network level and packages, to control the delivery of multimedia content with an acceptable level of quality over heterogeneous networks (Zapater & Bressan, 2007). However, recent advances in multimedia system have presented the benefits of QoE metrics in assessing the quality level of applications based on the user's perspective/human perception.

There are several subjective and objective methods to measure the quality level and detect impairments (blocking, blurring and color errors) of multimedia streaming. Subjective methods are performed to acquire information about the quality level of processed video based on human opinion score schemes, while objective methods are used to estimate the performance of multimedia systems, by using models that approximate results of subjective quality assessment.

Subjective Metrics

Subjective metrics assess how audio and/or video streams are perceived by users, i.e., what is their opinion on the quality of particular audio/video sequences, as described in ITU-T recommendation BT 500 (1990). The most popular subjective metric is called MOS. The quality level of a video (or audio) sequence based on MOS model is rated on a scale from 1 to 5, with 5 being the best possible score, as presented in Table 1.

The MOS values are achieved based on subjective tests and methodologies performed with a set of viewers. For instance, the *Single Stimulus Continuous Quality Evaluation* (*SSCQE*) test allows viewers to dynamically rate the quality of an arbitrarily long video sequence using a slider mechanism with an associated quality scale. The drawback of subjective metrics is the fact that they are neither practical nor scalable for real-time multimedia environments.

Objective Metrics

Several objective QoE metrics have been studied and developed to estimate the quality level of multimedia streaming applications according to the user's perception. Among them, the PSNR is a traditional objective metric used to measure, in *decibels* (*dB*), the video quality level based on original and processed video sequences. Typical values for the PSNR in lossy videos are between 30 dB and 50 dB, where higher is better. The PSNR of a video is defined through the *Mean Square Error* (*MSE*) metric; considering the luminance (Y) of the processed and original frames and assuming frames with MxN pixels, the MSE is obtained using *Equation 1*,

$$MSE = \frac{1}{MxN} \sum_{i=0}^{M-1} \sum_{j=0}^{N-1} \| Ys(i,j) - Yd(i,j) \|^2 \quad (1)$$

$$PSNR = 20 \log_{10} \left(\frac{255}{\sqrt{\frac{1}{MxN} \sum_{i=0}^{M-1} \sum_{j=0}^{N-1} \| Ys(i,j) - Yd(i,j) \|^2}} \right) \quad (2)$$

In *Equation 1*, while $Ys(i,j)$ designates the pixel in the position (i, j) of the original frame, the $Yd(i,j)$ represents the pixel located in the position (i, j) of the processed frame. Based on the MSE

Table 1. Mean Option Score

MOS	Quality	Impairment
5	Excellent	Imperceptible
4	Good	Perceptible but not annoying
3	Fair	Slightly annoying
2	Poor	Annoying
1	Bad	Very annoying

definition and on 8bits per sample, the PSNR, in a logarithmic scale, is obtained using *Equation 2.*

The MSE and PSNR metrics only provide an indication of the difference between the received frame and a reference signal, and do not consider any other important aspects that can strongly influence the video quality level, such as *Human Visual System* (*HVS*) characteristics. A detailed analysis of HVS can be found in Wang, Lu and Bovic (2004). The PSNR can also be used to map MOS values as described in Table 2.

SSIM improves the traditional PSNR and MSE, which are inconsistent with HVS characteristics, such as human eye perception (Wang, Bovik, Sheikh & Simoncelli, 2004). The SSIM metric is based on frame-to-frame measuring of three components (luminance similarity, contrast similarity and structural similarity) and combining them into a single value, called index. The SSIM index is a decimal value between 0 and 1, where 0 means no correlation with the original image, and 1 means the exact same image.

The VQM method defines a set of computational models that also have been shown to be superior to traditional PSNR and MSE metrics (Revés, Nafisi, Ferrús & Gelonch, 2006). The VQM method takes as input the original video and the processed video and verifies the multimedia quality level based on human eye perception and subjectivity aspects, including blurring, global noise, block distortion and color distortion. The VQM evaluation results vary from 0 to 5 values, where 0 is the best possible score.

Table 2. PSNR to MOS conversion

PSNR(dB)	MOS
> 37	5 (Excellent)
31 – 37	4 (Good)
25 – 31	3 (Fair)
20 – 25	2 (Poor)
< 20	1 (Bad)

The *Moving Picture Quality Metric* (*MPQM*) evaluates the video quality using HVS modeling characteristics (Lambrecht & Verscheure, 1996). The input to the MPQM metric is an original video sequence and a distorted version of it. The distortion is first computed as the difference between the original and the distorted sequences. The original and the error sequences are then decomposed into perceptual channels segmented using uniform areas, textures and contours classification. After that, HVS-based contrast sensitivity and masking parameters are considered for each perceptual channel in detection threshold calculation. Finally, data from channels are gathered to yield a single figure and to account for higher levels of perception, which is called pooling. Due to the MPQM's purely frequency-domain implementation of the spatio-temporal filtering process, this complex metric requires huge memory consumption. The final quality measure can be expressed either using a Masked PSNR equation or can be mapped to MOS scale as detailed in (Lambrecht & Verscheure, 1996).

The *Perceptual Evaluation of Video Quality* (*PEVQ*) (*PEVQ*, 2010) provides MOS values of the video quality degradation as a consequence of end-to-end communication. The PEVQ approach is based on the combination of spatial and temporal artifacts measurement with human visual system behavior. PEVQ provides MOS scores of the video quality; in addition, PEVQ also provides information about the perceptual level of distortion in luminance, chrominance and temporal aspects of the evaluated video.

The previous QoE methods are based on a set of user/service information about the original and processed video. In order to reduce the system complexity and the amount of available reference information, a packet-based method, called *Media Delivery Index* (*MDI*), was proposed in *Internet Engineering Task Force* (*IETF*) *Request For Comment* (*RFC*) 4445 by Welch & Clark (2006). The MDI metric is not the most accurate video quality level method and does not provide a good characterization of QoE, but can provide an indication of the video quality in a cost effective manner. The MDI scheme provides an indication of traffic jitter, a measure of deviation from nominal flow rates and a data loss at-a-glance measure for a particular multimedia service. According to MDI values, the overall video quality level through an end-to-end communication path can be estimated.

Regarding voice measurement, the E-Model is a non-intrusive *Voice over IP* (*VoIP*) metric based on the concept that impairments, which affect the voice call, are independent. Five factors are considered: the basic signal-to-noise ratio, which includes sources of noise as the circuit or environment, the impairments which occur more or less simultaneously with the voice signal, the impairments caused by delay, the impairments introduced by the equipment as losses, and the advantage factor (Bandung, Machbub, Langi & Supangkat, 2008; ITU-T G.107, 2005). The Advantage factor allows the compensation of impairment when there are other advantages of access to the user. A conventional wired access will have a smaller compensation compared to a wireless access in a remote area. Each parameter is calculated separately and combined to obtain the final result.

The *Perceptual Evaluation of Speech Quality* (*PESQ*) evaluates the voice QoE by comparing the signal sent and received in analog or digital networks. PESQ evaluation includes factors of distortion due to channel/encoder, losses and jitters and is recommended for speech quality assessment of narrow-band handset telephony and speech codecs. Effects of loudness loss, delay, side tone, echo, and other impairments related to two-way interaction are not reflected in the scores (ITU-T P.862, 2001).

The PESQ presents values from -0.5 (lower value) to 4.5 (best value), although for most cases the output range will be a listening quality score, such as the MOS scale, between 1.0 and 4.5. Although these results cannot directly be mapped to MOS, they can be approximated to it (Liang, Ke, Shieh, Hwang & Chilamkurti, 2006).

Figure 1 presents the main difference between objective and subjective QoE approaches. Objective metrics try to model the human perception based on network, transport and application factors, while subjective methods also include emotions, experiences and billing services.

In order to improve wireless control schemes, QoE quality metrics which have been designed and approved for video and voice quality assessment can be integrated into network/content management infrastructure, so that management operations can be proposed and selected according to user perceived quality.

QUALITY OF EXPERIENCE ASSESSMENT APPROACHES

Recent researches have focused on different ways of assessing the quality of video streaming considering the additional information used in the assessment process (by using subjective and objective metrics). These solutions are needed in wireless systems to measure the quality level of current and new applications, as well as, to provide input for optimization procedures in congestion situations or during failures in networking devices. The three main application-level approaches used to classify video quality assessment methods, based on reference-related video procedures are: *Full Reference* (*FR*), *Reduced Reference* (*RR*) and *No Reference* (*NR*).

The FR approach assumes unlimited access to the original video/multimedia sequence. This approach uses the video reference to predict the quality level (degradation) of the processed video, by comparing the difference of every pixel in each image of the distorted video with its corresponding pixel in the original video. As consequence, it provides, in general, superior quality assessment performance. The FR method is difficult to implement in real-time networking systems (QoE-aware

Figure 1. Subjective and objective QoE approaches

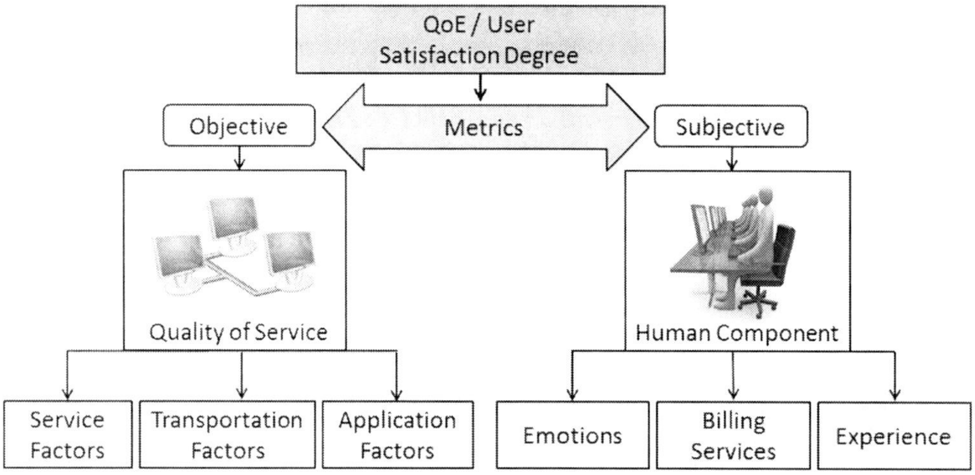

equipment/monitoring agent) because it always requires the original sequence during the evaluation process (common for offline experiments). Examples of metrics based on an FR approach are PSNR, SSIM and MPQM.

For in service video quality measurements, RR and approaches are generally more suitable. The RR approach differs from the FR approach, whereas only selected multimedia parameters (or characteristics) are required during quality evaluation process, such as motion information. The set of reference parameters can be transmitted piggy-backed with the multimedia flow or by using a secondary channel. The objective of RR is to be as accurate as the full reference model, although using less network and processing resources. An example of an RR scheme is the metric VQM, developed by the *National Telecommunications and Information Administrative* (*NTIA*) and reported by Pinson & Wolf (2004).

The NR approach tries to assess the quality of a distorted multimedia service without any reference to the original content. This approach is usually, employed when the encoding method is known. NR-based metrics can be used in in-service network monitoring/diagnostic operations, when the original multimedia sequence is not available. The drawbacks of NR metrics are the following: (i) low correlation with MOS; (ii) high CPU and memory consumption; (iii) time limitation. An example of NR schemes is the V-Factor model (V-Factor, 2010) that outputs MOS. One important challenge in wireless multimedia systems is the specification and implementation of well-defined and high-performance NR schemes.

In addition to the previously application-level measurements approaches (reference-based classification), and, due to the time and processing demands, as well as feasibility issues of content based assessments, multimedia quality prediction mechanisms can be used as a manner to evaluate the quality level of video sequences in wireless networks. These schemes predict the quality level that a specific content will have after the encoding process, based on the encoding parameters, packet inspections, and network conditions. Further processing of the original data is not required, minimizing the associated complexity and resource consumption.

Implementations of prediction mechanism are the utility function models that offer a user-layer extension to existing user-aware measurement schemes to better assess the requirements of multimedia applications as detailed by Mu, Mauthe and Garcia (2008). Impairment utility functions model the impact from each network QoS dimension (e.g., delay and packet loss) on the perceived quality. Utility values which are generated from all impairment functions are then aggregated as the application utility, which quantifies the user's experience on target multimedia streaming application. The utility value represents the impact of application requirements and network resources on the user's perception.

Network-based approaches are suitable alternatives to assessment controllers in emerging wireless systems, where they evaluate the quality of multimedia content, basically, by verifying all transmitted packets related with the application and network conditions, and no decoding processing is required. The main issue is that such approaches need to perform deep-packet inspection. They need to gather information about the current network conditions, such as packet loss rate and packet one-way delay, to be used in the evaluation process. The final quality level assessment decision can be taken based on previous information, together with information about the multimedia characteristics, such as frame-rate, *Group of Picture* (*GoP*), frame type and dependence, only available at application level. This approach is preferable for in-service (real time) multimedia applications since the computational complexity is reduced. The performance is low to medium, but the feasibility is high. For example, in a simplest scenario, quality indicators are only some QoS parameters such as packet loss ratio or bit error rate.

Finally, Hybrid Content Inspection and Network Measurement approaches have also been proposed (Romaniak, Mu, Mauthe, D'Antonio & Leszczuk, 2008). The main reason for the development of this kind of scheme is to allow network operators to combine the benefits of the previous approaches and adjust performance, complexity and feasibility, as well as to adjust operational cost issues according to different needs, multimedia content type, networks and equipments.

Figure 2 summarizes the main difference among NR, FR and RR approaches. FR metrics need unlimited access to the original video sequence. RR metrics requires some additional (extra) information during quality evaluation process. NR measures the quality of a distorted video without any reference to the original content. Notice that hybrid schemes are based on the combination of the previous one, but taking into accounting performance and feasibility issues.

QUALITY OF EXPERIENCE OPTIMIZATION IN WIRELESS NETWORKS

There are different QoE optimization approaches for emerging wireless networking, ranging from MAC layer improvements and channel adaptation to routing (Gomes, Junior, Cerqueira & Abelem, 2010) and packet control mechanisms (Rodrigues, Silva, Cerqueira & Monteiro, 2008). The last approach is very useful, optimizes the

Figure 2. Differences among NR, FR and RR approaches

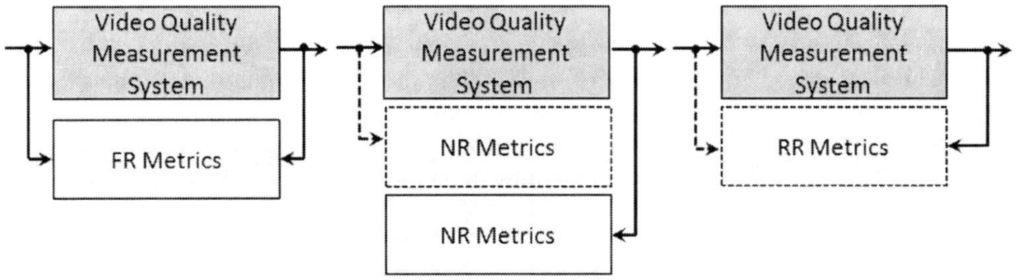

usage of network resources, maximizes the user's satisfaction and will be detailed in this section. A state-of-art analysis will be presented and a novel and efficient user-aware solution to control the quality level of multimedia applications on IEEE 802.11 wireless systems will be described. Since multimedia flows are different in terms of encoder parameters, intra-frame dependence, as well as other QoS and QoE requirements, one key challenge to keep applications with good quality levels during congestions periods is to implement an IEEE 802.11 packet controller mechanism to discard packet according to the impact of each frame on the user perception.

Lastly, advances in wireless networks IEEE802.11e aim to provide multimedia quality level assurance and are proposed and tested in (Ngoc, Tan, Lee & Oh, 2007). This proposal adds a second queue to *Distributed Coordination Function* (*DCF*) in order to give priority to multimedia streams, and therefore reduces their delay. However, this approach is obsolete and does not focus on the new *Hybrid Coordination Function* (*HCF*) operation mode introduced by the IEEE802.11e draft, which already includes several classes and queues that allow different priorities. Furthermore, few details are provided about performance evaluation issues and applicability scenarios. Thus, it is impossible to know how the multimedia flows were assessed and which scenarios were used. Finally, the conclusions are only based on network performance measurements, such as delay and losses, and no user level assessments were accomplished.

Another approach uses cross-layer architecture to map packets with different priorities to different *Enhanced Distributed Channel Access* (*EDCA*) queues of IEEE 802.11 (Haratcherev, Taal, Langendoen, Lagendijk & Sips, 2006). In order to setup the importance of each packet, flows are divided into *Parameter Set Concept* (*PSC*), *Instantaneous Decoding Refresh* (*IDR*) pictures and partitions. PSC packets contain information about each flow, such as the resolution or the encoding process. IDR packets transport intra-pictures, i.e., pictures that require no other pictures to be decoded. The remaining pictures are divided into slides and each slide is divided into partitions using H.264 *Data Partitioning* (*DP*). The DP uses 3 partitions types: A partitions include header information; *B* partitions have intra-predictions and *C* partitions inter-predictions.

Because PSC packets contain the most important information, they are mapped to the AC with highest priority, which is, AC3. Given the mean priority of partitions *A*, they are mapped to AC2. *B* and *C* partitions are placed in AC1. Finally, best-effort traffic is mapped to the category with lowest priority, i.e., AC0. However, this proposal is not in accordance with the recommendations of the IEEE 802.11 standard, as well as, additional traffic placed in AC3 may damages voice flows and no scenario related to this problem was analyzed. Moreover, the use of AC1 for video flows makes it impossible to be used for other traffic, and thus no distinction between various types of non-multimedia flows can be done. Furthermore, the mapping of packets from the same flow in different queues increases the jitter that can lead to losses in the buffer on the receiver side. Finally, simulations only take into accounting QoS metrics and no user-based measurements are performed.

Moid and Fapojuwo (2009) present a framework for streaming of H.264 video over an IEEE 802.11-based wireless network. A proposal was based on a cross-layer mechanism that jointly adapts the video transcoding parameters at the application layer and the video transmission parameters at the data-link layer to the network conditions defined by buffer length and wireless propagation channel. The validation of the proposal made by using the ns2 and Evalvid show that the model is adaptive to change according to network conditions and also in relation to the frame size of videos, with a gain of about 3 dB compared with the simulation model implemented. But the big gap left in this work is that it only deals with the metric, which despite being widely used has several limitations related to the *Human Behavior System*, as presented in section 2.

As discussed above, new QoE packet controllers for multimedia wireless systems are still challenges for both academy and industry. Novel approaches are required to assure the quality level support for video streaming applications based on user perception and also in mobility situations. Furthermore, the dependency of each frame of a sequence must be taken into accounting during the adaptation process. The remainder of this section will present the benefits and discuss an intelligent packet controller mechanism for IEEE 802.11.

The proposed optimization wireless mechanism is configured with different selective dropping levels (extends the MAC QoS classifier, meter and dropper with multimedia and QoE-awareness), where a percentage of discarding associated with video and non-video traffics can be assigned to be used during adaptation process. For example, in congestion situations, video traffic can be protected to be discarded last (concurrent traffic is dropped first) or the system can be configured to drop only 10% of all video packets.

This scheme has two operational modes as follows:

1. In its basic configuration, it adapts video sessions to the current network conditions, by dropping frames only according to their importance in order to keep the system as simple as possible (low processing and state stored).
2. In its enhanced configuration, it can adapt the video quality level also taking into account the dependency of intra-frame sequences and other relevant control information, such as audience size or cost.

Since video streaming are data flow containing application-level objects with special proprieties and dependence, the QoE optimization mechanism improves the packets dropping based on the dependency of a set of frames. The packet controller discards packets based on the importance of each frame. A typical MPEG-1/2/4 structure is presented in Figure 3, where *I* is the frame with highest priority and *B* the lowest ones.

The QoE wireless controller includes an advanced packet drop scheme that performs its multimedia quality level management based on the importance of each frame of the CODEC. Therefore, if a packet containing an *I* frame is marked to be dropped (in congestion periods during failures), it will be beneficial to check whether a packet corresponding to a *P* or *B* frame is currently in the buffer. Since the number of frames that depends on a *P* or *B* frame is reduced, the loss of these frame types will have a reduced impact on the final quality and, therefore, on the user perception. The same process happens when a *P* frame is selected to be dropped and a *B* frame is in the queue. It will be beneficial to drop the *B* frame that is already in the buffer and enqueue the incoming *P* frame.

Figure 3. A typical MPEG-1/2/4 structure

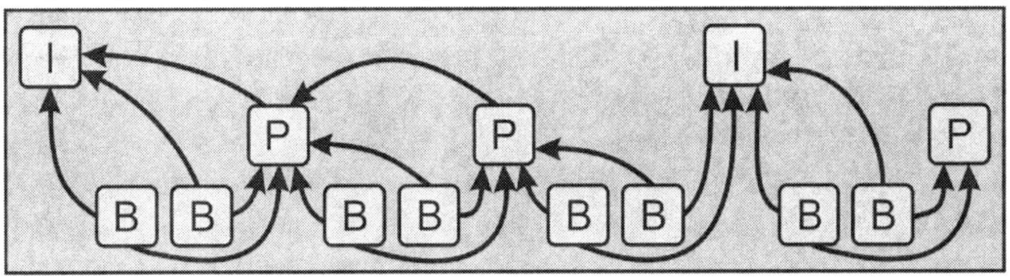

PERFORMANCE EVALUATION OF A QOE PACKET CONTROLLER

This section describes the benefits and impacts of the proposed QoE packet controller in IEEE802.11e wireless networks on the user perception based on well-known assessment metrics. The evaluation was carried out by using the NS2. The Evalvid platform was also implemented to assess the video streaming quality delivery and configured to support MPEG *I*, *P* and *B* frames.

The main objectives of the simulation experiments are the following:

1. Analyze the percentage of packet losses associated with frames of video streaming and non-video applications; and
2. Analyze the perceived quality of a video sequence by verifying PSNR, SSIM, VQM and MOS.

Four IEEE802.11e different configurations are implemented for the experiments, they are:

1. *Best-Effort* (no traffic differentiation),
2. *Pure* QoS (with traffic differentiation, but without QoE and multimedia support),
3. QoE control (the most important frames are protected in congestion periods) and
4. QoE control with 3% of Advanced Drop, called *QoEAdv* (the most important frames are protected and the percentage of non-video packets to be discarded is increased in 3%, in order to save more video packets).

For each approach, 10 experiments were performed with different congestion rates (from 0 up to 200% of congestion in a system).

The *Boston Representative Internet Topologies Generator* (*BRITE*) (Medina Lakhina, Matta & Byers, 2001) was used to generate a random topology with wired core nodes, wireless access points and users for the evaluation. Figure 4 illustrates the evaluated scenario, which is composed of 4 wireless nodes (2 sources and 2 receivers), 4 access points and 21 core routers. The wireless and core links have a bandwidth of 11Mb/s and 100Mb/s, respectively. The propagation delay of each link was assigned according to the distance between the edges of each link. Each source sends a real video sequence with average rate of 350Kb/s and *Constant Bit Rate* (*CBR*) traffic, in order to congest the links. The video sequence, denominated "News" (Evalvid, 2010), consists of 300 frames (30 fps) with YUV format, sampling 4:2:0 and dimension 352x288. The video sequence was compressed with a MPEG-4 CODEC. The GOP of the sequence is composed of 30 frames, using two *B* frames for each *P* frame. Frames are then fragmented in blocs with 1024*B*.

Frame Losses

This Section measures the percentage of frame losses: *I*, *P*, *B* and other frames (CBR traffic) for different congestion rates when a system is configured with *Best-Effort*, Pure QoS, QoE Control (the most important frames are protected in congestion periods) and *QoEAdv* (the percentage of non-video packets to be discarded is increased in 3%, in order to save more video packets) mechanisms.

As presented in Figure 5.a, when the system is implemented only with the *Best-Effort* approach, packets are discarded randomly. Hence, when the network's congestion increases, the percentage of packet losses will increase proportionally for all frame types (including packets associated with the CBR traffic).

Figure 5.b describes the system behavior when the pure QoS approach is being used. The results reveal that it provides packet differentiation and protect the discard of video packets (compared to the *Best-Effort* class), because they are accommodated in a most important class. However, all video packets are dropped in a random "black-box" way (without frame type differentiation).

Figure 4. Evaluated Wireless QoE-aware Scenario

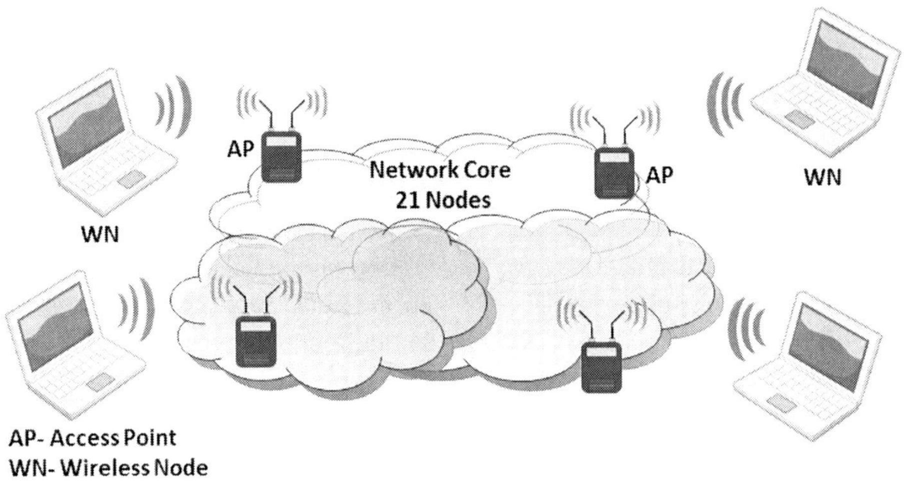

Figure 5. Frame lost for each mechanisms

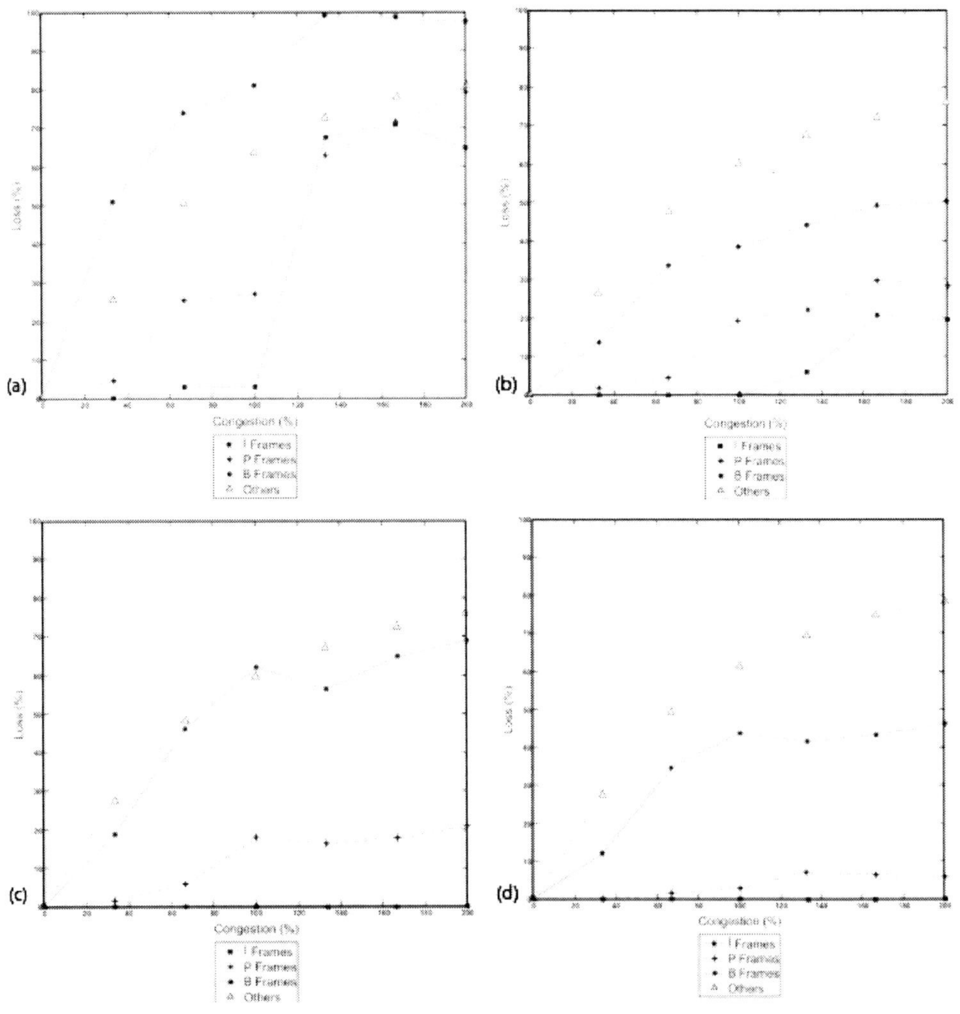

The benefits of the QoE control proposal on the video quality level are depicted in Figure 5.c. By adapting the video content according to the importance of each frame, the QoE controller aims to protect most important video frames in congestion periods, which aims to keep videos with better quality levels. Hence, *B* frames are dropped first and *I* frames are discarded last, increasing the QoE. Compared to *Best-Effort* and pure QoS approaches, the QoE solution reduces the percentage of *P* frame loss in 60% and 23% respectively, when the system overload is 100%.

Due to its frame protection schemes, *I* frames are not discarded during congestions periods when the QoE and *QoEAdv* approaches are configured in the wireless system. However, compared to a system with only QoE support, the percentage of *P* and *B* frame loss is decreased in 66% and 30% respectively, while the percentage of CBR packet loss is increased to a mere 3%,as presented in Figure 5.d. Notice that there are more CBR packets than video packets in the system and, consequently, more non-video packets are discarded during congestion periods.

PSNR, SSI, VQM and MOS

Since packet loss rate does not indicate the real impact on the video quality level, PSNR values of the video sequences in different congestion periods were analyzed. Figure 6 shows the average PSNR for the videos with different approaches.

The results reveal that when the *Best-Effort* approach is being used, the PSNR of the videos decreases as fast as the traffic increases, attaining a minimum value of 27dB. When the system is configured with only QoS, the PSNR of the videos is maximized in comparison with the *Best-Effort* (e.g., the PSNR is increased in 3% when the network is overloaded by 80%). Compared to

Figure 6. PSNR for each approach and congestion rates

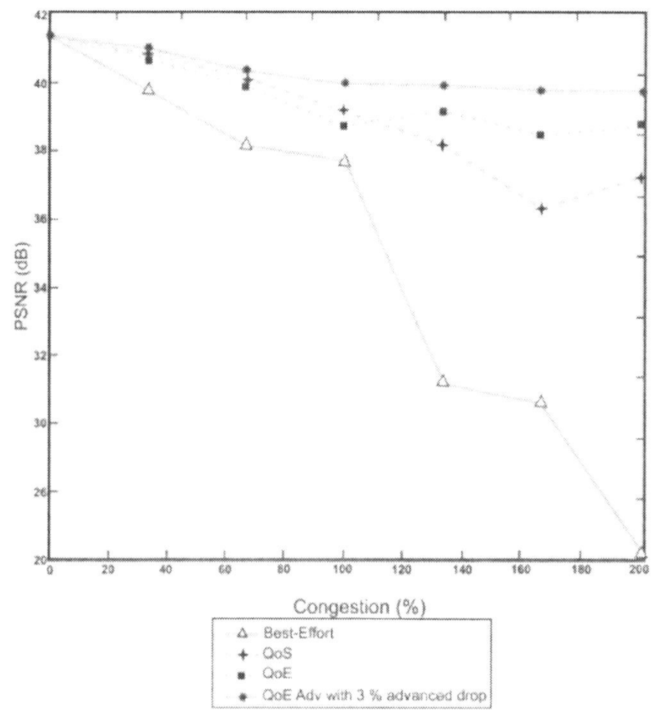

the pure QoS approach, the QoE control increases the quality level of video sequences in 3% and 6% when the system is overloaded in 130% and 150%, respectively.

The SSIM results give more detail about the video quality level taking human perception into account. Figure 7 illustrates the average SSIM of the video sequences when the system is configured with *Best-Effort*, QoS, *QoE* and *QoEAdv* drop mechanisms.

The results reveal that, when the *Best-Effort* approach is configured, the correlation between the original and the received video is poor after a congestion of 100%. Compared to the pure QoS, the QoE controller increases the SSIM of video sequences in 3% when the system is overloaded in 200%. On average, the *QoEAdv* approach increases in 0.5% the video SSIM for all experiments, when compared with simulations based only on the QoE configuration.

VQM is an important metric to verify the video quality level based on human eye perception and subjectivity. Figure 8 presents the average VQM results for each approach when the system has different congestion level.

Due to its packet differentiation scheme, the pure QoS approach increases the video VQM in 18% (compared to the *Best-Effort*) when the system load is 80%. Additionally, during a congestion period of 100%, QoE control increases the video VQM in 43% and 70%, when the system is configured with the QoS and *Best-Effort* approaches, respectively. Compared to a system with the QoE, the *QoEAdv* mechanism maximized the video VQM, on average,by15%, when the system is overloaded in 120%.

In order to present the user experience for each approach during congestion periods, the MOS was evaluated using Table 1 and illustrated in Figure 9.

The results revealed that, according to the MOS metric, the QoE and *QoEAdv* approaches kept videos with an excellent quality level during all congestion periods. In order to show the impact of the QoE control (compared to the pure IEEE802.11eQoS controller mechanism) from the user point of view when the wireless system

Figure 7. SSIM for each approach and congestion rates

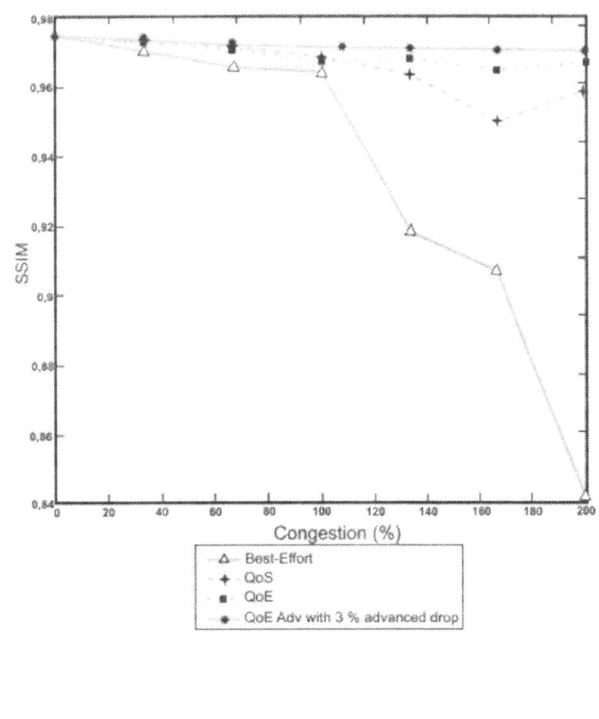

Figure 8. VQM for each approach and congestion rates

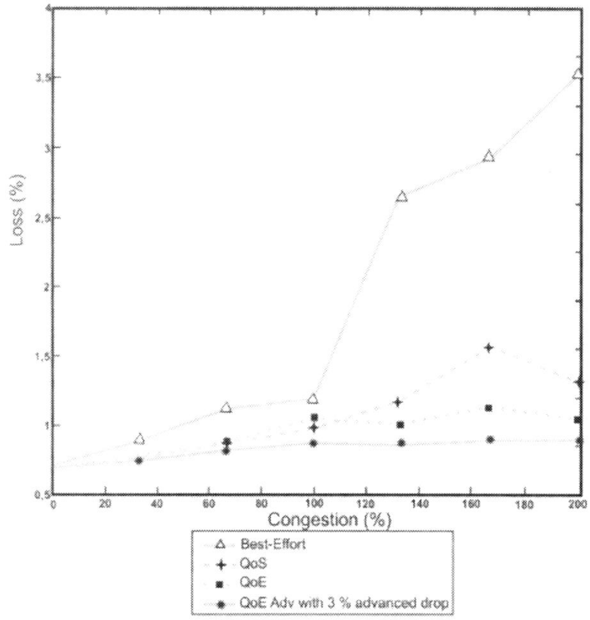

is experiencing 15% of congestion, some frames of the real video sequence, named News, were captured (Figure 10). The benefits of the QoE adaptation process are visible in the frames of the video, particularly in the ballet dancer.

FUTURE RESEARCH DIRECTIONS

Recently, solutions have been presented regarding multimedia measurement and optimization over heterogeneous wireless networks. However, there are still many important challenges that need to be addressed in future multimedia networks in

Figure 9. MOS variation when network load increase

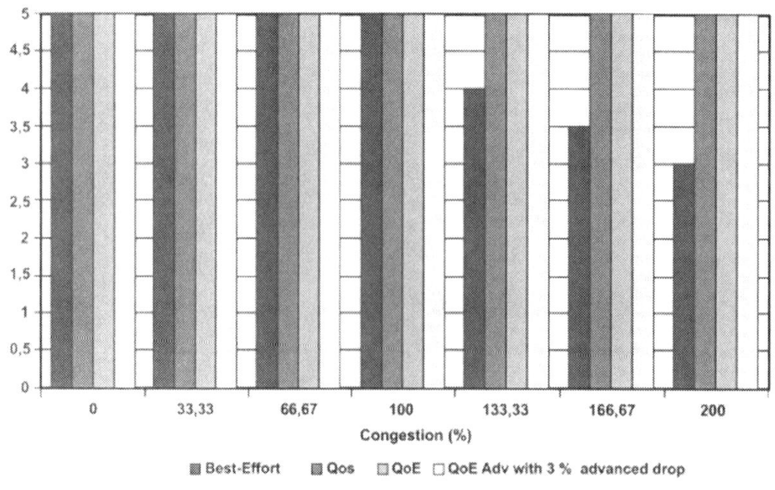

several areas. It is not the goal of this chapter to propose an integrated solution for QoE management, but rather to identify the main issues from application to network layers.

Packet/network inspection-based (or even hybrid approaches) can be used to predict and measure video quality based on information gathered from packet and network conditions without accessing the decoded video streaming. The results of these approaches are useful for pricing/billing, management and optimization operations in future wireless multimedia systems.

In addition to QoE-aware intelligent packet controllers presented in this chapter, new QoE-based application, transport and network level optimization mechanisms (whether a cross-layer approach is used or not) are still required in a near future, such as routing, inter/intra-session adaptation, resource reservation, traffic controller, seamless multimedia mobility and base station selection/user experience schemes. Moreover, the multi-homing capability of current devices can also provide an improved performance for multimedia streaming applications by taking advantage of the multiple connectivity levels from each wireless device.

As future work, the implementation and evaluation of the proposed solutions in networks are needed, as well as, the extension of QoE-aware intelligent packet controllers to operate in all types of wireless systems, such as IEEE 802.11s, IEEE 802.16, IEEE 802.15.4, 3G and Long Term Evolution are also required to optimize resources and increase the satisfaction of thousands of wireless-based users. The implementation of such QoE-aware schemes will create a new networking multimedia era and will change the live stile of users and multimedia markets, where wireless providers (or even user) will create and share their content according to the user needs and perception. Quality of Feeling, Quality of Smelling and other kind of new user experience approaches are expected to be implemented in the future in order to keep and attract new costumer.

CONCLUSION

Research in wireless multimedia area is envisioned to continue with various challenges emerging as a result of new applications, approaches, technologies, operational costs, changing user and terminal requirements, and highly heterogeneous

Figure 10. Some frames of "News" with different packet control mechanisms

networking infrastructures and devices. This chapter is intended to highlight some of the important topics in wireless and multimedia areas that need attention to address some of the most pressing challenges associated with them. The chapter is focused on two key areas, where the first one was on the assessment schemes and the second was on packet controller optimization schemes, including performance evaluation results.

QoE, although not always quantifiable, numerically, is the most important factor in assessing the user experience through metrics and techniques that can be categorized mainly into two categories, based on the subjectivity or objectivity. In the first category, in general, a group of viewers rate the quality of media, such as through the MOS, in its three axes, *Viewing (V)*, *Audio (A)*, *Interaction (C)*. The second category in turn, makes use of indexes that have qualities associated with mathematical calculations and / or measurement by specific equipment, such as PSNR, MPQM and MDI.

The benefits generated by QoE wireless optimization schemes contribute to the creation of a ubiquitous multimedia era, where the user experience is a key parameter during the development of new solutions to maximize the usage of network resources and the costumer´s satisfaction. QoE-aware packet controllers are examples of simples (and efficient) mechanisms to be implement in wireless multimedia systems.

We hope that this work will help improve the understanding of the issues and challenges that lie ahead in wireless multimedia networks and will serve as a catalyst for designers, engineers, and researchers to seek innovative solutions to address and solve those challenges.

ACKNOWLEDGMENT

Authors would like to thank the Pará State Government (Brazil), FAPESPA, REDE TIC, Federal University of Pará/PROPESP and CNPq (476202/2009-4,559994/2010-9,475312/2010-4, 304105/2010-4).

REFERENCES

Akyildiz, I. F., & Wang, X. (2005, September). A survey on wireless mesh networks. *IEEE Communications Magazine*, *43*(8), 23–30..doi:10.1109/MCOM.2005.1509968

Andrews, J. G., Ghosh, A., & Muhamed, R. (2007). *Fundamentals of WiMAX: Understanding broadband wireless networking*. Upper Saddle River, NJ: Prentice Hall.

Bandung, Y., Machbub, C., Langi, A. Z. R., & Supangkat, S. H. (2008). Optimizing voice over internet protocol (VoIP) networks based on extended e-model. In the *IEEE Conference in Cybernetics and Intelligent Systems* (pp. 801-805). doi10.1109/ICCIS.2008.4670898

ETSISTQ. (2010). *European Technical Committee for Speech, Transmission, Planning, and Quality of Service*. Retrieved from http://portal.etsi.org

Evalvid (2010). Retrieved from http:// ww.tkn. tu-berlin. de/ research/ evalvid/ cif

Gomes, R. L., Junior, W. M., Cerqueira, E., & Abelem, A. J. (2010, April). Using fuzzy link cost and dynamic choice of link quality metrics to achieve QoS and QoE in wireless mesh networks, *Journal of Network and Computer Applications*. ISSN 1084-8045.

Haratcherev, L., Taal, J., Langendoen, K., Lagendijk, R., & Sips, H. (2006, January). Optimized video streaming over 802.11 by cross-layer signaling. *IEEE Communications Magazine*, *44*(1), 115–121..doi:10.1109/MCOM.2006.1580941

http:// www.everest-ist. upc. es/ publicdocuments/ conferences/ Reves_ISTMS06.pdf

ITU-T. (2010). *International Telecommunication Union – Telecommunication Standardization Sector*. Retrieved on July, 2010, from http://www.itu.int/ITU-T/

ITU-T BT-500. (1990). *Methodology for the subjective assessment of the quality of television pictures*. Technical Report. Retrieved from http://www.itu.int/ITU-T/

ITU-T G.107 (2005). *The e-model: A computational model for use in transmission planning*. Retrieved from http://www.itu.int/ITU-T/

ITU-T P.862 (2001). *Perceptual evaluation of speech quality (PESQ)*: An objective method for end-to-end speech quality assessment of narrow-band telephone networks and speech codecs. Retrieved from http://www.itu.int/rec/T-REC-P.862/

Klaue, J., Rathke, B., & Wolisz, A. (2003). EvalVid - A framework for video transmission and quality evaluation. In the *13th International Conference on Modelling techniques and Tools for Computer Performance Evaluation* (pp. 255-272). doi10.1007/978-3-540-45232-4_16

Lambrecht, C., & Verscheure, O. (2006, January). Perceptual quality measure using a spatio-temporal model of the human visual system. *Proceedings of the Society for Photo-Instrumentation Engineers, 2668*, 450–461.

Liang, H. M., Ke, C. H., Shieh, C. K., Hwang, W. S., & Chilamkurti, N. K. (2006). Performance evaluation of 802.11eEDCF in the ad-hoc mode with real audio/video traffic. *International Conference on Wireless and Optical Communications Networks*, 1-6. doi10.1109/WOCN.2006.1666581

Lidinsky, W. P., Jeffree, T., Chambers, A., & Jeffree, T. (2001). *IEEE standard for local and metropolitan area networks: Overview and architecture*. Retrieved from http:// standards.ieee.org/getIEEE 802/ download/ 802-2001.pdf

Marks, R. B., Stanwood, K., Eklund, C., & Chang, D. (2004). *IEEE recommended practice for local and metropolitan area networks: Coexistence of fixed broadband wireless access*. Retrieved from http:// standards.ieee.org/ getIEEE 802/ download/ 802.16.2-2004.pdf

Medina, A., Lakhina, A., Matta, I., & Byers, J. (2001). BRITE: An approach to universal topology generation. In *Proceedings of the International Workshop on Modeling, Analysis and Simulation of Computer and Telecommunications Systems* (pp. 343-353). doi10.1109/MASCOT.2001.948886

Moid, A., & Fapojuwo, A. O. (2009). A cross-layer framework for efficient streaming of H.264 video over IEEE 802.11 networks. *Journal of Computer Systems, Networks, and Communications*. Retrivied from http://www. hindawi.com/ journals/ jcsnc/ 2009/ 682813.abs.html

Mu, M., Cerqueira, E., Boavida, F. & Mauthe, A. (2009). Quality of experience management framework for real time multimedia applications. *International Journal Internet Protocol Technology*, 54 - 64.

Mu, M., Mauthe, A., & Garcia, F. (2008). A utility-based QoS model for emerging multimedia applications. *IEEE International Workshop on Future Multimedia Networking* (pp. 521 – 528). doi10.1109/NGMAST.2008.24

ns2 (2010). *The network simulator – ns-2*. Retrieved from http://www.isi.edu/nsnam/ns/

Ngoc, D. M., Tan, P. A., Lee, J. S., & Oh, H. (2007, April). Improvement of IEEE 802.11 for multimedia traffic in wireless LAN. In *Wireless Telecommunications Symposium* (pp. 1 – 4). doi10.1109/WTS.2007.4563296

Perkis, A., Munkeby, S., & Hillestad, O. I. (2006, June). A model for measuring quality of experience. In *Proceedings of the 7th Nordic Signal Processing Symposium* (pp. 198-201). doi10.1109/NORSIG.2006.275209

PEVQ. (2010). *PEVQ Perceptual Evaluation of Video Quality*. Retrieved from http://www.pevq.org/

Pinson, H., & Wolf, S. (2004, September). A new standardized method for objectively measuring video quality. *IEEE Transactions on Broadcasting*, *50*(3), 302–311..doi:10.1109/TBC.2004.834028

Revés, X., Nafisi, N., Ferrús, R. & Gelonch, A. (2006, June). User perceived quality evaluation in a B3G network test bed. *IST Mobile Summit*. Retrieved from

Rodrigues, D., Silva, D., Cerqueira, E., & Monteiro, E. (2008). Quality of service and quality of experience in video streaming. *International Workshop on Traffic Management and Traffic Engineering for the Future Internet*. Retrieved from http:// eden.dei.uc.pt/ ~edmundo/ Conf%20 Internacionais/ CI120% 202008% 20FITraMEn% 20Rodrigues.pdf

Romaniak, P., Mu, M., & Mauthe, A. D'Antonio & S., Leszczuk, M. (2008, May). Framework for the integrated video quality assessment. *Specialist seminar on quality of experience*. Retrieved from http:// www.comp.lancs.ac.uk/ ~mum1/ files/ Framework Assessment.pdf

Standard, I. E. E. E. 802.11e (2010). *Medium Access Control (MAC). Enhancements for quality of service specifications*. Retrieved from http:// grouper.ieee.org/ groups/ 802/ 11/ Reports/ tge_update.htm

Takahashi, A., Hands, D., & Barriac, V. (2008, February). Standardization activities in the ITU for a QoE assessment of IPTV. *IEEE Communications Magazine*, *46*(2), 78–84..doi:10.1109/MCOM.2008.4473087

V-Factor. (2010). *V-factor quality of experience platform*. Retrieved from http://www.pevq.org/

VQEG. (2010). *Video quality experts group*. Retrieved from http://www.its.bldrdoc.gov/vqeg/

Wang, Z., Bovik, A. C., Sheikh, H. R., & Simoncelli, E. P. (2004, April). Image quality assessment: From error visibility to structural similarity. *IEEE Transactions on Image Processing*, *13*(4), 600–612..doi:10.1109/TIP.2003.819861

Wang, Z., Lu, L., & Bovik, A. C. (2004, January). Video quality assessment using structural distortion measurement. *Signal Processing: Image Communication. Special Issue on Objective Video Quality Metrics*, *19*(2), 121–132. doi:.doi:10.1109/ICIP.2002.1038904

Welch, J. & Clark, J. (2006). *A proposed media delivery index* (MDI). IEFTRFC 4445.

Zapter, M., & Bressan, G. (2007, January). A proposed approach for quality of experience assurance for IPTV. In *Proceedings of IEEE International Conference on the Digital Society*, 25 -31. doi10.1109/ICDS.2007.4

ADDITIONAL READING

Cerqueira, E. C., Veloso, L., Curado, M., Mendes, P., & Monteiro, E. (2008). Quality level control for multi-user sessions in future generation networks. In *Proceedings of IEEE Global Telecommunications Conference* (pp. 1 – 6). doi10.1109/GLOCOM.2008.ECP.273

Cerqueira, E. C., Zeadally, S., Leszczuk, M., Curado, M., & Mauthe, A. (2010). Recent advances in multimedia networking. *Multimedia Tools and Applications*, *52*, 1–13. doi:.doi:10.1007/s11042-010-0578-z

Engelke, U., & Zepernick, H. J. (2007). Perceptual-based quality metrics for image and video services: A Survey, *3rd Conference on Next Generation Internet Networks* (pp. 190-197). doi10.1109/NGI.2007.371215

Garcia, M., Canovas, A., Edo, M., & Lloret, J. (2009). *A QoE management system for ubiquitous IPTV devices. 3rd* International Conference Mobile Ubiquitous Computing, Systems, Services and Technologies (pp. 147-152). doi10.1109/UBICOMM.2009.31

Greengrass, J., Evans, J., & Begen, A. C. (2009). Not all packets are equal, part 1: Streaming video coding and SLA requirements. *IEEE Internet Computing, 13*(1), 70–75..doi:10.1109/MIC.2009.14

Grega, M., Janowski, L., Leszczuk, M., Romaniak, P., & Papir, Z. (2008). *Quality of experience evaluation for multimedia services* (pp. 142–153). Przegląd Telekomunikacyjnyi Wiadomości Telekomunikacyjne.

Jailton, J. J., Dias, K. L., & Cerqueira, E. C., Cavalcanti & Dave, C. (2009). Seamless handover and QoS provisioning for mobile video applications in an integrated WiMAX/MIP/MPLS architecture. *International Journal of Advanced Media and Communication, 3*, 404..doi:10.1504/IJAMC.2009.028710

Jain, R. (2004). Quality of experience. *IEEE MultiMedia, 11*(1), 95–96. doi:10.1109/MMUL.2004.1261114

Ke, C., & Chilamkurti, N. (2008). A new framework for MPEG video delivery over heterogeneous networks. *Elseiver Computer Communications, 31*(11), 2656–2668..doi:10.1016/j.comcom.2008.02.029

Kishigami, J. (2007). The role of QoE on IPTV services style. *9th IEEE International Symposium* (pp. 11-13). doi10.1109/ISM.2007.4412350

Monteiro, J. M. & Nunes, M. S. (2007). A subjective quality estimation tool for the evaluation of video communication systems. *IEEE Symposium on Computers and Communications* (pp. 75 –80). doi10.1109/ISCC.2007.4381645

MSU. (n. d.). *MSU video quality measurement tool*. Retrieved from http:// compression.ru/ video/ quality measure/ index en.html

Parker, A. (2006). *Addressing the cost and performance challenges of digital media content delivery* (p. 2P). Media Summit.

Pinson, H., & Wolf, S. (2004). A new standardized method for objectively measuring video Quality. *IEEE Transactions on Broadcasting, 50*(3), 312–322..doi:10.1109/TBC.2004.834028

Schwarz, H., & Marpe, H. (2008). Overview of the scalable video coding extension of the H.264/AVC standard. *IEEE Transactions on Circuits and Systems for Video Technology, 17*(2), 1103–1120. doi:.doi:10.1109/TCSVT.2007.905532

Siller, M., & Woods, J. (2003). Improving quality of experience for multimedia services by QoS arbitration on QoE framework. *Proceedings of the 13th Packed Video Workshop 2003.*

Video Traces Research Group. (n. d.). *YUV 4:2:0 Video sequences*. Retrieved from http://trace.eas.asu.edu/yuv/qcif.html

Yamada, H., Fukumoto, N., Isomura, M., Uemura, S., & Hayashi, M. (2007). A QoE based service control scheme for RACF in IP-based FMC networks. *E-commerce Technology and the 4th IEEE International Conference on Enterprise Computing, E-commerce, and E-services* (pp. 611-618). doi10.1109/CEC-EEE.2007.10

Zhao, H., Ansari, N., & Shi, Y. (2005). Layered MPEG video transmission over IP DiffServ. *Proceedings of the International Conference on Information Technology: Coding and Computing.*

Chapter 12
Routing and Coding Enhancements to Improve QoS of Video Transmissions in Future Ad Hoc Networks

Juan C. Guerri
UPV, Spain

Pau Arce
UPV, Spain

Patricia Acelas
UPV, Spain

Wilder E. Castellanos
UPV, Spain

Francisco Fraile
UPV, Spain

ABSTRACT

Video services are much demanded nowadays but bandwidth and delay requirements of this kind of services are very restrictive. Offering real-time video services in wireless ad-hoc networks is not an easy task because of the difficulty of guaranteeing certain quality in a shared medium. Practical solutions should try to improve communications at (and gathering information from) several layers of the protocol stack. Mobile Ad-hoc Networks are infrastructure-less wireless networks characterized by being very versatile, dynamic and self-organized but also by the difficulty to achieve a good Quality of Service in video transmissions due to packet losses and node mobility. On the other hand, the Wireless Mesh Network is presented as the next step in wireless networks. Wireless Mesh Networks have a hierarchical topology, clustered structure and static backbone, which all help to improve the network stability. In the way towards Wireless Mesh Networks, hierarchical routing protocols could transform an ad-hoc network in a more robust wireless network. Therefore, in this chapter, hierarchical routing protocols

DOI: 10.4018/978-1-61350-144-3.ch012

have been studied, particularly Hierarchical Optimized Link State Routing Protocol, and compared with a traditional flat routing protocol named Optimized Link State Routing. Furthermore, additional video coding techniques have been used in order to improve video quality in reception. At application layer, results show that Multi-description Coding achieves better quality on video transmissions when nodes have medium or high mobility, especially when using multipoint-to-point transmission or disjoint paths in a hierarchical structure. Video trace simulations have allowed us to perform subjective quality tests to assert the Quality of Experience improvements in video transmissions.

INTRODUCTION

Video transmission with *Quality of Service* (*QoS*) over IP networks has been discussed for decades and has become the main topic in many papers and research works. Studies carried out by researchers years ago such as Little (Little & Venkatesh, 1994), Bolot (Bolot & Turletti, 1994), Schulzrinne (Busse, Deffner & Schulzrinne, 1996) and Steinmetz (Steinmetz, 1990) triggered a process that allowed the publication and use of standards (which should be considered a great success) such as *Real-time Transport Protocol* (*RTP*) / *Real-time Control Protocol* (*RTCP*), *Real Time Streaming Protocol* (*RTSP*), *Session Initiation Protocol* (*SIP*), *Session Description Protocol* (*SDP*)… and, moreover, it provoked an evolution in compression systems more and more efficient such as *Scalable Video Coding* (*SVC*) included in the norm H.264 (H.264/SVC). Furthermore, the emergence of the concept of "*new content generator*" associated with any user (*youtube* being the best example of the new content generation model) has also led to the widespread use of pseudo streaming solutions based on *HyperText Transfer Protocol* (*HTTP*), proprietary solutions such as Flash Video or open solutions such as *HyperText Markup Language version 5* (*HTML5*) video, which initially appear to resolve the expectations of users. Moreover, the transmission using the *High Definition* (*HD*) standard, the imminent appearance of *3D television* (*3D-TV*) or enhanced immersive interactive user-based applications, will entail a change in the design of traditional networks, increasing the network intelligence in order to avoid over-provisioning and guarantee QoS.

However, there are some proposals for network topologies that are far away from supporting this kind of services. *Ad Hoc networks,* especially *Mobile Ad Hoc Networks* (*MANETs*), and its integration in *Wireless Mesh Network* (*WMN*) are examples of these networks. Currently, it is only possible to guarantee a minimum connectivity service between mobile nodes in an ad hoc network, despite the great amount of works that have been carried out about this topic. The possibility of offering a video streaming service with good QoS and a good *Quality of Experience* (*QoE*) to final users is still complicated. The support of a good QoS and QoE for HD or 3D-TV services is still a bigger challenge.

As a starting point we will use the typical definition of MANET, i.e. a group of independent nodes, interconnected through wireless links without using network infrastructure or centralized administration. Due to nodes mobility, the topology is dynamic and unpredictable. The network can work in isolated mode or being connected to a network infrastructure like the Internet. Routes between nodes may include multiple hops forming a multi-hop wireless ad hoc network.

The theoretical advantages that ad hoc networks bring are well-known. Among these advantages, we can highlight their easy configuration and their ability to establish a communication path -using the proper routing protocols- between any nodes of the network without using a central element. In many situations such as catastrophes or emergencies in which there are no infrastructures or they cannot be installed because of geographical or temporal reasons, this kind of networks may be the ideal solution. They are also an essential element in the future *4G* (*4th Generation*) networks.

However, the own definition of MANET based on wireless links and on the mobility of nodes, is the main reason of its biggest problems: reduction of the bandwidth available for transmission and a dynamic topology which causes continuous connection losses and a negative effect in TCP and UDP traffic throughput. Possible solutions to these problems would be: replacing the protocol 802.11b/g with a MAC protocol suitable for a multi-hop environment; modifying the routing protocols to consider the disconnections and multiple metric algorithms; or proposing *cross-layer* solutions. Moreover, another possible solution would consist on increasing the network coverage and reducing the mobility of part of the nodes by means of including a static backbone, like in the WMNs.

On the top of the protocols stack, at the application layer, some solutions have been proposed: from loss differentiation algorithms, which increase the functionality of applications to adapt themselves to the communication state, to the new video coding techniques, which allow video flows to adapt to network conditions, such as SVC o *Multi-description Coding* (*MDC*). Multi-description is a video encoding technique capable to enhance the error resilience of a video delivery system providing multiple video flows. This encoding technique is especially useful in multipoint to point transmissions due to the disjunction of packet routes. Among different solutions, multi-description appears as one of the most suitable solutions in multi-hop networks due to the possibility to use disjoint paths towards destination.

The rest of this chapter is organized as follows. Section 2 shows the quality degradation suffered by video transmissions due to the nature of wireless ad hoc networks and presents some of the most common parameters used to measure video quality (objectives and subjectives). Section 3 illustrates the performance improvements when a hierarchical structure is implemented in ad hoc networks following the philosophy of wireless mesh networks. In Section 4, an application layer solution to improve video quality using multi-description video coding is proposed. We present our further research directions in Section 5. Finally, conclusions are drawn in Section 6.

QUALITY DEGRADATION

In this section, we expect to show briefly the well-known results about the degradation that communications suffer in a multi-hop network and, particularly, the effect that this degradation causes on video transmissions (Figure 1). For this study, throughput and *Peak Signal to Noise Ratio* (*PSNR*) have been used as the measurement metrics.

The performance of the video transmission has been evaluated in terms of *Packet Delivery Ratio* or *Throughput*. This metric is defined as the percentage of video packets successfully delivered to the destination against total packets sent. Nevertheless, in terms of user QoE, every packet correctly received may not be used to decode the video flow due to previous errors in reception, so additional measurement such as PSNR is needed. When receiving video, a frame (depending on the codec used) can be decoded if all fragmented packets of this frame (and other packets that this frame depends on) are completely received and decoded.

The PSNR is one of the most common objective parameters used to evaluate video quality. PSNR is the ratio between the maximum possible power of a signal and the power of corrupting noise that affects the fidelity of its representation and it is defined via the *Mean Squared Error* (*MSE*) between an original and decoded frame as following:

Figure 1. Effect of degradation in mobile multi-hop networks

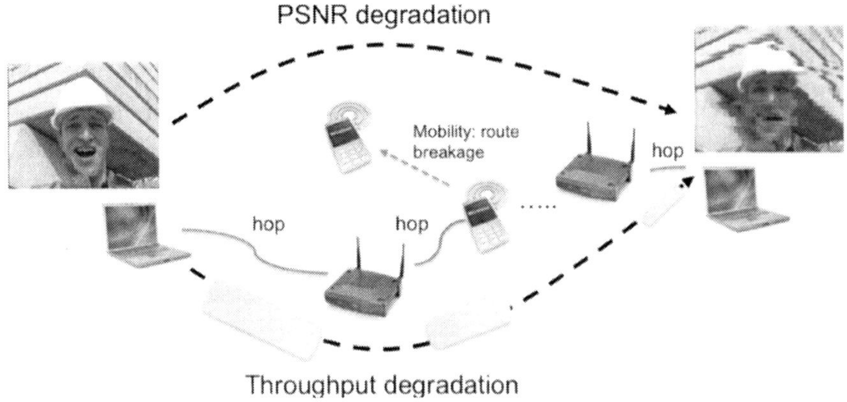

$$PSNR = 10\log_{10}\left(\frac{255^2}{MSE}\right)$$

PSNR has long been used as a comparative measurement by the image and video processing community. There are a number of reasons for the popularity of this metric. The formula for computing is as simple as to understand and implement and also it is easy and fast to compute. Despite its popularity, PSNR only has an approximate relationship with the video quality perceived by human observers, simply because it is based on a byte-by-byte comparison of the data without considering what they actually represent. Moreover, additional comparative parameters, more complicated and precise, have been designed to improve the consistency with human eye perception, such as the *Structural Similarity index* (*SSIM*) (Wang, Bovik, Sheikh & Simoncelli, 2004) or *Video Quality Metric* (*VQM*) (Xiao, 2000). Changes in brightness and contrast have high influence in PSNR but barely affect the SSIM index, which is the desired behavior. However, changes in hue or saturation (image color) have a slight effect on SSIM but have a high influence on human perception. VQM has a good correlation with subjective video quality assessment but the implementation is much more mathematical complex.

Therefore, another kind of subjective parameters can be measured instead. The subjective assessment consists in building a panel of human observers, which will evaluate sequences of video depending on their point of view and their perception. The output of the test is in terms of absolute scales or comparative scales. An example of an absolute qualitative scale is the *Mean Opinion Score* (*MOS*) scale (*Bad* (1), *Poor* (2), *Fair* (3), *Good* (4), *Excellent* (5)) which was initially standardized by the International Telecommunication Union (Table 1). On the other hand, in comparative scales, observers set mutual relations between sequences that eventually define objects quality hierarchy. Some standard methods for conducting subjective video quality evaluations exist such as the *International Telecommunication Union* (*ITU*)-R BT.500 (2002). Its variations are *Single Stimulus* (*SS*), *Double Stimulus Impairment Scale* (*DSIS*), *Double Stimulus Continuous Quality Scale* (*DSCQS*), *Single Stimulus Continuous Quality Evaluation* (*SSCQE*), *Simultaneous Double Stimulus for Continuous Evaluation* (*SDSCE*) and *Stimulus Comparison Adjectival Categorical Judgment* (*SCACJ*). All the variations are pretty much similar; differences are derived from the possibilities for constructing the subjective test methodology (Sat & Wah, 2009).

Table 1. Mean Opinion Score

MOS	Impairment
5 (Excellent)	Imperceptible
4 (Good)	Perceptible but not annoying
3 (Fair)	Slightly annoying
2 (Poor)	Annoying
1 (Bad)	Very annoying

However, subjective methods have some drawbacks to take into account, such as the impossibility to work in real time since they need both the original and the distorted video sequences received; and, moreover, they are time consuming and can only be interpreted under restricted conditions. Therefore, objective methods are more suitable when the aim of video quality measurements is to include them in adaptive rate control or video codec mechanisms. Consequently, both objective and subjective parameters should be considered whenever possible in order to perform a complete evaluation.

Apart from PSNR, throughput is another objective measurable parameter. With regard to the throughput, the medium access control algorithm used in *International Electrical and Electronic Engineers* (*IEEE*) 802.11 (i.e. *Carrier Sense Multiple Access with Collision Avoidance, CSMA/CA*) presents several mechanisms in order to avoid packet collisions, such as random back-off time or *Ready-to-Send/Clear-to-Send* (*RTS/CTS*) handshake mechanism. This kind of collision avoidance medium access entails some limitations in wireless ad hoc networks and, particularly, one of these limitations is the achievable transmission capacity. The maximum bandwidth is reduced due to the fact that nodes cannot simultaneously access the shared medium. More specifically, when a node is transmitting a packet, neighbor nodes within its *Interference Range* (*IR*) must not transmit. As a result an overall degradation of the data rate is produced.

Moreover, when a transmission is established, the nodes must cooperate to forward the packets through the network, which means that the available throughput on each host is limited not only by the access channel, but also by forwarding load. When the hop count is increased, the maximum throughput of one flow is decreased substantially and falls down because of the overhead of MAC layer and the mutual interference between packets of the same flow. This effect is called "*Intra-flow Contention*" (Sanzgiri, Chakeres & Belding-Royer, 2004) and represents the contention between packets of the same flow being forwarded at different hops along a multi-hop path, causing the actual bandwidth consumption. Figure 2 depicts this decrease of the overall throughput depending on the total amount of nodes that are taking part in the path to destination.

In addition, multiple-hop flows have to contend for the medium at each hop on its way to the destination node. Therefore, flows that extend over more hops will spend more time on competing with more nodes in the network. The packets forwarded on a longer path have more chance of being dropped than those of a shorter path. This end-to-end unfairness gets even worse when the

Figure 2. Throughput degradation in multi-hop networks

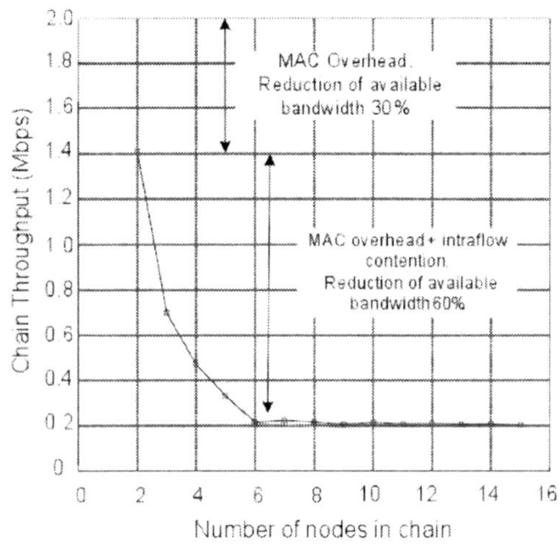

transport protocol cannot distinguish the reasons of packet drop such as network congestion, packet collision or broken route. Furthermore, the hidden node and the exposed node problems cause a local unfairness, which is partially solved with mechanisms such as RTS/CTS, but with a consequent reduction in the achievable throughput.

The nodes capacity to move freely can become a problem because of the link breakages induced and the necessity to discover new routes towards destination for data transmission. These facts cause packet losses and delay while new routes are being configured, with the consequent loss of quality at the receiver. Figure 3 shows the simulations results obtained for a video streamed over a static environment compared with a scenario where nodes are constantly moving. The video length is 110 seconds and it is sent over an IEEE 802.11b ad-hoc network where nodes are moving at a maximum speed of 15 mps. Figure 3 depicts the quality (PSNR) degradation suffered by a video transmission due to the packet losses caused by mobility and route breakage compared with a scenario formed by static nodes. Video packet losses cause a considerable reduction in PSNR, and consequently in video quality, during some transmission intervals. Therefore, maintaining an acceptable quality during the entire transmission can become a real challenge. This PSNR measurement includes the degradation due to packet losses and video encoding. Even when there are no losses, PSNR is not as high as expected because the video was encoded at medium quality because of the bandwidth limitations of the wireless channel. From the point of view of the subjective evaluation, it can be appreciated how the assessment using MOS regarding the quality experienced by the users is reduced from values MOS=4.7 to MOS=1.9 when mobility of nodes is increased. The MOS evaluation has been carried out by comparing the video received with the encoded video sent.

In order to minimize video quality degradation, several solutions can be introduced at the different layers of transmission stack. For instance, specific routing algorithms and video coding methods can be used as seen below.

HIERARCHICAL ROUTING IN AD HOC NETWORKS: TOWARDS WMN

Ad hoc networks have the ability of self organization and self configuration but the topology of

Figure 3. PNSR and MOS degradation in multi-hop mobile networks

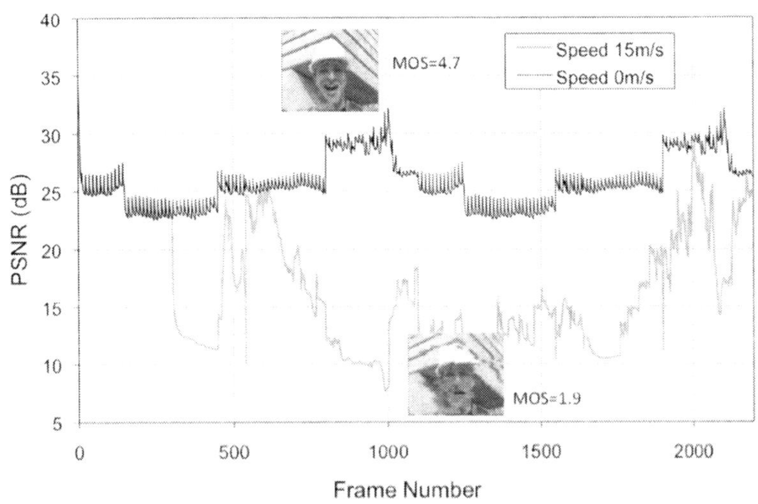

these networks is usually highly dynamic. An ad hoc wireless network can be set up at very low cost compared with those traditional *Wireless Local Area Networks* (*WLAN*) that relay on wired networks to carry out bridging functions and reach similar coverage. WMNs could become a trade-off between cost effort and the transmission quality offered. Figure 4 compares qualitatively the relationship between wireless ad hoc networks, emerging mesh networks and traditional wired access point networks regarding the cost to deploy and the achievable QoS.

Both in ad hoc networks and in the way towards WMNs, it is worth noting that routing protocols play an important role. Routing protocols are in charge of discovering and maintaining routes for packets from the source node to the destination. Traditionally, routing protocols have been classified as reactive, proactive or hybrid, depending on how their routing table is updated.

In reactive protocols, routes are discovered or updated on demand, i.e. route discovering process is only carried out when nodes start a new communication or a broken link is somehow detected, so there are no extra control packets for maintenance. On the other hand, this kind of routing protocol can cause high latency time when establishing the new route. Some reactive protocols for ad hoc networks are *Dynamic Source Routing* (*DSR*) (Johnson, Maltz, & Hu, 2007) and *Ad-hoc On-demand Distance Vector* (*AODV*) (Perkins & Royer, 2003).

Figure 4. Relationship between wireless networks

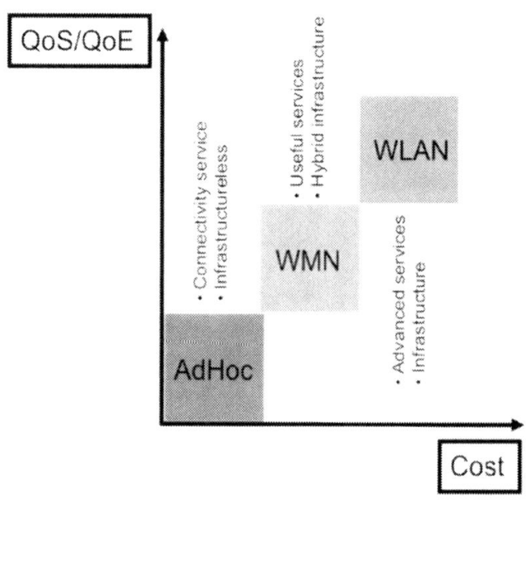

In proactive approaches, each node steadily keeps and updates a route to every node in the ad hoc network. Thus, routes are always available when they are needed. As a consequence, there is a constant overhead due to routing traffic, but there is no initial delay in data communications. This constant overhead could become a disadvantage in large ad hoc networks or with high mobility nodes. Some proactive routing protocols are *Optimized Link State Routing* (*OLSR*) (Clausen & Jacquet, 2003) and *Destination-Sequenced Distance Vector* (*DSDV*) (Perkins& Bhagwat, 1994).

Hybrid approaches try to combine the advantages of reactive and proactive protocols. The *Zone Routing Protocol* (*ZRP*) (Haas & Pearlman, 2001) is an example of this.

A lot of works has been done on the routing protocols in ad hoc networks, taking into account different scenarios and traffic conditions (Boukerche, 2004; Haerri, Filali & Bonnet, 2006; Kao, Ke, & Shieh, 2006). Moreover, other routing algorithms based on QoS parameters (such as packet delay or bandwidth estimation) have been proposed in order to improve the delay and the throughput of wireless transmissions (Zhang & Mouftah, 2005; Munaretto & Fonseca, 2007; Badis & Agha, 2005).

Most of the routing protocols proposed consider ad hoc networks as homogeneous; this means that all nodes have the same capabilities. This kind of protocols is known as flat protocols. Ad hoc networks are usually networks with a flat topology. When a route is broken for any reason, there is a gap in time until the new route is discovered, which causes packet losses and communication delays. In such a flat topology, when the network size increases, this route recovery time could increase considerably, which definitely deteriorates the user quality of experience in reception. This scalability issue is also noticeable because routing control traffic overhead is increased as well.

However, many ad hoc networks may be considered heterogeneous because they contain mobile nodes with different capabilities (bandwidth, transmission range, etc.). An important factor to bear in mind when designing routing protocols is scalability, which is the capacity of a network to maintain its performance when the number of nodes increases. To maintain scalability in these heterogeneous ad hoc networks, hierarchical routing protocols (Figure 5) should be considered as a good option (Villasenor-Gonzalez, Ge & Lamont, 2005; Zhao et al, 2003; Pei, Gerla & Chen, 2000).

A hierarchical routing protocol maintains a hierarchical network structure, which offers scalability and reduces the complexity in routing computation. Therefore, the network is organized as a hierarchy, consisting of certain number of clusters and a cluster leader is elected within each cluster. This hierarchical network structure reduces the number of participating nodes and the communication overhead for routing discovery and maintenance.

Due to mobility and the multi-hop nature of ad hoc networks, any node can move out of the coverage area during a short time, losing the packets routed to destination and even causing route breakages to other communications. Not relying only on ad hoc nodes but deploying new wireless nodes in order to increment the coverage area could be a trade-off to improve the overall performance of the network. These new nodes will work like the other devices in the network, taking part in the routing table creation and traffic forwarding. Somehow, these nodes help mobile nodes to ensure a traffic route to destination. This technique for maintaining some ad hoc nodes as if they work as an infrastructure is used in emerging WMN.

The aim of WMN is to guarantee connectivity. WMNs are multi-hop ad hoc networks consisting of a mesh backbone and mesh client nodes, and build a multi-hop wireless backbone to interconnect isolated local area networks and to extend backhaul access to users not within range of typical access points (Akyildiz, Wang & Wang, 2005).

WMN backbone routers use multi-hop communication similarly to ad hoc networks. On the other hand, mobile users connect to the backbone via these mesh routers that play the role of access points. The backbone routers typically are stationary. Therefore, the topology is relatively stable, which permits routing metrics to model link quality instead of simply using the number of hops.

The stationary nature of the backbone nodes should not be seen as a drawback but a kind of

Figure 5. Evolution from flat to hierarchical routing

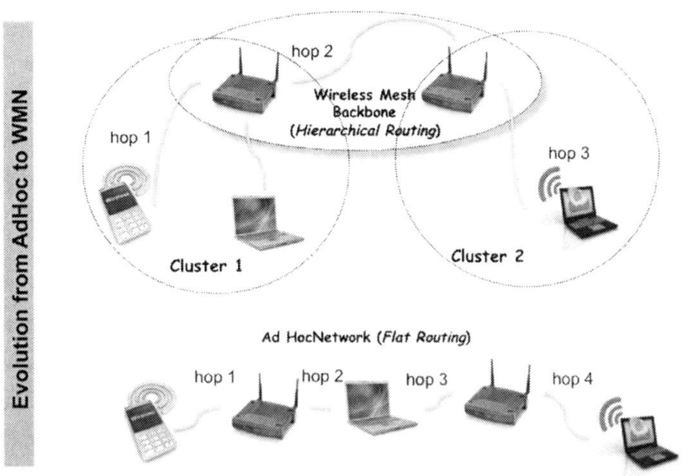

infrastructure support for MANETs. It contributes to extend the coverage area and reduces packet losses and route breakages. Moreover, these static nodes could be power supplied so energy problems of the network would be reduced. Multi-radio and multi-channel capability can be used in order to increase throughput and reduce interferences.

In addition, and taking into account the coverage that the static nodes in mesh networks provide, it could be possible to group the nearest nodes to the mesh routers in clusters. A backbone node serves as the cluster-head and advertises its reachability to other cluster periodically.

These mesh router nodes could have several wireless interfaces. On one interface, they can communicate intra-cluster, i.e. communication with the nodes inside the cluster. The other interface is used for inter-cluster communication, i.e. communication with the other nodes that form the static mesh network. Nodes of a cluster that want to communicate with nodes from other cluster use the cluster-head as an outgoing gateway towards the network. Cluster-heads, which form a higher-level network in turn, route the packets towards the cluster which destination node belongs to. Control packets are kept inside the cluster boundary and the cluster-head announces to the other cluster-heads which nodes are contained in its cluster. Address aggregation can be used and thereby the size and number of control packets would be reduced.

In a multi-channel network, all cluster-head nodes are equipped with multiple radio interfaces to operate on different orthogonal channels. The positioning of these cluster-heads could be one of the critical factors that determine the network performance, because a random or bad distribution could leave areas with poor coverage or saturated or even underused.

Therefore, hierarchical protocols could provide an important support to create more robust wireless networks and improve the quality of data or video transmissions. In order to quantitatively evaluate this assertion, a comparison study has been carried out. From the aforementioned hierarchical routing protocols, *Hierarchical Optimized Link State Routing* (*HOLSR*) proposal is very suitable to be implemented in a plain network because it is an extension of the current OLSR protocol as described by Villasenor-Gonzalez et al. (2005). This approach of HOLSR dynamically organizes nodes into cluster levels by means of new cluster announcement messages and transmits the membership information of a cluster with new hierarchical topology messages. Furthermore, there are even other implementations of HOLSR, which maintain the protocol performance but simplify its implementation and have backward compatibility (Arce, Guerri, Pajares & Lazaro, 2008). They use the existing OLSR messages for both inter-cluster and intra-cluster topology dissemination. As a drawback, cluster-heads cannot be dynamically selected and must be predefined.

A flat scenario using OLSR has been compared with a hierarchical structure running HOLSR routing protocol on NS-2. A 120-seconds *Quarter Common Intermediate Format* (*QCIF*) video has been sent through a network made up of 50 nodes that move according to the Random Waypoint Model at different maximum speeds (0 – 20 mps) within an area of 1200 x 600 m. In the hierarchical scenario, two clusters of 600 x 600 m were set with 25 nodes each one, connected through cluster-heads, which were situated at the center of each cluster and had a higher transmission power in order to gain range of coverage. Background traffic consists of 20 UDP connections with constant bitrate (1 KBps) between arbitrary nodes. The radio model used is based on the Two-Ray Ground Propagation Model and the standard IEEE 802.11b. The channel capacity was set to 11Mbps. Node transmission and carrier-sensing ranges are approximately 170 and 423 m respectively. Results obtained for PSNR, frame losses and packet delay are illustrated in the Figure 6, Figure 7, and Figure 8 respectively.

Results obtained from the simulations show that a hierarchical environment presents better

Figure 6. Average PSNR vs. node speed using hierarchical routing

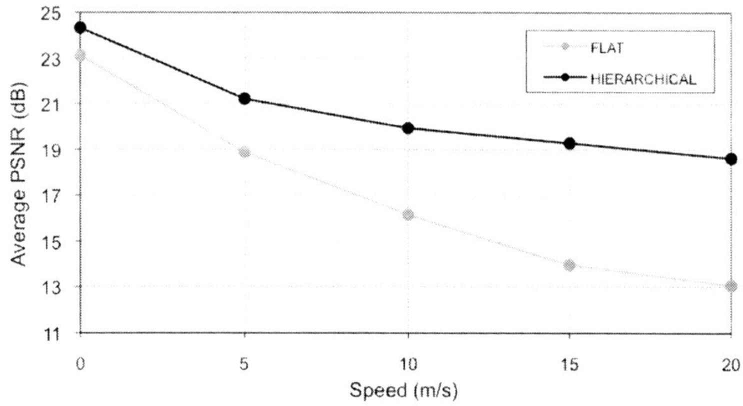

Figure 7. Frame loss vs. node speed using hierarchical routing

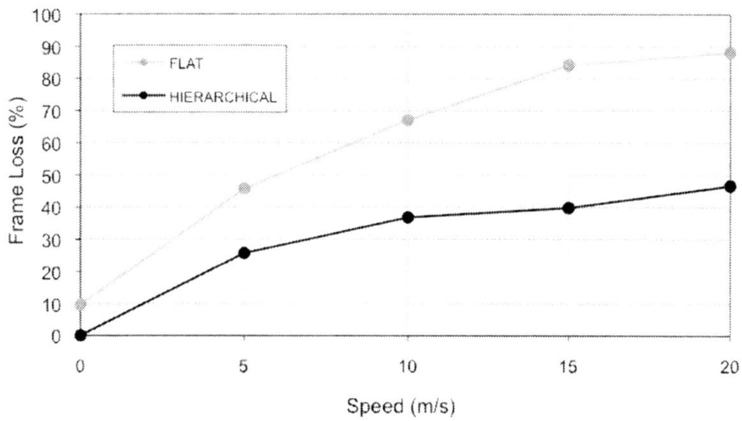

Figure 8. Packet Delay vs. node speed using hierarchical routing

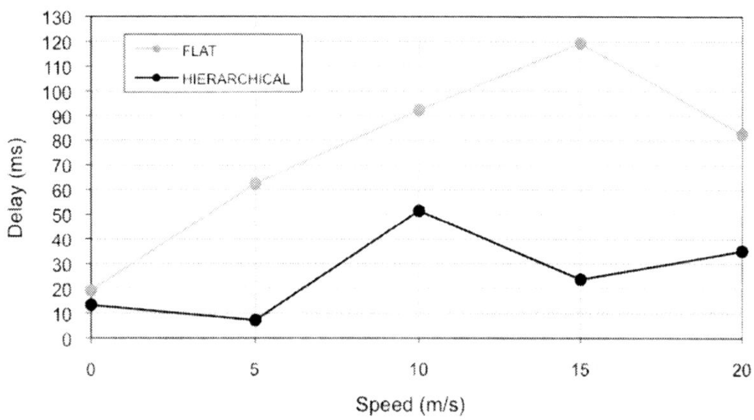

performance regarding the measured parameters. Average PSNR is improved in almost 6 dB for high speeds and frame loss and packet delay are considerably reduced (40% and 50 ms respectively). Better results are achieved with higher mobility of nodes. This is mainly due to the fact that when routes fall down, they are recovered faster in the hierarchical structure. Sometimes when a link breakes down inside a cluster, there is likely a section of the route that is still working between cluster-heads because these cluster-heads remain static. Therefore, only a part of the route has to be recalculated. Differences in performance between flat and hierarchical environments are emphasized as node speed grows.

As it has been shown, hierarchical or clustered routing introduces some advantages (Yu & Chong, 2005) but also presents some drawbacks that need to be solved, such as node movement between clusters or the overload that can be produced in the cluster-heads when forwarding inter-cluster packets. There are some proposals in the literature of suitable QoS algorithms to be implemented in a hierarchical scenario based on load balancing solutions and using multiple cluster-heads (Arce et al, 2008). They proposed new mechanism to include additional routes to other clusters when the current cluster-heads detect packet losses. When traffic load exceeds a certain threshold, nodes with cluster-head capabilities offer themselves as cluster-heads to reach a neighbor cluster. Access control could be performed if threshold is exceeded and no more cluster-heads are available.

Moreover, these hierarchical routing protocols can be used in combination with other quality improvement mechanisms such as video coding enhancement.

MULTI-DESCRIPTION EFFECTS IN MULTI-HOP NETWORKS

As a fourth point and regarding application layer solutions, the effects of MDC will be analyzed. Multimedia applications, especially real time applications, are very sensitive to packet delay and packet loss produced in the network. Therefore, it is worth introducing mechanisms that improve video transmission and provide better QoE.

The MDC has emerged as a promising mechanism to improve video error resilience and quality (Wang, Reibman & Lin, 2005). MDC is a video coding technique, which encodes several bitstreams from the same media source. Each bitstream is called description and every description can be independently decoded later, providing a useful visualization of the original stream. Moreover, the descriptions contain complementary information in order to improve the quality of the decoded video when the number of received descriptions increases. Different levels of reconstruction qualities can be obtained from different subsets of these descriptions. In contrast to *Layered Coding* (*LC*), there is no hierarchy among the descriptions so that each description may be independently decoded. The fact that each description is independent and equally important makes the use of MDC suitable on lossy systems. Moreover, an acceptable quality of video can be achieved in MDC without the need of retransmissions so it is appropriate for real-time applications, like video streaming.

Descriptions can be generated using several procedures. They are usually created by means of the division of the source video stream in different groups or sub-streams, in order to obtain flows independently encoded later. Segmentation of video sequence can be carried out in temporal, spatial or frequency domain. The most common technique is the temporal division. In this technique, descriptions are achieved by splitting the original video frames in subsets depending on their time position. Therefore, the raw video is separated into n descriptions with a frame rate n times lower than the original frame rate. Later, every sub-flow is independently encoded resulting in the video descriptions. The encoding process is shown in Figure 9.a.

On the receiver, each description is independently decoded and the reconstructed video flows are merged in order to generate the video with the same temporal resolution as the original sequence (Figure 9.b). At this point, the error compensation scheme replaces the lost or not decoding frames by the last successfully decoded frame, regardless the description it belongs to. Therefore, lost frames from a specific description are replaced with frames from the others descriptions so the perceived distortion is diminished. If losses occur in every description at the same time, no video reconstruction is obviously possible. Thus, the descriptions must be sent through disjoint paths.

The required bandwidth for multi-description transmissions is greater than the one required for *Single Description Coding* (*SDC*) video. This is due to the fact that the sub-streams are generated from non-consecutive frames causing lower compression efficiency. The more dynamic the video sequence, the bigger the mean frame size. It also increases when the amount of descriptions grows. The resulting average frame size limits the amount of descriptions that can be used. This effect has been studied by Chow and Ishii (Chow & Ishii, 2007) and results obtained suggest using 2 or 3 descriptions depending on the available bandwidth and the desired QoS. However, the use of 3 or more descriptions implies an increase of the system complexity while improvements in quality of service are not much significant.

The simulations were carried out under NS-2 with the Evalvid tool (Ke, Lin, Shieh & Hwang, 2006). The simulation environment consists of 50 wireless nodes random scattered in an area of 1200 x 600 m and OLSR is used. Two descriptions are transmitted from different nodes towards the same receptor performing a multipoint-to-point transmission. The radio model used for simulation is based on the Two-Ray Ground Propagation Model and the IEEE 802.11b standard at 11 Mbps. Node transmission range is 250 m approximately. Nodes move according to the Random Waypoint Model with maximum speeds of 0 mps (static), 2.5 mps, 5 mps, 7.5 mps, 10 mps, 12.5 mps and 15 mps. The video resolution used is 352x288 (*Common Intermediate Format* (*CIF*)) and two descriptions are generated based on it. The original video and the descriptions are encoded in MPEG-4 with the *GoP* (*Group of Pictures*) pattern IPPPPPPPPPPP.

Figure 9. Temporal Domain MDC Encoder (a) and Decoder (b)

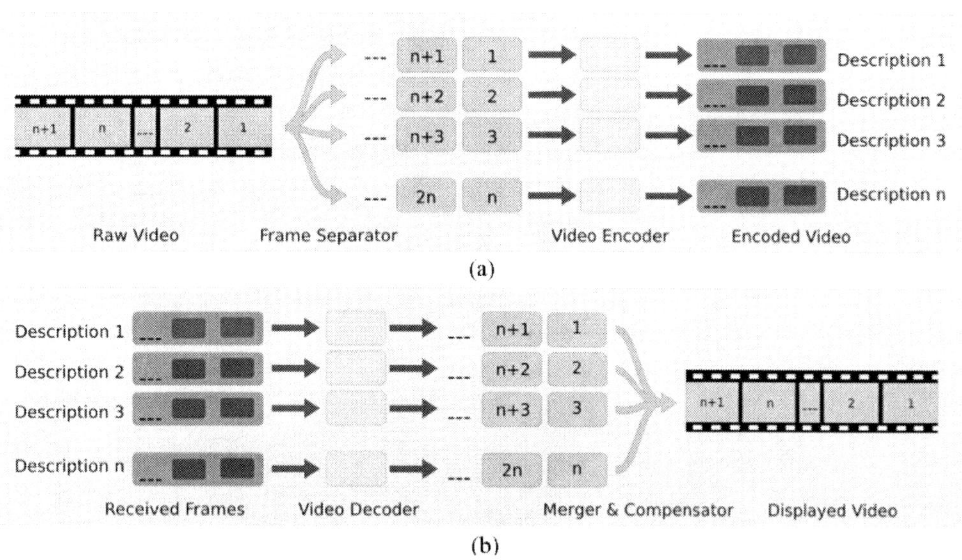

Original video source is used as SDC. Background traffic consists of 5 UDP connections with constant bitrate (1 KBps) between arbitrary nodes.

The results comparing the video evaluation using MDC and SDC (single flow coding) are shown below. Figure 10 shows how the average PSNR diminishes when node speed increases, as can be expected. For low speeds, SDC presents a better performance regarding to PSNR (approximately 0.5 dB). This is due to the fact that the temporal correlation between consecutive frames is higher in SDC. However, for higher speeds, MDC achieves better results (up to 3 dB) since multi-description technique increases the probability of receiving at least one of the descriptions.

Figure 11 illustrates the packet delivery ratio (throughput) and the decoding frame rate (Q) measured in the simulations. Results show that the percentage of correctly received frames decreases with higher node speeds. For medium and high speed, SDC suffers a greater amount of packet losses than MDC, according to the previous PSNR results. As it can be observed in Figure 11, the decoding frame rate Q is always below the throughput value. This occurs because not all the packets received are used when reconstructing the video, due to the hierarchical nature of MPEG-4 encoding. Q provides a measure of quality closer to the one perceived by the user because it takes into account only the packets that are actually used when playing the video.

Finally, as the aim of this study is to measure the degree of improvement in the quality of the resulting video using the MDC technique with respect to traditional coding system (SDC), several subjective quality tests have been realized using the SDSCE method. SDSCE is a method in which an observer evaluates a pair of short video sequences 1 and 2, simultaneously. The observers are requested to check the differences between the two sequences and to judge the fidelity of the video information of the impaired sequence with respect to the reference by marking on a continuous scale according to the Table 2. The observers are aware of which sequence is the reference and which sequence is the test condition. This method will be used to evaluate the improvements in QoE provided by MDC in the video transmission. Figure 12 shows the results of the test. In these tests, encoded videos using SDC are used as a reference (seq.1 in Table 2) and it is compared with videos encoded in MDC (seq.2 in Table 2). It can be observed that the obtained values are always positive, which means that for all speeds, users have experienced a better video quality using MDC with respect to SDC.

Nevertheless, there is still a long road ahead before high-quality video streaming could be achieved in wireless networks.

Figure 10. Average PSNR vs. node speed using Multi-description Coding

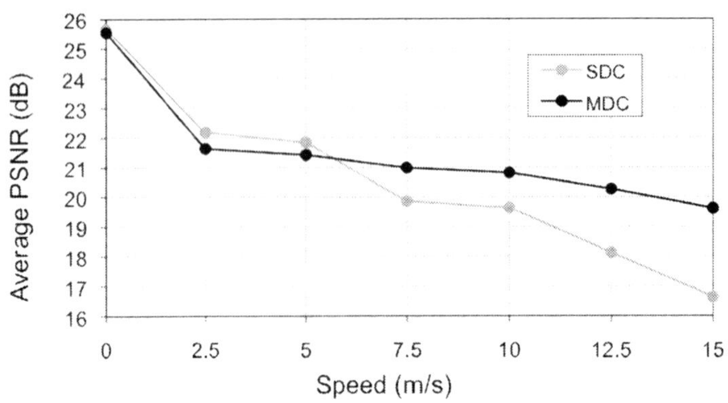

Figure 11. Throughput and decodable frame rate vs. node speed using Multidescription Coding.

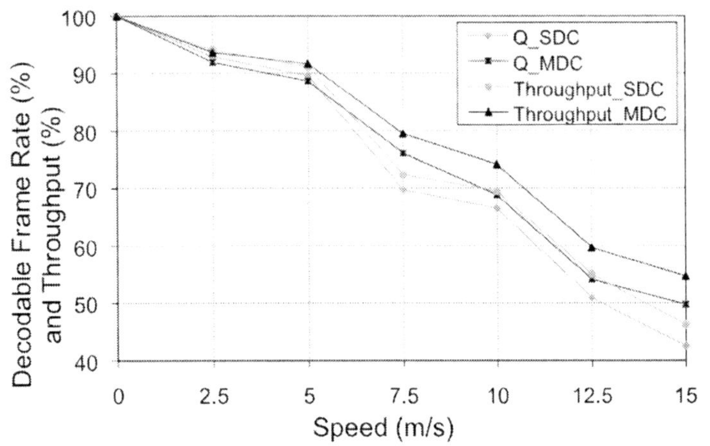

Figure 12. Subjective evaluation

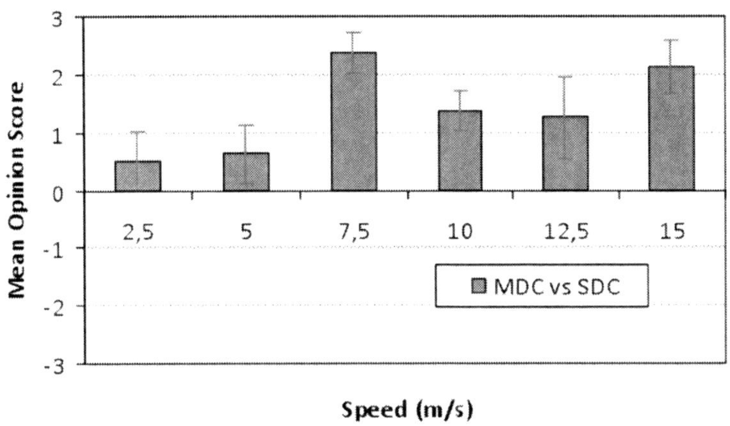

Table 2. SDSCE test

Value	Compare
3	seq.2 much better than seq.1
2	seq.2 better than seq.1
1	seq.2 slightly better than seq.1
0	seq.2 the same as seq.1
-1	seq.2 slightly worse than seq.1
-2	seq.2 worse than seq.1
-3	seq.2 much worse than seq.1

FUTURE RESEARCH DIRECTIONS

A realistic view about the future of MANETs must consider the integration in hierarchical networks, such as WMNs, taking advantage of the static nature of backbone nodes and the existence of multiple radio interfaces. These properties of a hierarchical structure will enable greater throughput and better QoE.

Nevertheless, there are still several open issues in order to provide successful multimedia services over multi-hop wireless networks. Among the hottest research topics, we are considering:

- *Load balancing* in order to prevent congestion situations.
- *Scalability* to accordingly deploy real world testbeds and carry out measurements with thousands of nodes.
- *Loss differentiation* between wired and wireless networks and between congestion situations and link breakages.
- *Loss protection* and error resilience.
- *Performance evaluation* of transport protocols over this kind of networks.

From our point of view, two ways to deal with these issues are feasible. On the one hand, it may be possible to develop new standards (adapted to multi-hop networks) to replace current MAC and routing protocols (e.g. IEEE 802.11s). IEEE 802.11s is an amendment of IEEE 802.11 standard, which introduces new MAC frame formats and a new optional medium access method in order to support multi-hop forwarding at the MAC layer, as well as other optimizations to improve performance and security of wireless mesh networks.

On the other hand, there is still the possibility to add new mechanisms to the available and widely spread standards (e.g. IEEE 802.11b/g, OLSR, UDP). This should be carried out by means of a *cross-layer* design (Liu, Gkelias, Hou & Leung, 2009) to make the most of the new coding systems and *Forward Error Correction* (*FEC*) schemes (Alay, Korakis, Wang & Panwar, 2010; Tsai, Chilamkurti, Zeadally & Vinel, 2010), as well as new routing proposals such as opportunistic routing (Biswas & Morris, 2005; Yan, Zhang, Zheng & Ma, 2010).

CONCLUSION

In wireless mobile ad hoc networks, because of the dynamic nature of these kinds of networks, it is a hard challenge to guarantee certain quality in real-time services such as video streaming. In this chapter, we have briefly looked through the main reasons video transmissions could suffer quality degradation in MANETs, such as the overhead of MAC layer, the intra-flow contention of packets of the same flow and, especially, the mobility of the nodes. Results have shown the impact of these effects in throughput, PSNR and the perceived video quality (MOS).

In order to improve this quality of user experience, feasible solutions have been proposed at different layers of the protocols stack. Following the structure of emerging wireless mesh networks, which have a hierarchical and clustered topology and some static backbone infrastructure, a hierarchical routing protocol has been evaluated and compared with a flat topology (HOLSR vs. OLSR). Regarding video measurements, a hierarchical environment presents lower delay, lower frame loss rate and better PSNR, which helps to achieve a better quality video reception.

In addition, multi-description coding technique has been used to improve the error resilience of video transmissions in wireless ad hoc networks. This application layer enhancement on video coding improves the perceived video quality and increments throughput and PSNR if it is used together with multipoint-to-point or disjoint path transmission.

The operation and characteristics of MANET (multi-hop path and mobile nodes) convert them in a good option for low scale networks. For real scenarios, however, integration in hierarchical networks and advanced coding techniques are needed. WMN and multi-description have been showed as valid solutions to improve services such as video transmission.

REFERENCES

Akyildiz, I. F., Wang, X., & Wang, W. (2005). Wireless mesh networks: A survey. *Computer Networks Journal (Elsevier)*, *47*(4), 445–487.

Alay, Ö., Korakis, T., Wang, Y., & Panwar, S. (2010). Dynamic rate and FEC adaptation for video multicast in multi-rate wireless networks. *Mobile Networks and Applications*, *15*(3), 425–434.

Arce, P., Guerri, J. C., Pajares, A., & Lazaro, O. (2008). Performance evaluation of video streaming over ad-hoc networks using flat and hierarchical routing protocols. *Mobile Networks and Applications*, *13*(3-4), 324–336.

Badis, H., & Agha, K. A. (2005). QOLSR, QoS routing for ad-hoc wireless networks using OLSR. *Wiley European Transactions on Telecommunications*, *15*(4), 427–442.

Biswas, S., & Morris, R. (2005). ExOR: Opportunistic multi-hop routing for wireless networks. *SIGCOMM*, *2005*, 133–144.

Bolot, J. C., & Turletti, T. (1994). A rate control mechanism for a packet video in the Internet. In *Proceedings IEEE INFOCOM'94*, Toronto, Canada (pp. 1216-1223).

Boukerche, A. (2004). Performance evaluation of routing protocols for ad-hoc wireless networks. *Mobile Networks and Applications*, *9*(4), 333–342.

Busse, I., Deffner, B., & Schulzrinne, H. (1996). Dynamic QoS control of multimedia applications based on RTP. *Computer Communications*, *19*(1), 49–58.

Chow, C.-O., & Ishii, H. (2007). Enhancing real-time video streaming over mobile ad-hoc networks using multipoint-to-point communication. *Elsevier Computer Communications*, *30*, 1754–1764.

Clausen, T. & Jacquet, P. (2003). *Optimized link state routing protocol (OLSR)*. Request for Comments 3626.

Haas, Z. J., & Pearlman, M. R. (2001). *ZRP: A hybrid framework for routing in ad-hoc networks, ad-hoc networking*. Boston, MA: Addison-Wesley Longman Publishing Co., Inc.

Haerri, J., Filali, F., & Bonnet, C. (2006). Performance comparison of AODV and OLSR in VANETs urban environments under realistic mobility patterns. In *Proceedings 5th IFIP Mediterranean Ad-Hoc Networking Workshop*, Lipari, Italy.

ITU-R Recommendation BT.500-11. (2002). Methodology for the subjective assessment of the quality of television pictures.

Johnson, D. B., Maltz, D. A., & Hu, Y.-C. (2007). The dynamic source routing protocol (DSR) for mobile ad-hoc networks. Request for Comments 4728.

Kao, K. L., Ke, C. H., & Shieh, C. K. (2006). Video transmission performance evaluation of ad-hoc routing protocols. In *Proceedings of the International Conference on Intelligent Information Hiding and Multimedia Signal Processing (IIH-MSP'06)*, Pasadena, California, USA (pp. 181-184).Washington, DC., USA: IEEE Computer Society.

Ke, C. H., Lin, C. H., Shieh, C. K., & Hwang, W. S. (2006). A novel realistic simulation tool for video transmission over wireless network. In *Proceedings of the IEEE International Conference on Sensor Networks, Ubiquitous, and Trsutworthy Computing (SUTC2006)*.

Little, T. D. C., & Venkatesh, D. (1994). Prospects for interactive video-on-demand. *IEEE MultiMedia*, *1*(3), 14–24.

Liu, C. H., Gkelias, A., Hou, Y., & Leung, K. K. (2009). Cross-layer design for QoS in wireless mesh networks. *Springer Wireless Personal Communications, Special Issue on ". Cross-Layer Design for Future Generation Networks*, *51*(3), 593–613.

Munaretto, A., & Fonseca, M. (2007). Routing and quality of service support for mobile ad hoc networks. *Computer Networks*, *51*(11), 3142–3156.

Pei, G., Gerla, M., & Chen, T. W. (2000). Fisheye state routing in mobile ad-hoc networks. *Proceedings of IEEE ICDCS Workshop.*

Perkins, C. E., & Bhagwat, P. (1994). Highly dynamic destination sequenced distance vector routing (DSDV) for mobile computers. In *Proceedings of ACM SIGCOMM'94*, London, UK (pp. 234–244).

Perkins, C. E. & Royer, E. M. (2003). *Ad-hoc on-demand distance vector (AODV) routing.* Request for Comments 3561.

Sanzgiri, K., Chakeres, I., & Belding-Royer, E. (2004). Determining intra-flow contention along multi-hop paths in wireless networks. In *Proceedings Broadnets 2004 Wireless Network Symposium* (pp. 611–620).

Sat, B., & Wah, B. W. (2009, October). Statistical scheduling of offline comparative, subjective evaluations for real-time multimedia. *IEEE Transactions on Multimedia, 11*(6), 1114–1130.

Steinmetz, R. (1990). Synchronization properties in multimedia systems. *IEEE Journal on Selected Areas in Communications, 8*(3), 401–411.

Tsai, M.-F., Chilamkurti, N. K., Zeadally, S. & Vinel, A. (2010). Concurrent multipath transmission combining forward error correction and path interleaving for video streaming. *Computer Communications, Article in Press*, Available online, February 10, 2010.

Villasenor-Gonzalez, L., Ge, Y., & Lamont, L. (2005). HOLSR: A hierarchical proactive routing mechanism for mobile ad-hoc networks. *IEEE Communications Magazine, 43*(7), 118–125.

Wang, Y., Reibman, A. R., & Lin, S. (2005). Multiple description coding for video delivery. *Proceedings of the IEEE, 93*(1).

Wang, Z., Bovik, A. C., Sheikh, H. R., & Simoncelli, E. P. (2004). Image quality assessment: From error visibility to structural similarity. *IEEE Transactions on Image Processing, 13*(4).

Xiao, F. (2000). *DCT-based video quality evaluation. MSU graphics and media lab.* Video Group.

Yan, Y., Zhang, B., Zheng, J., & Ma, J. (2010). CORE: A coding-aware opportunistic routing mechanism for wireless mesh networks. *IEEE Wireless Communications, 17*(3), 96–103.

Yu, Y. J. & Chong, P. H. J. (2005). A survey of clustering schemes for mobile ad-hoc networks. *IEEE Communications Surveys & Tutorials (1st Quarter 2005), 7*(1), 32-48.

Zhang, B., & Mouftah, H. T. (2005, October). QoS routing for wireless ad-hoc networks: Problems, algorithms, and protocols. *IEEE Communications Magazine*, 110–117.

Zhao, S., Tepe, K., Seskar, I., & Raychaudhuri, D. (2003). Routing protocols for self-organizing hierarchical ad hoc wireless networks. *IEEE Sarnoff Symposium*, Trenton, NJ.

ADDITIONAL READING

Akyildiz, I. F., & Wang, X. (2009). *Wireless mesh networks.* Wiley.

Basagni, S. (2004). *Mobile ad-hoc networking.* IEEE Press and Wiley.

Cheng, M. X., & Li, D. (2008). *Advances in wireless ad-hoc and sensor networks.* Springer Series on Signals and Communication Technology.

Hanzo, L. (2007). *Video Compression and Communications.* IEEE Press and Wiley.

Hiertz, G. R., Denteneer, D., Rakesh, S. M., Cardona, J., Berlemann, L., & Walke, B. (2010). *IEEE 802.11s: The WLAN Mesh Standard* (pp. 104–111). IEEE Wireless Communications.

Ilyas, M. (2003). *The handbook of ad-hoc wireless networks*. CRC Press.

Misra, S., Misra, S. C., & Woungang, I. (2009). *Guide to wireless mesh networks*. Springer.

Paul, S. (2010). *Digital video distribution in broadband, television, mobile and converged networks: Trends, challenges and solutions*. Wiley.

Steinmetz, R., & Nahrstedt, K. (1995). *Multimedia: Computing, communications and applications*. Upper Saddle River, NJ: Prentice Hall.

Winkler, S. (2005). *Digital video quality: Vision models and metrics*. Wiley.

Chapter 13
Resilient P2P Video Streaming over an Urban VANET

Nadia N. Qadri
University of Essex, UK

Martin Fleury
University of Essex, UK

Mohammed Ghanbari
University of Essex, UK

ABSTRACT

This Chapter proposes the concept of resilient multi-source streaming using a peer-to-peer overlay network over an urban vehicular ad-hoc network. Peer-to-peer overlay networks with multiple sources have proven to be robust, distributed solutions to multimedia transport, including streaming. To achieve video streaming over a vehicular ad-hoc network overlay, the Chapter introduces spatial partitioning of a video stream based on flexible macroblock ordering. Flexible macroblock ordering is an error resilient feature of an H.264 video codec in which macroblocks can be assigned to slices in non-raster scan order. Furthermore, the Chapter examines the impact of differing traffic densities and road layouts upon a peer-to-peer overlay network's performance when streaming video. The research demonstrates that the vehicles' mobility pattern and their drivers' behaviors need to be carefully modeled to determine signal reception. The Chapter also considers the impact of the wireless channel, which also should be more realistically modeled, compared to the usual two-ray path loss model.

INTRODUCTION

Video streaming within an automotive *Vehicular Ad-hoc NETwork* (*VANET*) is a challenging prospect because of the lack of network infrastructure, the motions of the vehicles, and the wireless channel conditions leading to unreliable wireless links. A VANET will typically be formed by a group of people moving around on vehicles. Direct inter-vehicle communication (Yousefi, 2006) can then be an aid both to the passengers' comfort and to road safety. Compared to a cellular network, a VANET may be toll free, avoids the delay in setting up a long communication circuit, and will

DOI: 10.4018/978-1-61350-144-3.ch013

operate on a highway when there are coverage gaps in a cellular network. A scenario explored in this Chapter is a team of early responding emergency vehicles converging upon an emergency scene within a city. Real-time video communication allows early responders approaching an incident to better understand the nature of the problem at the scene of an emergency (Park et al., 2006) but the lack of centralized routing and network resource management must be overcome. Transmission of crash scenes, views of fleeing vehicles or burning buildings are some illustrative applications of video streaming, while there is also now a strategic incentive to provide coverage during a more serious, general emergency such as a terrorist attack.

Video communication in a VANET will be of considerable benefit in an urban emergency and may be preferable to communication over a cellular network. A novel aspect of the scheme developed by the authors in this Chapter is that the closed group formed by the emergency vehicles must take advantage of other vehicles to relay the video streams. Otherwise the numbers of vehicles and their dispersion throughout the urban area will make forming routing paths difficult to accomplish. Therefore, this Chapter highlights the concept of an embedded emergency vehicle-streaming group to aid video transfer within a more general VANET operating within a city.

To achieve video streaming over a VANET overlay, resilient multi-source streaming using a *Peer-to-Peer* (*P2P*) overlay network (Buford et al., 2009) over a VANET can be considered. P2P overlay networks with multiple sources (Jurca et al., 2007) have proven to be robust, distributed solutions to multimedia transport, including streaming as well as simple file download. A multi-source P2P streaming system as an overlay network can be organized in such a way that a number of destination vehicles receive the same streaming data from multiple peers within the overlay. The presence of multiple sources means that if one source fails during the streaming process another can step in. This is because, in a VANET, a source could easily fail if a vehicle is parked or moves out of the area. Multiple sources are also required if the video stream is split into several streams in order to exploit path diversity within the VANET using *Multiple Description Coding* (*MDC*) (Chakareski et al., 2005). In MDC, any single description or partial video stream is sufficient for a watchable version of the video but the quality is enhanced when one or more other descriptions are delivered. In contrast, in layered video coding, the base layer must be received because otherwise enhancement layers cannot be utilised.

Another important issue is to what extent video error resilience measures (Stockhammer & Zia, 2007) can protect video streaming across these networks. The H.264/*AVC* (*Advanced Video Coding*) codec (Wenger, 2003) has introduced a number of such measures, some of which are new to this codec, an example being *Flexible Macroblock Ordering* (*FMO*) (Lambert et al., 2006). The Chapter analyses the response of the error resilience measures to isolated and burst error patterns. In fact, MDC based on the spatial decomposition of FMO can counter the effect of packet loss on video streams. By employing checkerboard or dispersed FMO, each frame can be separated into two or more slices. These slices can aid the reconstruction of adjacent macroblocks (a macroblock is the unit of block-based motion compensation, whereby only the difference between a block in the current frame and a matching area in the reference frame is encoded, as the energy of the residual frame is reduced compared to the current frame) in another slice in the event of the packet bearing that slice being lost. Checkerboard FMO is the only one of the H.264/AVC FMO types that has this property. An error resilience strategy adopts FMO for high packet loss regimes but includes distributed intra-refresh to reduce the sudden data bursts that would otherwise arise from transmitting periodic intra-coded I-frames.

This Chapter also compares different VANET mobility models (Härri, 2010) and their impact on different traffic densities upon a P2P overlay network's performance. The work models the emerging IEEE 802.11p standard (Jiang & Delgrossi, 2008) for wireless VANETs. Some mobility software incorporates an ideal model of road layouts and as such does not account for traffic signals at intersections or other road 'obstacles' that give rise to queuing. While these models may be perfectly adequate for generic comparisons, clustering of vehicles within wireless range results in increased wireless channel access contention. Vehicular mobility can be split into macro-mobility effects such as road layout, number of lanes, and speed limits, and micro-mobility effects, especially driver behavior, which should account for the presence of other vehicles, both nearby and due to traffic congestion. This implies that the random waypoint mobility model (Broch et al., 1998) that is common in *Mobile ad-hoc Network* (*MANET*) simulations is no longer applicable to VANETs, whereas vehicle density has a more important role to play. Moreover, choice of routing algorithm (Lee et al., 2003), especially position-aware algorithms, is also important.

In a VANET, wireless path loss through diffraction and multipath (Tan & Bahai, 2010) (Matolak, 2008) is a serious threat to transmissions and consequently it needs to be modeled accurately. General-purpose VANET simulations tend to restrict modeling of the wireless channel to either free-space or two-ray. That is they use *line-of-sight* (*LOS*) modeling and do not account for wireless fading (fast or slow) through multi-path, as can occur, especially in a built environment. Moreover, it is possible that the simulator may not properly account for inter-node interference. Modeling of ad-hoc routing protocols at higher layers of the protocol stack but neglecting physical-layer modeling may, therefore, result in misleading results.

The remainder of this Chapter is organized as follows. Section 2 outlines the emergence of VANETs as a feature of intelligent transportation systems. To achieve resilient video streaming, two possibilities are to use P2P streaming and to apply error resilience to the video streams. Section 2 outlines the problems and issues that exist in those respects. Before a solution can begin to be applied, one requires greater modeling fidelity. This Chapter has two main recommendations: improve mobility modeling and include more realistic channel models. Section 3 considers these recommendations. Section 4, is the tentative solution of the authors, which is multiple description coding combined with P2P streaming. As we believe prior modelling has been too general, Section 4 presents a carefully controlled simulation in which to evaluate the merits of the proposal. Section 5 presents future research development, while Section 6 makes some concluding remarks.

BACKGROUND: APPLICATIONS OF VANETS

VANETs are a cornerstone of the envisioned *Intelligent Transportation System* (*ITS*). By enabling vehicles to communicate with each other via inter-vehicle communication, as well as with roadside base stations via roadside-to-vehicle communication, vehicular networks will contribute to safer and more efficient roads by providing timely information to drivers and the relevant authorities. The interesting research area of vehicular networks (Biswas et al., 2006) is where ad-hoc networks can be brought to their full potential.

The main goal of a VANET is providing safety and comfort for passengers. To this end a wireless transceiver inside each vehicle provides ad-hoc network connectivity for the passengers. This network tends to operate without any infra-structure or legacy client and server communication. Each vehicle equipped with VANET transceiver will be a node in the ad-hoc network and can receive and relay others' messages through the wireless network. Collision warning, road sign alarms and in-place traffic view will give the driver essential tools to decide the best path along the way. Automatic payment for parking lots and toll collection are other examples of possibilities enabled by VANETs.

Rather than moving at random, vehicles tend to move in a co-ordinated fashion. The interactions with roadside equipment can likewise be characterized fairly accurately. Most vehicles are restricted in their range of motion, for example by being constrained to follow a paved highway. In general, wireless VANETs, just like general ad-hoc networks can: relieve congested cells; extend coverage; and service dead-spots within cellular networks, assuming dual cellular and ad-hoc interfaces in-vehicle transceivers or act as relay vehicles for those vehicles not so equipped.

There can also be multimedia and Internet connectivity facilities for passengers, all provided within the wireless coverage of each car. Furthermore, VANETs can facilitate in the exchange or sharing of personal video sequences, as may occur in a social network. They may also have a role in reporting traffic conditions, as seen by roadside cameras, and likewise have a role in reporting crisis scenes during an emergency. An example scenario is shown in the Figure 1, in which a video sequence captured at the site of an accident is relayed over the multi-hop network to approaching emergency vehicles.

It is the transience of the video sources in a VANET that poses a problem to video communication on a highway (Guo et al., 2005), as the sources can quickly be isolated with a network partition (Wu et al., 2004). On the other hand, even at fast car speeds (up to 104.6 kmh) it is possible (Bucciol et al., 2008) to achieve a 1 Mbps data rate with an IEEE 802.11b system operating with a range of 250 m. In Guo et al. (2005) multiple car-borne video sources were modeled traveling on a 4-5 lane highway in Atlanta. Video was collected by sending a request trigger from a car approaching a destination to a camera on a remote vehicle passing a destination region. Video transport back to the requestor was by a store-carry-and-forward sub-system, though the method was not detailed in Guo et al. (2005). The main analysis in Guo et al. (2005) was of delay characteristics, whereas in fact delay can be absorbed by a large on-board buffer.

In Xie et al. (2007), a similar scenario to Guo et al. (2005) was again simulated but video quality was now considered. It was observed that the delivered video quality was good (30 dB *Peak Signal-to-Noise Ratio* (*PSNR*) at medium traffic density with sufficient buffer size (300 packets and above)) but deteriorated at lower densities. In

Figure 1. VANET with captured video relayed by cars to first responder emergency vehicle

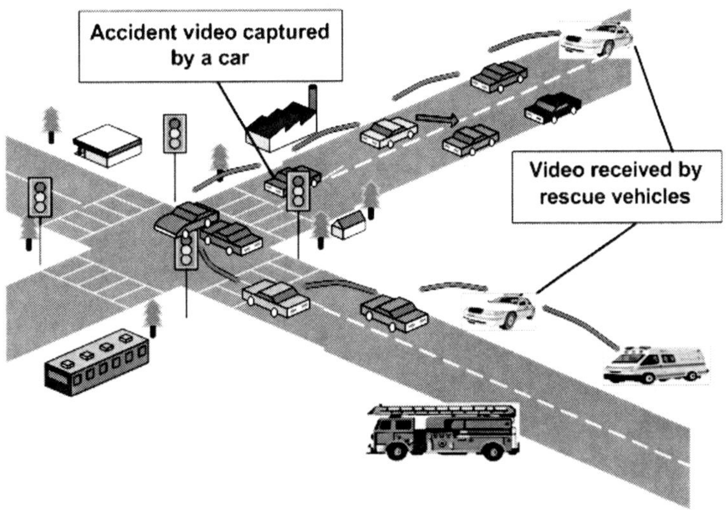

the event of buffer overflow, an early drop policy was better than the normal drop-tail policy but the buffer drop test was only conducted for a relatively small buffer size of 100 packets, which was already known to result in poorer quality delivered video. Wireless range was 400 m with a data-rate of 6 Mbps but the propagation model was not given and may have been free space (see Section 5.2).

Research in Park et al(2006), extending the work in Guo et al. (2005), simulated a two-ray wireless propagation model (see Section 5.2) and imposed an application-layer *Forward Error Correction* (*FEC*)-based solution through network coding (Ho & Lun, 2008). Though network coding of FEC and in particular rateless error coding (MacKay, 2005) is an effective means of limiting the impact of packet erasures upon streamed video, it depends on action by intervening nodes. When these nodes are not possible destinations and consequently may not be expected to make special provision for video data, then network coding is not feasible.

VANET VIDEO STREAMING: ISSUES AND PROBLEMS

In this section, we consider the main issues associated with P2P streaming within a VANET, namely how to organize the P2P overlay and how to protect the video stream in the hostile transmission conditions within a VANET.

P2P Streaming in an Ad-Hoc Network

Decentralized P2P streaming systems such as mesh-P2P streaming (Yong et al., 2008) have been deployed to provide live and on-demand video streaming services on the Internet and the same ideas may be useful in providing real-time video streaming in ad-hoc networks. The P2P streaming concept has nowadays been developed into several trial P2P streaming systems, such as Joost, Sopcast, Zattoo, and PPlive (Hei et al., 2007). The online broadcasting arena is evolving in response to the clear commercial interest for these new technologies in support of IPTV (TV over IP-framed networks). P2P streaming architectures can be categorized according to their distribution mechanisms. The various approaches to P2P streaming have been surveyed by Liu et al. (2008). Two main topologies have emerged, i.e. tree-based and mesh-based P2P.

In an ad-hoc network, the network topology changes randomly and unpredictably over time. Hence, an application that can easily adapt to the dynamic behavior of the ad-hoc network will be an effective solution. Mesh-P2P streaming is flexible and can be managed easily in comparison to a tree-based topology. Moreover, it is not affected by the churn of peers or the effects of handover. A mesh-based topology can also overcome the bandwidth heterogeneity present in an ad-hoc network. Consequently, a mesh-based solution to P2P streaming in an ad-hoc network can effective. Mesh-based distribution is becoming more widespread than tree-based distribution and has been adopted by most successful P2P streaming systems (Reza, 2006).

In fact, little prior research has apparently considered the possibility of running P2P streaming applications across ad-hoc networks, though P2P file download has been actively investigated, as will be apparent from the survey in Yan (2005) which mentions eight contributions to file sharing across ad-hoc networks. However, streaming applications have additional latency constraints. P2P overlay networks allow physical networks to better cope with the higher bitrates required for video streaming by increasing the number of peers and Gürses & Kim (2008) investigated ways of facilitating peer selection to optimise throughput across an ad-hoc network. To do so, congestion information across the underlay network (the physical network underlying the P2P overlay network) was utilised to assign rates between peers. Network-wide optimisation of throughput was achieved through the method of Lagrange multipliers. However, there was no investigation of node speed and limited investigation of the effect of network size and distribution protocol. In contrast, a naturally-inspired peer search algorithm, based on the behaviour of ant colonies, was investigated in Zuo et al. (2007). A simulation with 1600 nodes supported the idea, though clearly this work was at a preliminary stage in the investigation. It should also be mentioned that unless in a stadium or other crowded venue it is unlikely that there will be so many nodes available. Similarly, in Liu et al. (2006) a hierarchical arrangement of MANETs was introduced to reduce distribution latency. P2P streaming was given a practical investigation Giordano et al. (2008), when vehicles were driven around a building in a university campus to check the feasibility of video streaming. The main finding was that provided the number of hops was limited then streaming was possible across a vehicular network. Like other work in this field, research in Giodano et al. (2008) is at an exploratory stage but it is certainly likely that there will be many cars able to act as peers. Recently P2P video streaming over an ad-hoc networks using MDC have been studied in Niraula et al. (2009), proposing *Cross-layer and P2P based Solution* (*CLAPS*) using the *Multicast Overlay Spanning Tree* (*MOST*) protocol. However, configuration and performance details are unavailable.

Error Resilience for VANETs

Error resilience source coding (Stockhammer & Zia, 2007) is a way of protecting compressed video with reduced overhead compared to FEC. The two can be combined, though herein we concentrate on error resilience alone because of the demands on bandwidth from adding FEC to every packet and the risk of high packet loss in the VANET, which might require excessive overhead to protect packets. Error control through various forms of *Automatic Repeat reQuest* (*ARQ*) is another form of protection but, across a multi-hop VANET, the added latency involved in sending an ARQ packet back to the video source is of serious concern to a real-time application. The decoder can also employ error concealment (Wang & Zhou, 1998), whereby in one way or another previous video information serves to reconstruct missing information. However, error concealment is an implementation-dependent feature of a decoder, as there can be a significant computational cost involved in picture reconstruction if simple previous frame replacement is not used.

There is a variety of error resilience methods available some of which serve to recover from corrupt data and some of which serve to recover from lost packets. Prior work on error resilience coding in an ad-hoc network, e.g. Mau et al. (2003), has largely taken advantage of the path diversity available through MDC. In MDC, a video stream is split into two or more descriptions or flows in such a way that if packet loss from either description still allows reconstruction at the decoder, albeit at a reduced quality. For practical reasons, the original video may be split temporally (Apostolopoulos, 2002) in a frame-wise fashion, though with additional overhead due to the need to preserve the *Group of Pictures* (*GOP*) structure (encoded video pictures or frames are normally introduced by a reference frame that is purely spatially or intra-coded. Other frames within a GOP are predictively coded until the next reference frame that is normally inserted after about 12 or 15 frames.) in both descriptions. This is because in general, MDC is difficult to apply and computationally complex (Wang, 2005), because it requires synchronization between encoder and decoder to reduce motion estimation error drift.

As an example of MDC, the research in (Setton et al., 2004) examined point-to-point *Constant Bit-Rate* (*CBR*) video streaming in a 15-node ad-hoc network in a 1000 × 1000 m^2 area. The reference *Foreman* video clip was encoded at *Quarter Common Intermediate Format* (*QCIF*) (176 × 144 pixel/frame) at 30 fps. The sending rates simulated ranged from around 50 kbps up to 350 kbps. GOP sizes were varied with playout buffer settings equivalent to 350 ms and 500 ms of video. The paper showed that optimal regimes did exist but that simple formulas to achieve the optimal quality of delivered video require full network knowledge. Research in Setton et al. (2004) considers more general ad-hoc networks than VANETs. It also concerns itself with the effect of playout buffer size, which is not an issue in a VANET. A variety of other diversity schemes for video over ad-hoc networks have been considered (Mau et al., 2003) though again without the restrictions arising in a VANET.

Even if MDC were to be used in a VANET, the emergency scenario we have sketched in Section 1 does not allow a single MDC destination to be readily identified, because an early responder to an emergency may not necessarily be aware of what other emergency vehicles were in the vicinity. It is also not possible for a video receiver to trigger a video (Guo et al., 2005) from an early responder, as it may be unaware that an emergency has occurred. Instead, as mentioned in Section 1, we suppose that the emergency vehicles constitute a closed group to which video is distributed.

Turning to FMO error resilience, compressed frame data is normally split into a number of slices each consisting of a set of macroblocks. In the long established MPEG-2 codec, slices could only be constructed from a single row of macroblocks. Slice resynchronization markers at either end of the slice bit-stream ensure that if a slice is lost then the decoder is still able to continue. Therefore, a slice is a unit of error resilience and it is normally assumed that one slice forms a packet, after packing into a *Network Adaptation Layer unit* (*NALU*) at the output of an H.264/AVC codec. In H.264/AVC, by varying the way in which the macroblocks are assigned to a slice (or rather group of slices), FMO gives a way of reconstructing a frame even if one or more slices are lost. Within a frame up to eight slice groups are possible.

A simple FMO method is to continue a row of macroblocks to a second row of the picture but it is also possible to form disjoint slice groups (Lambert et al., 2006). Checkerboard or dispersed slice group selection allows one slice group to aid in the reconstruction of another slice group (if its packet is lost) by temporal (using motion vector averaging) or spatial interpolation. Assignment of macroblocks to a slice group can be general (type 6) but the other six types pre-define an assignment formula, thus reducing the coding overhead from providing a full assignment map. Katz et al. (2007) proposes a combined spiral and checkerboard pattern of macroblock selection. Because there is approximately at least a 10 to 15% overhead from slice group mapping, depending on choice of entropy coding method for Common *Intermediate Format* (*CIF*) (352 × 288 pixels per frame) frames, even with just two slice groups in a checkerboard pattern, the work herein does not use more than two slice groups for this pattern, which is feasible for the QCIF-sized frames used. To reduce overhead, it is also preferable to choose the H.264/AVC option preventing reference outside the slice group (Lambert et al., 2006), though at some cost in coding efficiency.

Assume that a set of macroblocks within a slice need to be reconstructed. For motion-copy error concealment in H.264/AVC (Vars & Hannuksela, 2002), the motion vectors of correctly received slices are computed in order to reconstruct an average motion vector for a missing macroblock if the average motion activity is sufficient (more than a quarter pixel). It is also possible to select the intra-coded frame method of spatial interpolation to reconstruct a missing macroblock. In the experiments in Section 5, though experience favors a motion vector-based method, we employed both methods and selected the superior result in terms of average PSNR across the video sequence. In a live situation, it is possible to choose the method that best reduces 'blockiness' at macroblock boundaries.

To enhance error robustness in H.264/AVC, the encoder may create duplicate copies of some or all parts of a picture, which can be used for the purposes of error resilience (Baccichet et al., 2006). Redundant frames (or strictly redundant slices making up a frame) are coarsely quantized frames that can avoid sudden drops in quality marked by freeze frame effects, if a complete frame (or slice) is lost in transmission. As redundant frames are predictively coded they require fewer bits than *Instantaneous Decoder Refresh* (*IDR*) frames. Hence, they are more efficient than including additional intra-coded IDR frames. It is also possible to combine redundant frames with MDC. Consequently, redundant frames also offer a way to stream across a VANET or other ad-hoc network using MDC. The redundant frames can be discarded if not required. A weakness of this scheme is the delay in encoding and transmitting redundant frames, making it more suitable for one-way communication. If the redundant frame/slice replaces the loss of the original frame/slice, there will still be some mismatch between the encoder and decoder; the encoder will assume the original frame/slice was used. However, the effect will be much less than if no substitution took place.

RECOMMENDATIONS FOR GREATER FIDELITY

In early investigation of the possibility carried out by members of the video coding community, rather general scenarios were chosen. However, car manufacturers are more interested in detailed simulation of VANETs. This section discusses two ways in which greater fidelity can be achieved, the first by mobility models that include driver behavior and the second by including the effects of wireless channel models other than the two-ray ground reflection propagation model. Though the two-ray model is perfectly appropriate for generalized modeling with line-of-sight communication, it is less appropriate for VANETs because of the presence of buildings.

Mobility Models

In the context of P2P streaming, greater realism is needed in modeling road layouts and driver behavior. Some mobility modeling software incorporate an ideal model of road layouts and as such do not account for traffic signals at intersections or other obstacles that give rise to queuing. While these models may be perfectly adequate for generic comparisons, clustering of vehicles within wireless range results in increased access contention. Interestingly, behavior at bottlenecks (Treiber et al., 2006) in the presence of obstacles of whatever form[1] is similar in road systems across the globe. Such behavior results in synchronized traffic, evidenced by clusters of vehicles passing along a road.

Five different clustering patterns (Treiber et al., 2006) have been identified from mathematical modeling. Vehicular mobility can be split into macro-mobility effects such as road layout, number of lanes, and speed limits, and micro-mobility effects, especially driver behavior, which should account for the presence of other vehicles, both nearby and due to traffic congestion. This implies that the fixed speed simulations that are common in ad-hoc network modeling are no longer applicable to VANETs, whereas vehicle density has a more important role to play. It is difficult to conduct repeated live experiments and because the complex vehicle mobility models that arise are unlikely to be represented analytically, simulation is the main tool for research on VANETs (Tonguz et al., 2009). The recent trend amongst research supported by the manufacturers is for detailed simulations (Liu et al., 2009) to represent the large number of variables (vehicle density, car speeds, driver behavior, road obstacles, road topologies …) that can occur.

In (Bai et al., 2003), two models relevant to vehicular mobility were described, namely Freeway Mobility and Manhattan Mobility. The Freeway model limits vehicles to 1-D motion in either direction. Vehicles are tied to one of several lanes; the speed is dependent on a vehicle's previous speed; and in the 'car-following' restriction, a following vehicle cannot exceed the speed of a preceding vehicle to avoid approaching within a safety distance. The Manhattan model, an extension of the Freeway model, restricts the number of lanes in either direction to just one, but introduces a turning probability to give greater mobility. Both Freeway and Manhattan are related in that they should result in high spatial and temporal dependency. This layout is characteristic of many North American city centers but differs from older cities such as London with more irregular street layouts or more recent North American urban areas.

More complex models are supported by VanetMobiSim, which is an extension of the *CANU Mobility Simulation Environment* (*CanuMobiSim*) (Fiore et al., 2006), a flexible framework for user mobility modeling. CanuMobiSim is Java-based and can generate movement traces in different formats, supporting different simulation tools for mobile networks. The VanetMobiSim extension focuses on vehicular mobility, and features new realistic automotive motion models at both macroscopic and microscopic levels. The micro-mobility models presented by VanetMobiSim are of increased sophistication in driver behavior. In VanetMobiSim a variety of urban road layouts can be generated of increasing density by means of the random backbone mode using Voronoi tessellations. It also adds support for multi-lane roads, separate directional flows, differentiated speed constraints and traffic signs at intersections. The real advantage of VanetMobiSim is the variety of driver behavior models.

The *Constant Speed Motion* (*CSM*) model does not produce realistic motions (as it is possible for vehicles to overlap during motion) and is included simply for comparison with earlier results in previous research. In this model, the speed of each vehicle is determined on the basis of the local state of each car and any external effect is ignored. A vehicle follows a random movement (within the road topology) in the sense that a destination is selected and vehicle moves towards it, possibly pausing at intersections. While this model may work for isolated cars, it fails to reproduce realistic movements of groups of vehicle.

In the *Fluid Traffic Model* (*FTM*), the traffic density affects a vehicle's speed. In this model, the speed of car considers the presence of nearby vehicles. However, the car mobility is considered for single lanes only and does not take into account the interaction between cars on multiple lanes. This model describes the vehicle's speed as a monotonically decreasing function of the traffic density. When traffic congestion reaches a critical state, then the node speed is constrained by Equation (1).

$$s = \max\left[s_{\min}, s_{\max}\left(1 - \frac{k}{k_{jam}}\right)\right] \tag{1}$$

where S is the output speed, S_{min} and S_{max} are minimum and maximum speed constraints respectively, k_{jam} is the vehicular density at which a traffic jam is declared, and k is the current vehicle density of the road the vehicle of interest is moving on. $k = n/l$, where n is the number of vehicles on that road and l is the length of the road (Xu et al., 2004).

The *Intelligent Driver Model* (*IDM*) accords with the car-following model and is based on live observations. The instantaneous acceleration of a vehicle is computed by Equation (2).

$$\frac{dv}{dt} = a\left[1 - \left(\frac{v}{v_0}\right)^4 - \left(\frac{s^*}{s}\right)^2\right] \tag{2}$$

where a is the maximum acceleration of the vehicle, υ is the current speed of the vehicle, υ_0 is the desired velocity, s is the distance from the preceding vehicle and s^* is a desired dynamical distance, as given by the following equation:

$$s^* = s_0 + vT + \frac{v\Delta v}{2\sqrt{ab}} \tag{3}$$

where s_0 is the minimum desirable bumper-to-bumper distance, and T is the minimum safe time between two vehicles. The speed difference with respect to front vehicle velocity is Δv, and the normal acceleration and de-acceleration rate is a and b respectively.

VanetMobiSim modifies IDM by modeling intersection management (IDM-IM) as a lead vehicle approaches an intersection. Whenever a vehicle approaches an intersection with stop signs or traffic lights, the parameters in Equation (4) are used in IDM-IM to set its speed.

$$\begin{cases} s = \sigma - S \\ \Delta v = v \end{cases} \tag{4}$$

where σ is now the current distance to the intersection and S is a safety distance, as the vehicle does not stop exactly at the intersection. Once the vehicle halts at an intersection a decision is made on whether it should pass over the intersection or wait at the intersection. This decision is based on several criteria. These criteria are principally: the number of other vehicles on other roads leading into the intersection; and the state of any traffic lights at the intersection.

The IDM-IM includes lane change and overtaking behaviour in the IDM-LC model. At intersections, if a lane ceases to exist on the other side of the intersection, a vehicle will wait until a gap appears. More generally, for overtaking using lanes, a game-theoretical model (Kesting et al., 2007) is applied. A lane change will occur if the advantage of a vehicle changing lanes to a new lane is greater than the disadvantage to following vehicles in the current lane and the new lane. This is measured in terms of acceleration, a, by the following inequalities:

$$a^l - a \pm a_{bias} > p(a_{cur} - a_{new} - a^l_{cur} - a^l_{new}) + a_{thr} \tag{5}$$

$$a^l_{new} > -a_{safe}$$

Thus, $(a^l - a)$ is the gain in acceleration from moving over to lane l, with similar expressions for the loss in acceleration for a following vehicle in the current lane and the new lane (when the lane swapping vehicle joins it). Thus, a_{cur} is the current acceleration of the following vehicle, while a^l_{cur} is the acceleration of the following vehicle should a lane change occur and a_{new} is the acceleration of a following vehicle in the new lane, while a^l_{new} is the following vehicle's acceleration in the new lane should a lane change occur. p is a driver politeness factor; a_{bias} is a lane bias factor encouraging lane changes to a particular side; and a_{safe} restricts movements to avoid a driver in the new lane having to decelerate too quickly as a result of the lane change. Thus, a^l_{new} is the de-acceleration required in the new lane by the following vehicle, which needs to be below a safe limit of de-acceleration. The term a_{thr} is a minimum acceleration below which lane changing has limited value.

Wireless Channel Models

General-purpose VANET simulations tend to restrict modeling of the wireless channel to either free space or two-ray, LOS modeling and do not account for wireless fading through multi-path, as can obviously occur, especially in a built environment. Moreover, it is possible that the simulator may not properly account for inter-node interference. Consequently, modeling of ad-hoc routing protocols at higher layers of the protocol stack but neglecting physical layer modeling (Takai et al., 2001) may result in misleading results. In fact, the well-known ns2 simulator may have deficiencies (Wellens et al., 2005) in that respect, which is one reason why GloMoSim (Zeng et al., 1998) is selected herein for network simulations. Simulations may also mislead if they take no account of packet length by thresholding the *Signal-to-Interference-Noise Ratio* (*SINR*) level rather than determining the *Bit Error Rate* (*BER*). Notice that signal reception thresholding is acceptable if all packets are long. However, the latter is not normally the case for video communication.

Both Rayleigh and Rician models (Sklar, 1997) (Lindsey, 1964) are used to model wireless fading caused by multi-path interference. The Rayleigh distribution represents a situation in which nodes are highly mobile. The restrictions of an urban topology make highly mobile nodes unlikely. In communication between vehicles in a city there will often be a LOS component, as vehicles will be aligned along road segments and mobility is often restricted. Here Rician fading channel is appropriate, as the Rician distribution includes a LOS component.

The free space propagation model predicts received signal strength when the transmitter and receiver have a clear and unobstructed line-of-sight between them. This model predicts that transmission power is attenuated in proportion to the square of the distance. According to this model, the free-space equation for non-isotropic antennas is the following:

$$P_r = P_t \left(\frac{\lambda}{4\pi d} \right)^n G_t G_r \tag{6}$$

where P_r is the received power, P_t is the transmitted power (in Watts or milli-Watts), λ is the carrier wavelength (in meters), d is the distance between transmitter and receiver (in meters), n is the path loss coefficient, G_t is the antenna gain at the transmitter and G_r is the antenna gain at the receiver.

The antenna gain is a measure of the directionality of an antenna. Antenna gain is defined as the power output, in a particular direction, compared to that produced in any direction by a perfect omni-directional antenna or isotropic antenna. For example, if an antenna has a gain of 3 dB, that antenna improves upon the isotropic antenna in that direction by 3 dB, or a factor of 2. The increased power radiated in a given direction is at the expense of other directions. Therefore for isotropic or omni-directional antenna the gain will be 0. The decibel gain is given by Equation (7), with P_{in} being the input power and P_{out} being the output power:

$$G_{dB} = 10 \log_{10} \frac{P_{out}}{P_{in}} \tag{7}$$

For the ideal omni-directional antenna, the free space loss equation at a distance d is:

$$\frac{P_t}{P_r} = \frac{(4\pi d)^2}{\lambda^2} = \frac{(4\pi f d)^2}{c^2} \tag{8}$$

where c is the speed of light (3 x 10^8 mps) and f is the frequency (in hertz).

With this path loss model, even nodes far from the transmitter can receive packets, which can result in fewer hops to reach the final destination. Therefore, simulation results with the free space path loss model tend to be better than with other path loss models but unfortunately they are unrealistic. However, as signal propagation with little power loss may cause stronger interference for concurrent transmissions, it does not necessarily yield the best performance under all scenarios.

The two-ray path-loss model is suited to LOS channels in urban environments. The two-ray model considers both the direct path and a ground-reflected propagation path between transmitter and receiver. The model predicts that received power decays with distance raised to the fourth power, or at a rate of 40 dB per decade. This is a much more rapid path loss than is experienced in free space but it is more realistic. The following equation is used to calculate the receiver power for the two-ray model.

$$P_r = P_t \frac{h_t^2 h_r^2}{d^4} G_t G_r \quad (9)$$

where h_t and h_r are the height of the transmitter and receiver antennas respectively. G_t and G_r is the antenna gain at the transmitter and the receiver respectively. d is the distance between the receiver and transmitter and P_t and P_r are the transmitter and receiver power respectively.

Fading is a variation of signal power at receivers, caused by the node mobility that creates varying path conditions from transmitters. In a Rician model, there is one LOS signal. In Rayleigh fading, there is no LOS component and, consequently, Rayleigh fading represents a worst-case scenario (provided there is no frequency selective fading.) Specifically, the Rayleigh distribution of the signal envelope is given by the following function:

$$f(z) = \frac{z}{\sigma^2} \exp\left[-\frac{z^2}{2\sigma^2}\right], z \geq 0 \quad (10)$$

where σ^2 is the variance of the zero-mean Gaussian distributed in-phase, r_I, and quadrature, r_Q, components of the signal envelope:

$$z(t) = \sqrt{r_I^2(t) + r_Q^2(t)} \quad (11)$$

The Rician distribution is given by

$$f(z) = \frac{z}{\sigma^2} \exp\left[\frac{-(z^2 + s^2)}{2\sigma^2}\right] I_0\left(\frac{zs}{\sigma^2}\right), z \geq 0 \quad (12)$$

where the in-phase and quadrature components are no longer Gaussian, as there is a fixed LOS component. I_0 is the modified Bessel function of zeroth order. The average power in the non-LOS components is given by $2\sigma^2$ and s^2 is the power of the LOS component. The fading parameter $K = s^2/2\sigma^2$ is the ratio of these powers. In fact, the right-hand-side of this equation can be rewritten in terms of K and the average received power. If $K = 0$ then this equation reverts to a Rayleigh distribution and with $K=\infty$ there is only a LOS component. Setting $K = 3$ (or 4.7 dB) by default results in a moderate departure from the Rayleigh distribution and, in fact, in many cases K does not exceed 7 dB (Doukas & Kalivas, 2006).

Notice that the two-ray model considers path loss and as such it is deterministic. Fading models are stochastic and represent an additional attenuation as well as existing path loss. Thus, in terms of dB it is possible to accumulate the effects of these impacts, when calculating the mean effect within a simulation.

A POTENTIAL SOLUTION: MDC-BASED P2P VIDEO STREAMING

The authors' tentative solution to VANET video streaming (Qadri et al, 2010), in the context of an emergency response, is through P2P streaming using multiple description coding. To avoid a general solution, a controlled and more precisely defined simulation was undertaken, as a means of evaluating the proposal.

Simulation Scenario

In this Section, a mesh architecture with all-to-all connectivity (Long et al., 2008) is simulated, Figure 2. In the example scenario illustrated, seven nodes form a mesh within which two nodes are the original source nodes for the same video. A likely way that two nodes in a VANET could acquire the same video clip is that both could have passed a roadside unit offering informational/advertising video clips. However, it is also possible that these two nodes previously could have acquired the same video from a single node, prior to the distribution process illustrated in Figure 2. Three nodes (node C, node D and node E) receive video from these two source nodes. The three nodes in turn also act as sources to two further nodes (node F and node G). Hence, nodes C, D and E download and upload at the same time i.e. act as receivers and sources at the same time. All receiving nodes, C to G, are served by multiple sources. Moreover in Figure 2, nodes F and G can receive data from an alternative source if a source ceases to be available.

Peer selection for streaming purposes can either be achieved in one of three ways. Firstly, it is achieved in a hybrid fashion, by including some server nodes in the manner of commercial P2P streaming. Secondly, a structured overlay can be organized that allows quicker discovery of source peers than in an unstructured overlay. However, structured overlays require the source data to be placed in particular nodes, which may not be practical in a VANET. Therefore, this work assumes an unstructured and decentralized overlay such as Gnutella, KaZaa, or GIA.

Simulation Model

GloMoSim (Zeng et al., 1998) was employed to generate the simulation results. For video transport, IP framing was employed with UDP transport. The default network configuration consisted of seven nodes forming an overlay network, as arranged in Figure 2, with these nodes selected from up to a total of 100 nodes (vehicles) forming the complete VANET. GloMoSim was altered so that nodes start at random locations rather than at the origin, to avoid biased results. As a default a two-ray path loss model was used and other channel models were subsequently tested. IEEE 802.11p was the selected wireless technology. In IEEE 802.11p transmission is at 5.9 GHz, when the two-ray crossover distance is about 556.5 m. Receiver sensitivity in the simulation was set to -92 dBm. IEEE 802.11p's *Binary Phase Shift Keying* (*BPSK*) modulation mode with 1/2 coding rate was simulated. Accordingly, the data-rate was set to 3 Mbps. *Request to Send / Clear to Send* (*RTS/CTS*) signaling was turned on.

IEEE 802.11p is an amendment to the IEEE 802.11 standard to add *Wireless Access in Vehicular Environments* (*WAVE*) (Jiang & Delgrossi, 2007). It defines enhancements to 802.11 required to support ITS applications. This includes data exchange between high-speed vehicles and between the vehicles and the roadside infrastructure in the licensed ITS band of 5.9 GHz (5.85-5.925 GHz). The standard (Xu et al., 2004) operating in ad-hoc mode will facilitate VANET development, taking advantage of spectrum allocated both in Europe (30 MHz) and the USA (75 MHz) in

Figure 2. Mesh-based overlay topology sending video streams from sources to receivers

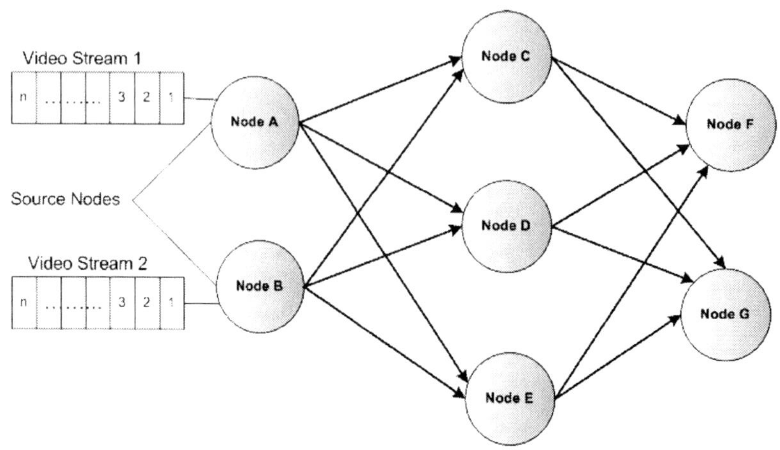

the 5.9 GHz range using 10 Mbps channels (the choice of a 10 MHz scaling of IEEE 802.11a for IEEE 802.11p ensures that the guard interval is sufficient to cover the root mean square Doppler delay spread expected in high mobility highway settings) operating at up to 27 Mbps, depending on modulation mode.

The *Location-Aware Routing* (*LAR*) (Ko & Vaidya, 2000) routing protocol was used at the routing layer. The LAR protocol is able to restrict the area for route propagation by virtue of *Global Positioning System* (*GPS*) information gathered from the vehicles in the VANET. The advent of satellite navigation systems has shown the benefits of GPS provision in vehicles and if a *Wireless Local Area Network* (*WLAN*) transceiver is available within the vehicle, GPS will also most likely be present. The result is that LAR will incur less control packet overhead compared to other well-known protocols such as *ad-hoc On-Demand Distance Vector* (*AODV*) (Perkins & Royer, 1999).

MDC exploits path diversity. In order to achieve diversity was decided to force disjoint paths (ones in which separate paths were taken). This was achieved by modifying the way that LAR chooses its initial routes. The authors have proposed (Rofoee et al., 2010) a family of multipath routing algorithms, which can be exploited for the purposes of MDC. However, consideration of multi-path routing algorithms would take the discussion away from video streaming.

In the simulations, a square 1000 m^2 area was defined and nodes (vehicles) were initially randomly placed within the area. Other spatial settings to do with road clusters, intersection density, lanes (2) and speeds are given in Table 1. The number of clusters refers to the number of rectangular areas with a road density of 2 obstacles per 100 m^2 (2e-4) for the Downtown model (to create randomly a non-homogeneous simulation terrain), while the remainder of the terrain is at the minimum road density, which was the same as that for a residential area, i.e. 2e-5. The number of traffic lights (at intersections) and time intervals between changes are also defined. In the following paragraph, the equation variable annotations in Table 1 are identified, with some uncommented terms explained soon. The values in Table 1 were the defaults for VanetMobiSim.

Table 1. VanetMobiSim configurable settings with road layout and mobility model

Global Parameters	
Terrain dimension	1000 m^2
Graph type	Space graph (Downtown model)
No. of road clusters	4
Min. intersection density	2e^{-5}
Max. traffic lights	6
Time interval between traffic lights change	10 s
Number of lanes	2
Min. stay	10 s
Max. stay	100 s
Nodes (vehicles)	50, 100
Min. speed (s_{min})	3.2 mps (11.27 kmph)
Max. speed (s_{max})	13.5 mps (48.28 kmph)
CSM model	
Min. and max. pause time	0 s
FTM Model	
Density of traffic jam	0.2 cars per m
Recalculation of traffic parameters interval	0.1 s
IDM Model, IDM-IM Model, IDM-LC Model	
Length of vehicle	5 m
Max. acceleration (a)	0.6 mps^2
Normal deceleration (b)	0.5 mps^2
Traffic jam distance	2 m
Node's safe time headway (T)	1.5 s
Recalculation of movement parameters interval	0.1 s
Other Parameters of IDM-LC Model	
Safe deceleration (a_{safe})	4 mps^2
Politeness factor of drivers when changing lane (p)	0.5
Threshold acceleration for lane change (a_{thr})	0.2 mps^2

In Table 1, 'Recalculation of movement parameters interval' is the interval after which the simulator recalculates parameters and makes a decision on those parameters. This time interval does not have a direct impact on the simulation time but indicates the update frequency. In the CSM model, pause time is set to zero because, unlike in a general ad-hoc network, it is not expected that a vehicle will malinger for any reason. The minimum and maximum stay times are times spent parking or stopping for any reason. These are set to low values by default.

The results were generated using two test video sequences to allow for content variation. Both sequences were encoded at QCIF video resolution (176×144 pixels per frame) @ 15 Hz (fps). Encoding was with the reference JM 15 software for an H.264/AVC codec. The *Foreman* sequence, with 300 frames, is a short, well-known sequence typical of one taken by a handheld camera. The sequence has jerky motion and a rapid camera pan towards the end. The *Paris* sequence consists of 1065 frames taken from a TV studio set. The background has significant spatial coding complexity. A trace file for the longer *Paris* clip served to simulate the impact of VANET transport. Both sequences were used to ascertain the video quality experienced by the user.

Evaluation Results

Averaged (arithmetic mean) results were recorded at the destination nodes in the overlay network of Figure 2. All simulations were repeated 50 times and the average taken. In repeating each run, it is important to realize that the road layout topology changes each time. Therefore, an average is across a set of road topologies, as well as node placements. Of course, for packets to reach a destination vehicle from the source vehicles, those packets will be routed across multiple hops to reach their destination. The effective wireless range was set to 625 m with a transmission power of 19.3 dBm (On a highway using IEEE 802.11p, 33 dBm with a potential range of 1300 m is more appropriate.) In most urban settings, the presence of buildings will impede transmission (Oishi et al., 2006) through reflections, diffraction and absorption of signals. The reduced power level is a concession to that fact.

In Figure 3, the packet loss ratios (ratio of packets sent to packets lost) would make unprotected streaming of video feasible in the denser network

Figure 3. Packet loss ratio compared by mobility model for different network sizes with no fading with routing by the LAR protocol

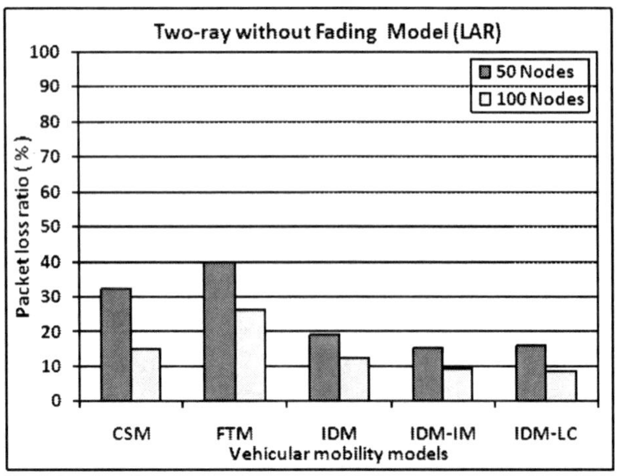

if lane changing was permitted, as losses below 10% occur in that scenario. It is possible that, for packet loss rates up to 20% but no more than 30%, inclusion of error resilience or application-layer forward FEC could compensate for high packet loss rates. Figure 3 shows that more careful modeling of driver behavior can actually decrease the predicted level of packet loss (going from CSM to IDM-LC). However, lane-changing (IDM-LC), which allows increased mobility, is not always possible within a city.

A feature of Figure 4a and 4b is that more careful wireless channel modeling results in a reduction in packet loss. However, the introduction of fading models causes the simulator to report greater packet losses in the denser network (100 nodes). This is because, when path loss only is modeled, then the distance between nodes is the principal effect and the sparser network (50 nodes) results in greater losses. However, if the nodes are closer together then interference at road obstacles due to traffic queuing increases the number of packet losses. In addition, both sparse and dense networks suffer from additional packet loss from fading. Figure 5 varies the fading parameter or factor *K* arising from the Rician distribution for reasonable values of *K* (Doukas & Kalivas, 2006) with the IDM-IM mobility model. When *K* is lower than 3, then an increase in packet loss occurs and the packet loss behavior begins to resemble that for the Rayleigh distribution, Figure 4b. When *K* is higher than the default value of 3, the increasing dominance of the LOS component means that the packet loss regime depends more on the LOS component. As *K* reaches 6 (7.8 dB) then the sparser network begins to suffer more packet losses, perhaps because of a relative increasing impact of path loss rather than fading.

As might be expected, when the node density increases, then end-to-end-delay increases. The number of hops traversed by a packet increases on average, resulting in longer delay. This is illustrated by the differences between the results for different network sizes, Figure 6. However, an interesting feature of these results is that more realistic modeling actually indicates that the delay for the sparser networks is less than that predicted by the coarser mobility models. That is the effect of network sparseness is equalized to some extent when the effects of driver behavior are taken into account. For the denser network of 100 nodes, delay is forecast to be less under a two-ray model alone. When fading is taken into account, Figure 7a, b, then predicted delay increases.

At the levels of delay reported, only one-way video streaming is possible[2] but this is not a problem as overlay network applications are generally not interactive. Therefore, the main purpose of estimating the delay is to assist in buffer dimensioning, which for the *Paris* video stream at 15 Hz obviously requires at least a 45 frame buffer. The size of the buffer will affect the start-up delay experienced by the user. If the user consciously selects a clip then delay will be perceptible, as it would need to be less than 20 ms for that to occur, which it clearly is not. If delivery is automatic, e.g. in a congestion monitoring application, then the user will not be aware of the start time of streaming.

The impact of cross-traffic on packet loss was also assessed. Four nodes that were not part of the overlay were introduced again with the IDM-IM mobility model, with each sending 100 packets of packet length 400 bytes as a CBR stream. Such messages could represent text safety alerts. From Table 2 in which the basic two-ray ground propagation model is used, it is apparent that moderate cross traffic does not greatly affect the packet loss ratios compared to the default values, though it might prove important to video quality as the losses reach above 20% (as it is difficult to reconstruct even with error resilience once losses reach above 20%). However, varying the overlay size certainly does have a significant effect on packet loss ratios. In the 'More Peers' scenario of Table 2, a total of 14 peers were part of the overlay, compared to the default scenario previously used.

Figure 4. Packet loss ratio compared by mobility model for different network sizes for a) Rician fading, and b) Rayleigh fading, with routing by the LAR protocol

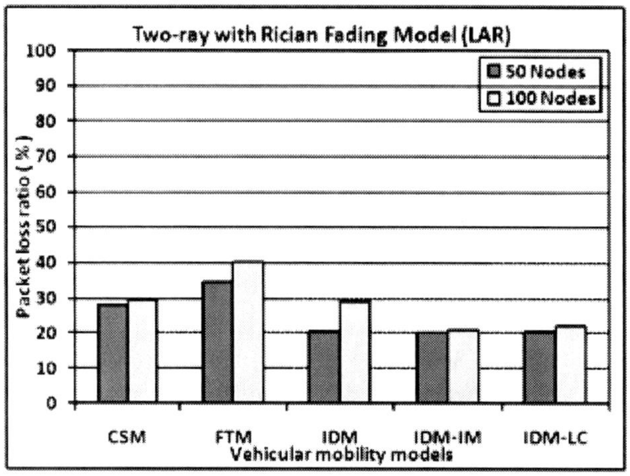

(a)

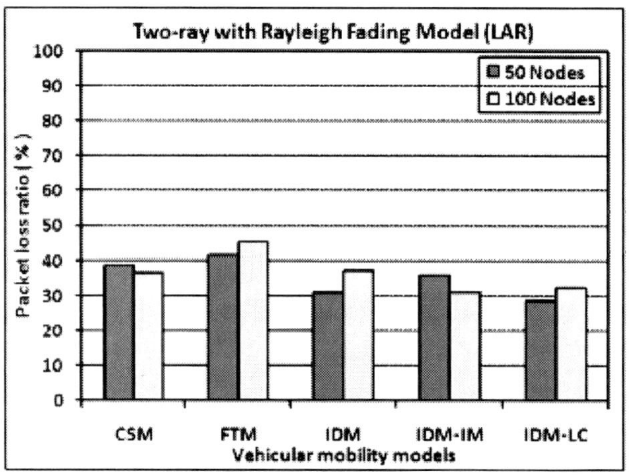

(b)

Two peers were source peers initially and six peers acted as receivers. These same six peers served four more peers. These four peers sent streams to two more peers. This means that 12 peers acted as source and 12 peers were receivers. Amongst these, ten peers simultaneously acted as receivers and senders. The mobility model was again IDM-IM and the two-ray channel model was used. As can be seen from the Table 2 the packet loss ratio increases significantly with this amount of traffic passing across the overlay. Therefore, in a VANET the size of an overlay network must be tightly controlled. This contrasts with an Internet overlay, when there are few or no restrictions on joining the overlay network.

The consequences of applying spatial MDC to the dual streams were considered. For comparison purposes, one simple form of spatial 'MDC' is to employ slicing (without FMO) in which the top part of the encoded frame forms one slice and the bottom half forms the other slice. Then each set of slices (top and bottom) form a

Figure 5. Packet loss ratio with Rician fading and varying K factor for an IDM-IM mobility model and LAR routing

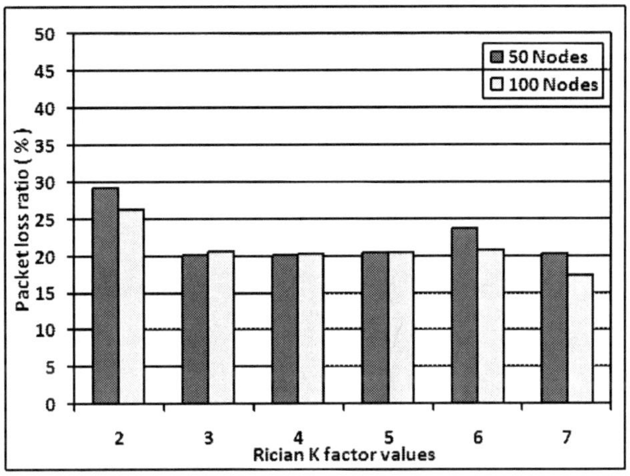

Figure 6. End-to-end delay compared by mobility model for different network sizes with no fading with routing by the LAR protocol

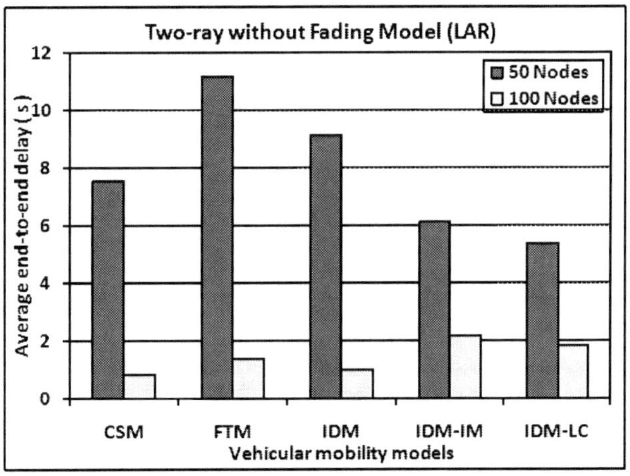

description but these cannot be used to reconstruct the other if the packet bearing that slice is lost. Instead, previous frame replacement must be employed by the decoder.

In checkerboard FMO, the macroblocks equivalent to the white squares of a checker or chess board form one slice, while the remaining macroblocks form the other slice. Error concealment is a non-normative feature of H.264/AVC. As described in Section 3, the method of concealment (spatial or temporal) can be selected depending on the smoothness of reconstructed macroblock border transitions.

As the lower complexity H.264/AVC Baseline Profile was employed the GOP frame structure was IPPPPP… To reduce error propagation both MDC schemes employed intra-refresh macroblocks by insertion of an intra-coded row of macroblocks into each encoded frame, cycling the replaced row through the video sequence. It is possible to

Figure 7. End-to-end delay compared by mobility model for different network sizes for a) Rician fading, and b) Rayleigh fading, with routing by the LAR protocol

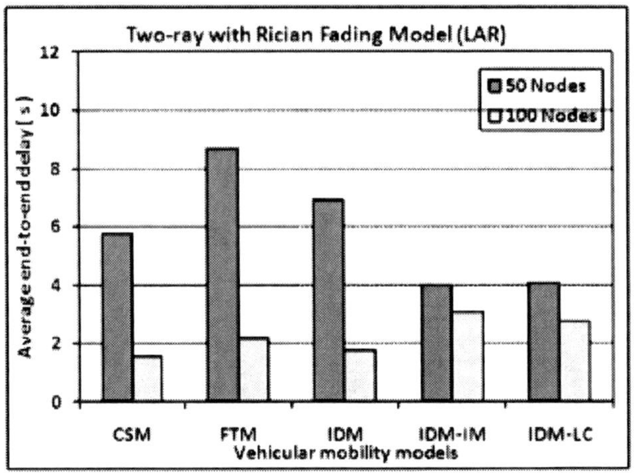

(a)

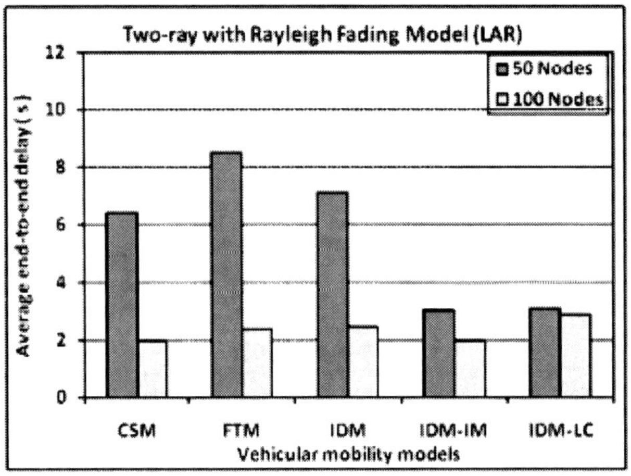

(b)

refine this scheme (Schreier & Rothermel, 2006) by adapting intra-refresh macroblock replacement through motion detection.

Table 2. Packet loss ratios when either introducing cross-traffic or more peers into the overlay for an IDM-IM mobility model, two-ray channel model, and LAR routing

Scenarios	Default	Cross Traffic	More Peers
50 Nodes	15.26%	22.22%	45.75%
100 Nodes	9.42%	21.45%	46.43%

In Figure 8, packet losses at rates up to 30% were generated from a Uniform distribution and the PSNR was calculated. From the Figure it is apparent that apart from zero packet loss, simple slicing (MDC) fares badly in comparison to checkerboard FMO with MDC (MDC with Error Res.). At zero packet loss the extra overhead from including the FMO macroblock mapping in a packet results in a drop in quality (if the datarate is the same as for simple slicing). However, from 8% packet loss onwards, the video quality for both sequences using simple slicing would be unacceptable. For *Paris*, the video quality using checkerboard FMO in conjunction with error concealment results in reasonable video quality. Though the PSNR drops below 30 dB, it is generally the case that users will tolerate PSNRs above 25 dB for mobile applications. In fact, it is possible (Chow & Ishii, 2007) to equate PSNR to the ITU-R's Mean Opinion Scores, when the range 25 to 31 dB inclusive is approximately equivalent to a score of 3 or "fair" (from a range 1 to 5 with 5 being "excellent"). For *Foreman*, the drop in quality is steeper, reflecting the more complex source coding task. However, below 20% packet loss ratio, the quality would be tolerable. From Figure 8b, this occurs under Rician channel conditions and IDM models when there are fifty vehicles in the network.

FUTURE RESEARCH DIRECTIONS

This section presents a summary of possible avenues of future research. Compared to the Internet environments that P2P streaming has generally

Figure 8. Applying MDC with and without error resilience to dual path video streaming of a) the Foreman sequence, b) the Paris sequence

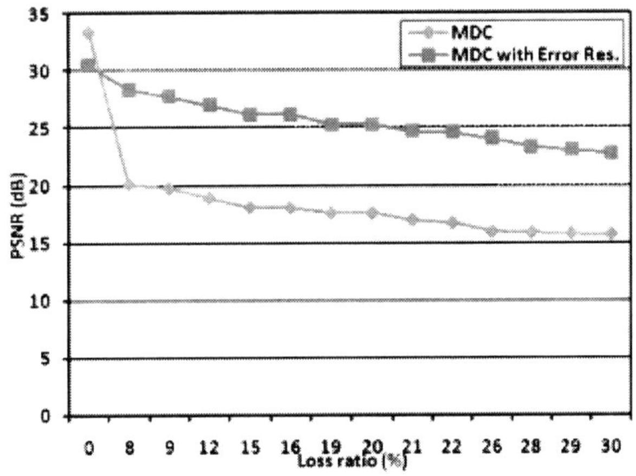

(a)

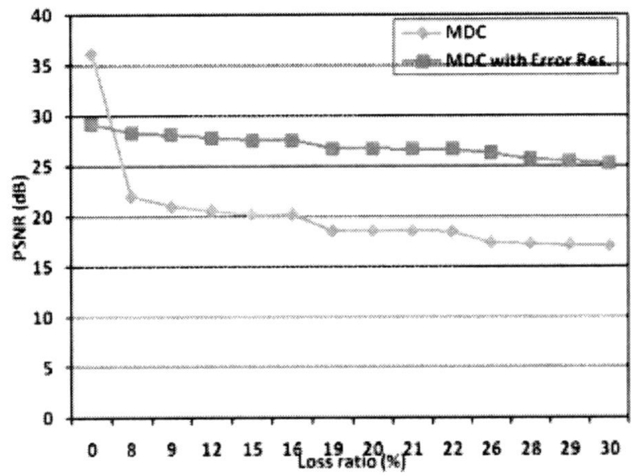

(b)

been applied, VANETs have lower bandwidth and higher error rates. Node mobility leads to a continually changing network topology. These are not the assumptions upon which P2P overlays have been built upon. Consequently, before P2P streaming can be properly implemented, it is necessary to adapt the overlay structure to VANETs.

Location of data in Internet P2P overlays is commonly via a *Distributed Hash Table* (*DHT*). However, this may be too much of an overhead in a small P2P overlay, as described in this work. Therefore, broadcast and even flooding may be an alternative loop-up mechanism (Buford et al., 2009) in these circumstances. In MADPastry (Zahn & Schiller, 2008), landmark nodes self-select themselves by comparing their identifiers with a set of pre-defined landmark identifiers. Nodes in the wireless vicinity of the landmark nodes adopt the prefix of the landmark node's address in the overlay for their address. P2P overlay routing tables are organized by prefix. Routing within the overlay is via a landmark node or any node with the same landmark prefix. At the physical network level, routing is by AODV. To go between any two nodes in the overlay network may take several hops. Furthermore each of the overlay hops might consist of several hops in the physical network. However, MADPastry has not been evaluated for high node speeds. For evaluation with higher speeds, one can consider Ekta (Pucha et al., 2004), FastTrack over AODV (Tang et al., 2008), and Dynamic P2P Source Routing (Hu et al, 2003). As most of these are concerned with file download and not video streaming, clearly there is scope for further investigation, in order to improve the provision of overlay functionality within ad-hoc networks.

This Chapter has concentrated on urban VANETs in which there are high densities of vehicles within a built-up area, with the vehicles moving at slower speeds than they would in open highways. However, initial investigation (Guo et al., 2005) into video streaming within a VANET was for highways. These networks assume a linear form, and unsurprisingly consist of vehicle traffic moving in two directions with much higher speeds than are possible within a city. Because of these characteristics, network fragmentation is more likely. Whether a P2P network can be established under these circumstances is worthy of investigation. It could be that roadside units could act as super-peers able to boot-strap the P2P overlay. To cope with network fragmentation a recent development in MANET routing has been concerned with delay-tolerant networks. While obviously video streaming is anything but delay tolerant, the insights into coping with network fragmentation from these routing mechanisms may be applicable to highway VANETs.

P2P overlays exist independently of the underlying physical network, though routing algorithms such as LAR, as mentioned in the previous section, are position aware. A network-aware P2P selection algorithm for ad-hoc networks based on least hop count and reliable source peers can be developed for effective delivery of video chunks. The algorithm can select a peer (to download video) based on its physical location and shortest path. This information can be determined from routing tables. The approach could reduce the packet loss experienced by nodes within an overlay created without prior knowledge of peer locations. Along with this, an efficient querying mechanism can also be developed for ad-hoc applications.

Reliability in video streaming over an ad-hoc network is one of the important challenges facing the technical community. Multi-path routing algorithms reduce the impact of packet loss, delay, and control overhead, which all impact upon video streaming. In (Rofoee et al., 2010), reliability was one of the criteria used to select paths. Perhaps, further improvements can be made by considering a reliability factor for route selection alongside the need to reduce the hop count. For example, this might be achieved through a feedback mechanism which collects networks statistics required to calculate the reliability of the routes.

CONCLUSION

As the integration of wireless networks takes place, it is likely that multimedia exchange will spread to VANETs. Some high-end cars already include WLAN capability and the advantages that the emerging IEEE 802.11p standard will bring for traffic safety will encourage manufacturers further in that direction. Video streaming has advantages over download in terms of regulated bandwidth usage, reasonable start-up delay for longer sequences, and commercial confidentiality. Multi-source P2P streaming across an overlay networks is potentially a robust form of video delivery. However, the size of such overlays will need to be strictly controlled in a VANET to prevent an increase in competing traffic amongst the overlay nodes, leading to unacceptable increases in packet loss ratios. This work has shown that more realistic mobility and driver behavior modeling, along with probabilistic fading models causes considerable differences in reported packet loss ratios, even with a reasonable number of nodes. Nevertheless, the spatial MDC method introduced are able to counter adverse conditions present in a VANET.

REFERENCES

Apostolopoulos, J. (2001). *Reliable video communication over lossy packet networks using multiple state encoding and path diversity* (pp. 24–26). SPIE, Visual Communications and Image Processing.

Baccichet, P., Shantanu, R., & Girod, B. (2006). Systematic lossy error protection based on H.264/AVC redundant slices and flexible macroblock ordering. *Journal of Zhejiang University. Science*, *A*(7), 900–909.

Bai, F., Sadagopan, N., & Helmy, A. (2003). A framework to systematically analyze the impact of mobility on performance of routing protocols for ad-hoc networks. In *IEEE INFOCOM* (pp. 825–835). IMPORTANT.

Biswas, S., Tatchikou, R., & Dion, F. (2006). Vehicle-to-vehicle wireless communication protocols for enhancing highway traffic safety. *IEEE Communications Magazine*, *44*(1), 74–82.

Broch, J., Maltz, D. A., Johnson, D. B., Hu, Y.-C., & Jetcheva, J. (1998). A performance comparison of multi-hop wireless ad-hoc network routing protocols. *4th Annual ACM/IEEE International Conference on Mobile Computing and Networking* (pp. 85-97).

Bucciol, P., Ridolfo, F., & de Martin, J. C. (2008). Multicast voice transmission over vehicular ad-hoc networks: Issues and challenges. In 7th Int. Conference on Networking (pp. 746-751).

Buford, J. F., Yu, H., & Lua, E. K. (2009). *Networking and applications* (p. 2P). Burlington, MA: Morgan Kaufmann.

Chakareski, J., Han, S. & Girod, B. (2005). Layered coding vs. multiple descriptions for video streaming over multiple paths. *Multimedia Systems*, online journal.

Chow, C.-O., & Ishii, H. (2007). Enhancing real-time video streaming over mobile ad-hoc networks using multipoint-to-point communication. *Computer Communications*, *30*(8), 1754–1764.

Doukas, A., & Kalivas, G. (2006). Rician K-factor estimation for wireless communication systems. In *IEEE International Conference on Wireless and Mobile Communications* (pp. 69-73).

Erceg, V., Ghassemzadeh, S., Taylor, M., Li, D., & Schilling, D. (1992). Urban/suburban out-of-sight propagation modelling. *IEEE Communications Magazine*, *30*(6), 56–61.

Fiore, M., Härri, J., Filali, F., & Bonnet, C. (2007). Vehicular mobility simulation for VANETs. In *40th Annual Simulation Symposium* (pp. 301-307).

Giordano, E., Ghosh, A., Pau, G., & Gerla, M. (2008). Experimental evaluation of peer-to-peer applications in vehicular ad-hoc networks. In *1st Annual International Symposium on Vehicular Computing Systems.*

Guo, M., Ammar, M. H., & Zegura, E. W. (2005). V3: A vehicle-to-vehicle live video streaming architecture. In *3rd IEEE International Conference on Pervasive Computing and Communications* (pp. 171-180).

Gürses, E., & Kim, E. (2008). Maximum utility peer selection for P2P streaming in wireless ad-hoc networks. In *IEEE GLOBECOM* (pp. 1-5).

Härri, J. (2010). Vehicular mobility modeling for VANET. In Hartenstein, H., & Laberteaux, K. P. (Eds.), *VANET: Vehicular applications and inter-networking technologies, Chichester, UK* (pp. 107–156). Hoboken, NJ: Wiley & Sons.

Hei, X., Liang, C., Liang, J., Liu, Y., & Ross, K. W. (2007). A measurement study of a large-scale P2P IPTV system. *IEEE Transactions on Multimedia, 9*(8), 1672–1687.

Ho, T., & Lun, D. S. (2008). *Network coding.* Cambridge, UK: Cambridge University Press.

Hu, Y. C., Pucha, H., & Das, S. M. (2003). Exploiting the synergy between peer-to-peer and mobile ad-hoc networks. In *9th Workshop on Hot Topics in Operating Systems.*

Jiang, D., & Delgrossi, L. (2008). IEEE 802.11p: Towards an international standard for wireless access in vehicular environments. In *IEEE Vehicular Technology Conference* (pp. 2036-2040).

Jurca, D., Chakereski, J., Wagner, J., & Frossard, P. (2007). Enabling adaptive video streaming in P2P systems. *IEEE Communications Magazine, 45*(6), 108–114.

Katz, B., Greenberg, S., Yarkoni, N., Blaunstein, N., & Giladi, R. (2007). New error-resilient scheme based on FMO and dynamic redundant slices allocation for wireless video transmission. *IEEE Transactions on Broadcasting, 53*(1), 308–319.

Kesting, A., Treiber, M., & Helbing, D. (2007). General lane-changing model MOBIL for car-following models. *Transportation Research Record: Journal of the Transportation Research Board, 1999*, 86–94.

Ko, Y. B., & Vaidya, N. H. (2000). Location Aided Routing (LAR) in mobile ad-hoc networks. *Wireless Networks, 6*, 307–321.

Lambert, P., de Neve, W., Dhondt, Y., & van de Walle, R. (2006). Flexible macroblock ordering in H.264/AVC. *Journal of Visual Communication, 17*, 358–375.

Lee, S. J., Hsu, J., Hayashida, R., Gerla, M., & Bagrodia, R. (2003). Selecting a routing strategy for your ad-hoc network. *Computer Communications, 26*(7), 723–733.

Lindsey, W. (1964). Error probabilities for Rician fading multichannel reception of binary and N-ary signals. *IEEE Transactions on Information Theory, 10*(4), 339–350.

Liu, B., Khorashadi, B., Du, H., Ghosal, D., Chuah, C., & Zhang, M. (2009). VGSim: An integrated networking and microscopic vehicular mobility simulation platform. *IEEE Communications Magazine, 47*(5), 134–141.

Liu, Y., Guo, Y., & Liang, C. (2008). A survey on peer-to-peer video streaming systems. *Journal of P2P Networking and Applications, 1*(1), 18-28.

Liu, Y.-S., Nuang, Y.-M., & Hsieh, M.-Y. (2006). Adaptive P2P caching for video broadcasting. In *Joint Conference on Information Sciences, 2006.*

MacKay, D. J. C. (2005). Fountain codes. *IEE Proceedings. Communications, 152*(6), 1062–1068.

Mao, S., Lin, S., Panwar, S. S., Wang, Y., & Celebi, E. (2003). Video transport over ad-hoc networks: Multi-stream coding with multipath transport. *IEEE Journal on Selected Areas in Communications, 21*(4), 1721–1737.

Matolak, D. W. (2008). Channel modeling for vehicle-to-vehicle communications. *IEEE Communications Magazine, 46*(5), 76–83.

Niraula, M. B., Kanchanasut, K., & Laouiti, A. (2009). Peer-to-peer live video streaming over mobile ad-hoc network. In *International Conference on Wireless Communications and Mobile Computing* (pp. 1045-1050).

Oda, Y., Tsunekawa, K., & Hata, M. (2000). Advanced LOS path-loss model in microcellular mobile communications. *IEEE Transactions on Vehicular Technology, 49*(6), 2121–2125.

Oishi, J., Asukura, K., & Watanabe, T. (2006). A communication model for inter-vehicle communication simulation systems based on properties of urban area. *International Journal of Computer Science and Network Security, 6*, 213–219.

Park, J.-S., Lee, U., Oh, S. Y., Gerla, M., & Lun, D. (2006). *Emergency related video streaming in VANETs using network coding (Technical Report, TR-070016)*. UCLA CSD.

Perkins, C. E., & Royer, E. M. (1999). Ad-hoc on-demand distance vector routing (AODV). In *2nd IEEE Workshop on Mobile Computing Systems and Applications* (pp. 90-100).

Pucha, H., Das, S. M., & Hu, Y. C. (2004). Ekta: An efficient DHT substrate for distributed applications in mobile ad hoc networks. In *6th IEEE Workshop on Mobile Computing Systems and Applications* (pp. 163–173).

Qadri, N. N., Fleury, M., Altaf, M., & Ghanbari, M. (2010). Multi-source video streaming in a wireless vehicular ad hoc network. *IET Communications, 4*(11), 1300–1311.

Reza, R. (2006). Anyone can broadcast video over the Internet. *Communications of the ACM, 49*, 55–57.

Rofoee, B. R., Fleury, M., Qadri, N. N., & Ghanbari, M. (2010). (in press). A flexible family of multi-path routing protocols over a MANET. *International Journal of Ad Hoc and Ubiquitous Computing*.

Rustako, A. J. Jr, Amitsay, N., Owens, G. J., & Roman, R. S. (1991). Radio propagation at microwave frequencies for line-of-sight microcellular mobile and personal communications. *IEEE Transactions on Vehicular Technology, 40*(1), 203–210.

Schreier, R. M., & Rothermel, A. (2006). Motion adaptive intra refresh for the H.264 video coding standard. *IEEE Transactions on Consumer Electronics, 52*(1), 249–253.

Setton, E., Zhu, X., & Girod, B. (2004). Congestion-optimized multipath streaming of video over ad-hoc wireless network. In *IEEE International Conference on Multimedia and Expo* (pp. 1619-1622).

Sklar, B. (1997). Rayleigh fading channels in mobile digital communication systems part I: Characterization. *IEEE Communications Magazine, 35(7)*, 90-100. Retrieved from http://en.wikipedia.org/ wiki/ Digital_object_identifier

Stockhammer, T., & Zia, W. (2007). Error-resilient coding and decoding strategies for video communication. In Chou, P. A., & van der Schaar, M. (Eds.), *Multimedia in IP and wireless networks* (pp. 13–58). Burlington, MA: Academic Press.

Takai, M., Martin, J., & Bagrodia, R. (2001). Effects of wireless physical layer modeling in mobile ad-hoc networks. In *2nd ACM International Symposium on Mobile ad-hoc Networking & Computing* (pp. 87-94).

Tan, I., & Bahai, A. (2010). Physical layer considerations for vehicular communications. In Hartenstein, H., & Laberteaux, K. P. (Eds.), *VANET: Vehicular applications and inter-networking technologies, Chichester, UK* (pp. 157–217). Hoboken, NJ: Wiley & Sons.

Tang, B., Zhou, Z., Kashyap, A., & Chiueh, T. (2005). An integrated approach for P2P File sharing on multi-hop wireless networks. In *IEEE International Conference on Wireless and Mobile Computing, Networking and Communication* (pp. 268-274).

Tonguz, O., Viriyasitavat, W., & Bai, F. (2009). Modeling urban traffic: A cellular automata approach. *IEEE Communications Magazine*, *47*(5), 142–150.

Treiber, M., Hennecke, A., & Helbing, D. (2000). Congested traffic states in empirical observations and microscopic simulations. *Physical Review E: Statistical Physics, Plasmas, Fluids, and Related Interdisciplinary Topics*, (62): 1805–1824.

Vars, V., & Hannuksela, M. N. (2001). *Non-normative error concealment algorithms, ITU-T SGI6 Doc., VCEG-N62*. CA: Santa Barbara.

Wang, Y., Reibman, A. R., & Lee, S. (2005). Multiple description coding for video delivery. *Proceedings of the IEEE, 93*(1), 57–70.

Wang, Y., & Zou, Q. F. (1998). Error control and concealment for video communication: A review. *Proceedings of the IEEE, 86*(5), 974–997.

Wellens, M., Petrova, M., Riihijarvi, J., & Mahonen, P. (2005). Building a better wireless mousetrap: Need for more realism in simulations. In 2nd Annual Conference on Wireless On-demand Network Systems and Services (pp. 150-157).

Wenger, S. (2003, July). H264/AVC over IP. *IEEE Transactions on Circuits and Systems for Video Technology*, *13*(7), 645–656.

Wu, H., Fujimoto, R. M., & Riley, G. (2004). Analytical models for information propagation in vehicle-to-vehicle networks. In *Vehicular Technology Conference* (pp. 4548-4552).

Xie, F., Hua, K. A., Wang, W., & Ho, Y. H. (2007). Performance study of live video streaming over highway vehicular ad-hoc networks. In *Vehicular Technology Conference* (pp. 2121-2125).

Xu, Q., Mak, T., Ko, J., & Sengupta, R. (2004). Vehicle-to-vehicle safety messaging in DSRC. In *1st ACM International Workshop on Vehicular Ad-hoc Networks* (pp. 19-28).

Yan, L. (2005). Can P2P benefit from MANET? Performance evaluation from users' perspective. In *International Conference on Mobile Sensor Networks* (pp. 1026-1035).

Yong, L., Guo, Y., & Liangl, C. (2008). A survey on peer-to-peer video streaming systems. *Peer-to-Peer Networking and Applications*, *1*, 18–28.

Yousefi, S., Mousavi, M. S., & Fathy, M. (2006). Vehicular ad-hoc networks (VANETS): Challenges and perspectives. In *6th International Conference on ITS Telecommunications* (pp. 761-766).

Zahn, T., & Schiller, J. (2008). Designing structured peer-to-peer overlays as a platform for distributed network applications in mobile ad hoc networks. *Computer Communications*, *31*(3), 643–654.

Zeng, X., Bagrodia, R., & Gerla, M. (1998). GloMoSim: A library for parallel simulation of large-scale wireless networks. In *Workshop on Parallel and Distributed Simulation* (pp. 154-161).

Zuo, D.-H., Du, X., & Yang, Z.-K. (2007). Hybrid search algorithms for P2P media streaming distribution in ad-hoc networks. In *7th International Conference on Computational Science* (pp. 873-876).

ADDITIONAL READING

Goldsmith, A. (2005). *Wireless Communications*. Cambridge, UK: Cambridge University Press.

Hartenstein, H., & Laberteaux, K. P. (Eds.). (2010). *VANET: Vehicular applications and inter-networking technologies*. Chichester, UK: Wiley & Sons.

Rappaport, T. (2001). *Wireless communications: Principles and practice*. Upper Saddle River, NJ: Prentice-Hall.

Schaar, Van der M., & Chou, P. (2007). *Multimedia over IP and wireless networks*. Amsterdam, Netherlands: Academic Press.

Shen, X., Yu, H., Buford, J., & Akon, M. (2010). *Handbook of peer-to-peer networks*. New York, NY: Springer.

Stallings, W. (2000). *Wireless communications and networks*. Upper Saddle River, NJ: Prentice-Hall.

ENDNOTES

1 Obstacles include lane closures, uphill gradients, and potholes.

2 For interactive video one-way delays up to 250 ms are possible, though delays of around 100 ms are ideal.

Chapter 14
Multi-Platform Video Streaming Implementation on Mobile Terminals

Elsa Macías
ULPGC, Spain

Alvaro Suarez
ULPGC, Spain

Francisco Espino
ULPGC, Spain

ABSTRACT

Video streaming technique is at present the key factor of some multimedia services. The efficient implementation of multimedia services in mobile telephones is currently a major challenge. Today, the deployment of multi-platform multimedia software in high-end mobile telephones is also a challenge. In this chapter, we will review current and upcoming mobile multimedia services and programming frameworks for video streaming software deployment in mobile telephones. We will show that the multi-platform implementation of video streaming clients in mobile telephones remains a fallacy. However, we did make innovative progress regarding the critical challenge of mitigating the adverse effects of long lasting service disruptions impacting a video streaming service. For instance, if a service disruption lasts a long time, the Server will terminate the video streaming session; subsequently, such an event usually leads to a user re-initiating the terminated session from scratch which provokes a huge loss of performance for the streaming technique; furthermore, these types of disruptions tend to be sporadic and impossible to predict. We used JADE-LEAP for programming a multi-platform software solution for the video streaming service disruptions' challenge by utilizing the internal service of JADE-LEAP which automatically controls the availability of the mobile telephone (that is, the service disruption). We implemented a novel mechanism (protocol) using that internal service to mitigate the adverse effects of video streaming service disruptions. It improved the user experience spectacularly reducing the jitter and minimizing video communication latency.

DOI: 10.4018/978-1-61350-144-3.ch014

INTRODUCTION

Multimedia communication (Pagani, 2009) tends to be delay sensitive, bandwidth intense and loss tolerant; furthermore, it must also satisfy the user experience. A multimedia system processes and communicates with independent pieces of information (video, audio and text) in order to eventually synchronize them. The overlapping of these processing and communication tasks is a technique whose objective is to achieve an efficient use of bandwidth and to hide latency. Ming (2000) named *Media Streaming* as the application of streaming technique to multimedia communications; it consist on the overlapping of the processing of the information pieces in the hosts (terminals) and the communication of their associated streams in the network. An up to date overview of Media Streaming have been edited by Zhu et al. (2011).

Mobile devices has experimented an important evolution in the last recent years. On the one hand, mobile manufacturers offer high-end mobile telephones with on chip *Wireless Network Interface Cards* (*WNIC*) like Bluetooth, *Wireless Fidelity* (*WiFi*), *Global Positioning System* (*GPS*) and *Third Generation* (*3G*) Wireless networks. On the other hand, mobile telco operators have improved their networks to facilitate the access to *Video on Demand* (*VoD*) servers (Qadeer & Ahmad, 2009) and to satisfy the user requirements (Pagani & Schipani, 2005). Savo (2006) identified some general challenges that must be taken on to satisfy those user requirements. On the one hand, the dynamic adaptation of the wireless channel and *Medium Access Control* protocol to the number of active users in a wireless *Access Point* (*AP*). On the other hand, the design of routing algorithms, transport protocols, security procedures and mobility management. Apostolopoulos and Trott (2005) remarked that the efficient implementation of wireless video streaming service presents some important differences as compared to other services.

There are recent and interesting research works in mobile media streaming. A critical challenge in mobile media streaming is to mitigate the adverse effects of video streaming service disruptions (hence referred to as "service disruptions"). For instance, if a mobile telephone suffers a long lasting loss of communication, i.e. a service disruption, the Server will terminate the video streaming session; thus the user starts a new video streaming session from scratch and the Server has to repeat again the delivery of the video from the beginning which implies a huge loss of performance. Different authors have identified potential scenarios and reasons for service disruptions. Video and signaling packets are frequently delayed, lost or even discarded (Tse & Viswanath, 2005; Ippolito, 2008). The mobile telephone disconnects while it is roaming (Gupta & Williams, 2006), it is moving in a boundary of the coverage area (WiFi) or it is stopped in a radio degraded zone (Suárez & Menza, 2007). The mobile telephone changes its *Internet Protocol* (*IP*) address (Bolla & Mangialardi, 2009) while it is receiving video frames: they proposed the adaptation of *Real Time Streaming Protocol* (*RTSP*) to avoid the VoD Server will terminate the streaming session and starts a new session from scratch when the mobile device changes its IP. This is a very interesting work but its implementation in commercial VoD servers is improbable.

We implemented an automatic and multi-platform software mechanism to mitigate the adverse effects of service disruptions. One important obstacle was the absence of practical multi-platform *Software Development Kits* (*SDK*). Specifically, we tried to minimize the adverse effects of the *Jump distance* and *Time off* problems presented by Costa and Cunha (2004). We considered a test scenario in which the mobile telephone was connected to a WiFi AP or a 3G mobile network *Base Station* (*BS*) and the VoD Server was inside wired Internet. We used the pure *RTSP* (Schulzrinne & Rao, 1998) because it has been the official standard protocol for video streaming. We also used pure network and transport Internet protocols. Our aim was to prevent the Server from terminating the session, thus the Client would not loss video frames during a service disruption. A general approach to avoid video frames loss was to design proxy architecture (see (Gao & Zhang, 2003) for definitions on proxy architecture). The proxies could implement a proactive buffering mechanism similar to that used by Bellavista and Corradi (2005) or Wu and Hsu (2007). We used a standard software agent platform named *Java Agent DEvelopment framework-Lightweight Extensible Agent Platform* (*JADE-LEAP*) (Bellifemine & Caire, 2003; Caire & Pieri, 2009) to specify the behaviour of the proxies which implement proactive buffering. JADE-LEAP controlled the physical disconnections and reconnections of the mobile telephone raising appropriated events. We took into account these events to specify the control and buffering actions in the behavior of the agents. The main reasons we used JADE-LEAP was that it can be used in current mobile telephones; furthermore, it is a completely open *Java* software platform which uses *Foundation for Intelligent Physical Agents* (*FIPA*) for messaging; moreover, it has been a consolidated project that is partnered by enterprises like: *Telecom Italia*, *Motorola*, *Whitestein Technologies AG*, *Profactor GMBH* and *France Telecom R&D*.

The rest of this book chapter is organized as follows. In Section 2 we review the highly heterogeneous and not interoperable streaming software frameworks for programming video streaming clients in current high-end mobile telephones. In section 3 we discuss some effects of media streaming service disruptions. In section 4 we present our agent based multi-platform to mitigate those effects and experimental results. We present our further research directions in section 5. Finally, we summarize our conclusions.

DESIGNING MULTI-PLATFORM MOBILE STREAMING SOFTWARE

The multi-platform implementation in mobile telephones of the old and the above new services based on video streaming is a challenge. The major companies of software and mobile telephones have not agreed on a common SDK to design a full interoperable Server, Client and protocol for video streaming. In this section we first present implementation problems of current clients, servers and protocols and we second sum up some conclusions.

The implementation of video streaming clients in mobile telephones requires the understanding of mobile telephones hardware evolution, their operating system and their programming platforms.

Present high-end mobile telephones have 0.5 GHz processors, up to 32 GB memory, heterogeneous WNIC, up to 12 M Pixels cameras and touch screen. Upcoming mobile telephones (named *ultra-small-mobile-computers*) could include the *SnapDragon* architecture: 1 GHz processor, 0.6 GHz *Digital Signal Processor* (*DSP*), multiple audio and video *codec* support, 1280x720 screen resolution and direct Mobile *Television* (*TV*) support. These mobile telephones accept powerful operating systems (with the same characteristics of the laptop version): W*indows Phone 7*, *Linux embedded*, *MAEMO*, *Symbian*, *Mac OS* (*iPhone*), *Android*, *Web OS* (*Palm Pre*) and *Blackberry* Java based operating system. The *Open Handset Alliance* around *Android* will produce new mobile telephones like *A1, Ophone mini3*, *H7*, *O1 Ophone*, *Xperia X3 Rachael...*

The programming frameworks of existing mobile telephones are mainly proprietary and they are not fully compatible. Each operating system (even each mobile telephone series) includes its particular SDK. Examples of different programming frameworks are *.NET compact framework*, *BREW*, *Java 2 Micro Edition* (*J2ME*) and *Objective C* (*Cocoa*). The aim of J2ME SDK is to generate multi-platform clients. But different implementations of the *Java Virtual Machine* produce non-compatible clients. An interesting tool named *alcheMo* translates standard applications compliant with J2ME to *BREW*, *Mac OS* and. *NET*. To the best of our knowledge, it does not work in *Android* and *Web OS*. The *alcheMo* tool provides J2ME equivalent functionality support for automatic garbage collection, multi-threading, extensive subsets of *Connected Limited Device Configuration* (*CLDC*) 1.1, *Mobile Information Device Profile* (*MIDP*) 2.0 and *Java Specification Requests* (*JSRs*).

The *Mobile Media Application Programming Interface* (*MMAPI*), compliant with RealNetworks media Server, is a library used by J2ME to generate multi-platform players but also present interoperability problems (Goyal, 2006). Belda (2006) adapted *Java Media Framework* (*JMF*) to an old series of Nokia mobile telephones with relatively low success. The *Freedom for Media in Java* (*FMJ*) does not consider mobile telephones architectures. A very promising scripting based framework over *J2ME* named *JavaFX 1.2 for mobiles* (Weaver & Gao, 2009) can work with *Groovy*, *Grails* and *Android*. But it does not work in commercial high-end mobile telephones.

Major software companies provide powerful media streaming servers for commercial use, for example: *Windows Media Server* (generally included in the *Windows Server* operating system), the *RealNetworks Media Server*, the *Darwin Streaming Server* (*Apple*) and the *Adobe Flash Media Server* (Sanders, 2008). These companies release both free limited and full (of payment) versions. Interesting examples of free open source video streaming servers are *VideoLAN* and *RED5* project (Allen & Arnold, 2008) that is part of the *openflash* project. They have been implemented in desktop versions of *Windows*, *Linux* and *Mac* operating systems using the *ffmpeg* library.

Normally the proprietary media servers work better with their compliant Client software. *Windows Media Server* does it with *Windows Media Player* (usually included in the *Windows Mobile* operating system with *.NET compact framework*). This Server should deliver information compatible with Microsoft proprietary *Silverlight* technology for mobile telephones. But this is still in its infancy. The *RealNetworks* Server is well intended only for clients compatible with MMAPI under *Symbian*. *Darwin Streaming Server* is well adapted to *Quick Time* Client. The *Flash Player 10* (and its compliant *Flash Media Server Version 10*) will be available for a wide range of mobile operating systems except for *iPhone*. The VideoLAN Client is not well-supported by *Android* and its performance on current mobile telephones is low. The RED5 client is not implemented in present high-end mobile telephones.

The implementation of the RTSP also varies in present mobile telephones. *Windows Media Server* implements a proprietary *Microsoft© Media Services* (*MMS*) protocol. *RealNetworks Media Server* also implements a proprietary protocol including the *Real Data Transport* (*RDT*) for video communication. Both RDT and MMS can use an implementation of *TCP Friendly Rate Control* (*TFRC*) (Handley & Floyd, 2003) for signaling and video communication. *Darwin Streaming Server* uses a pure RTSP implementation. *Flash Media Server* and RED5 use *Real Time Messaging Protocol* (*RTMP*) running over TCP. Lastly, *VideoLAN Server* uses several protocols: MMS, RDT and pure RTSP. Recently *Apple* has done a revival of *Hotmedia* project (Kumar & Lipscomb, 2001) defining *HiperText Transfer Protocol* (*HTTP*) *Live Streaming* (Pantos, 2009).

We have sniffed traffic of several video streaming protocols. Next we present parts of some protocol conversations in tables 1 (for iPhone), 2 (for Android simulator) and 3 (for MMS) analyzing different work cases for each platform.

We reached the following conclusions:

- A given proprietary video streaming Server usually requires a compliant Client. Open source video streaming servers do not run efficiently in any mobile telephone.

Table 1. HTTP for iPhone mobile telephone. This protocol show a versatile behavior for live sessions and VoD

The iPhone mobile telephone issues an HTTP connection and closes it. The Server response contains multimedia data broadcasting. - Client: GET /15834-33555 HTTP/1.0 ; … ; Connection: Close. - Server: HTTP/1.0 200 OK; Content-Type: application/vnd.ms.wms-hdr.asfv1 ; Pragma: no-cache, client-id=2277275327, xResetStrm=1, features="broadcast"; Multimedia data file.
The iPhone mobile telephone issues an HTTP connection and closes it. The Server responds sending partial content of the media file and closing the connection. - Client: GET /videoplayback?id=002f1f4964cadfba & itag=18 & ip=0.0.0.0 & ipbits=0 & expire=1247964667 & spara ms=ip%2Cipbits%2Cexpire%2Cid%2Citag & signature=827FD33A87C7826053415815C3F134ED80570D38. AEBF6E5C815FF659DB82AE25691AD579F9300800 & key=dk & el=standard & client=ytapi-apple-iphone & devKey =YdvQWEjeVwCvghTFBAYp6GD9LlbsOl3qUImVMV6ramM & app=youtube_gdata HTTP/1.1; … ; Connection: Close. - Server: HTTP/1.1 206 Partial Content ; Content-Disposition: attachment; filename="video.mp4" ; Connection: close ; Server: gvs 1.0.
The iPhone mobile telephone issues an HTTP connection and keeps it alive. The Server responds sending a multimedia data file. - Client: POST /gateway.php HTTP/1.1 ; … ; Connection: keep-alive. - Server: HTTP/1.0 200 OK; Content-Type: application/x-amf ; Pragma: no-store; Multimedia data file.

Table 2. Simulator of Android mobile telephone. The simulator can use several video streaming protocols

The Android mobile telephone issues an HTTP connection and closes it using the Android.Mediaplayer Application Programming Interface (API). The Server responds sending a multimedia data file and closing the connection. - Client: GET /sample_100kbit.mp4 HTTP/1.0 ; … ; Connection: Close. - Server: HTTP/1.1 200 OK; Connection: close ; Multimedia data file.
The Android mobile telephone and the Server maintain a pure RTSP conversation. - Client: OPTIONS rtsp://…/sample_100kbit.mp4 RTSP/1.0 ; CSeq: 0. - Server: RTSP/1.0 200 OK; CSeq: 0. - Client: DESCRIBE rtsp://…/sample_100kbit.mp4 RTSP/1.0 ; CSeq: 1. - Server: RTSP/1.0 200 OK; CSeq: 1. - Client: SETUP rtsp://… /sample_100kbit.mp4/trackID=3 RTSP/1.0 ; CSeq: 2. - Server: RTSP/1.0 200 OK; CSeq: 2. - Client: PLAY rtsp://193.145.145.177/sample_100kbit.mp4 RTSP/1.0 ; CSeq: 3. - Server: RTSP/1.0 200 OK; CSeq: 3. - Client: TEARDOWN rtsp://193.145.145.177/sample_100kbit.mp4 RTSP/1.0 ; CSeq: 4. *- Server: RTSP/1.0 200 OK; CSeq: 4.*

Table 3. Type of sessions of the MMS protocol. In full implementation, Microsoft Media Server and Windows Media Player work together; otherwise a slight version is used. MMS protocol distinguishes between VoD and live sessions

VoD session: - MMS protocol uses TCP for signaling and video communication. - The PAUSE command stops the session.
Live session: - MMS protocol uses *UserDatagram Protocol* (*UDP*) for signaling and video communication. - The PAUSE command terminates the session.

- Video streaming clients are programmed in particular SDK and the transformation of software among different SDK is not easy.
- The source code of video streaming protocols is not released. This makes difficult or impossible the connection between current servers (and clients) and proxies programmed by independent third parties.

MOBILE VIDEO STREAMING BASED SERVICE DISRUPTIONS

The above conclusions present some obstacles to the designing of a multi-platform application level mechanism to control service disruptions. In this section we introduce a simple model for video streaming technique and analyze some parameters influencing the service disruptions.

A Simple Performance Model for Video Streaming

As we mentioned in the introduction, video streaming technique tries to efficiently overlap the displaying, storage and communication of video frames. No lasting information is stored in the Client device (only a small buffer). The main components of Client-Server architecture for video streaming are:

Media frames. Pieces of video, image or text are encapsulated into data packets. They are communicated independently and should be eventually synchronized in the Client.

Server. It delivers the data packets from any sink of information to the Client. It normally consists of a communication protocol thread and a sink receiving thread (trans-coding threads can be also implemented). Important implementation issues are the scaling, load balancing (Frey & Hasler, 2009), the reduction in disk utilization rates and power consumption (Le Moal & Moralo, 2009).

Client. It receives the data from the Server and displays them. Normally it consists of a communication protocol thread and a displaying thread (trans-coding threads are not usually implemented).

Network. It can be abstracted as a set of nodes (hosts) and communication links. We suppose a heterogeneous network. The Server is allocated in a wired node and the last mile is a wireless link in which a mobile telephone allocates the Client. We also suppose the nodes are passive, there is no multipath communication, no re-routing and that the programmer uses sockets for video streaming.

Buffering system. The Client and the Server must arrange a buffering system to temporally store the media frames in order to avoid the jitter in the network and minimize the latency of communication.

Sessions. A session is allocated for each stream using an individual IP address and ports on the Client and Server. A data structure containing IP, ports, buffering data, etc. are allocated in the Server for each session. We suppose that the session initialization and the activation of its associated data structure take a non depreciable amount of time.

Application protocol. The RTSP Client establishes a session sending a SETUP command (additional commands can be used). After that it sends a PLAY command to start the video streaming session. We suppose that the delay of the session starting has a cost we name *Session Starting Cost* (*SSC*).

User terminal. It needs a very small buffer: it can be from a high-end mobile telephone to a gadget like an MP3, MP4 or a gaming device. We are interested in high-end mobile telephones.

Sink of multimedia information. The sink can be from a camera of a mobile telephone to a TV digitizer NIC. Curious sinks are: Unmanned Aerial Vehicle (Zhou, 2009), and Distributed Smart Cameras (Doblander & Zoufal, 2009).

In Figure 1 we present an illustrative graphic to introduce a simple mathematical performance model for service disruptions. In Figure 1.a we show a schematic picture of the Client (on the left), the Network (on the center) and the Server (on the right). Also with numbered square blocks (*1* to *n*) we indicate the direction of data packets communication from the Server to the Client. In Figure 1.b we show a temporal diagram of data and signaling communication (vertical axis) in front of the resource utilization (horizontal axis). First of all, the Client sends a *SETUP* and the Server will respond opening the session (a 200 OK command). The Client sends a *PLAY* and data transmission will start. Let us suppose that one data packet per unit of time arrives at the Client's buffer, after *n* units of time all the *n* packets will arrive at the Client's buffer. Let us observe the overlapped actions. At one particular unit of time, the communication of packets, buffering, visualization and reception of data in the Server actions are overlapped. The first fill of the Server's buffer and the flush of the Client's buffer actions do not overlap. We suppose the first fill of the Server and the Client buffers are included in *SSC*. Therefore, the simplified calculation of time (*T*) needed to finish the displaying of the *n* packets is:

Figure 1. a) Main components of a Client-Server media streaming system. b) A temporal data and signaling communication and overlapped actions

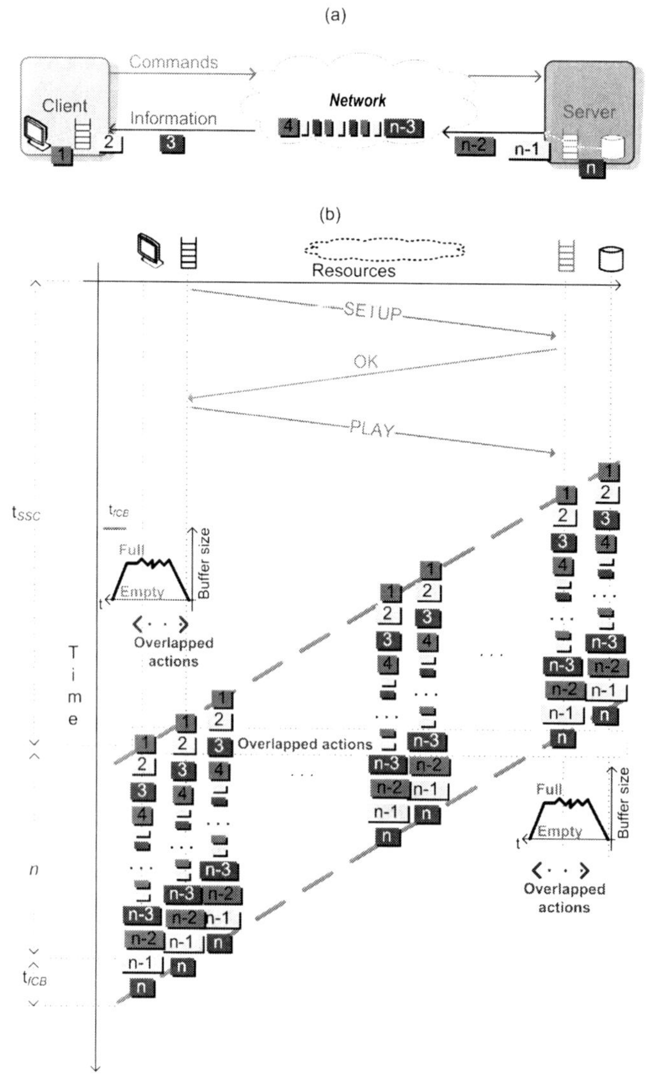

$$T = T_{SSC} + n + T_{fCB} \quad (1)$$

where, T_{SSC} is the unit of time the Client displays the first data packet and T_{fCB} is the time the Client needs to flush its buffer.

This model is a linear function where the dominant value is T_{SSC} (in an existing Client the displaying starts after its buffer is full). Our aim is only to show, at very high level of abstraction, the adverse effects of service disruptions. But in practical cases this simplistic model encloses additional complexity. For instance, the network behavior is not so simplistic and different data packets suffer different delays (variable jitter). A more complex model must be designed to mitigate the jitter effect and to control the *Client buffer duration*. Chou (2007) presented an excellent introduction to the modeling of Client buffering control. Firstly, he introduced a simplistic case of VoD *Constant Bit Rate* traffic. Secondly, he complicated the model to consider *Variable Bit Rate* traffic. Lastly, he treated the case of live broadcasting traffic. In all cases he analyzed the different parameters that influence the buffer sizes. Our aim in this chapter is not the formulation of a complete streaming mathematical model. But those interested in a completed mathematical modeling should read the integrated and very detailed multi-level (physical, link, network, transport and application) model formulated by Gomez (2009). For each level Gomez considered a multi-parameter model and composed the global model to adjust parameters of the different levels (he did not use a cross layer modeling).

Analysis of Video Streaming Based Services Disruptions

In this section we first analyze the causes of service disruptions at Physical, Link, Network, Transport and Application levels. We second present the adverse effects of service disruptions and our conclusions.

At physical level the main cause that influences the streaming service disruptions is the radio disconnections of the mobile telephone. In a given geographical point, the radio wireless channel eventually varies along time sporadically and intermittently. If a mobile telephone is moving in the boundaries of the coverage area, it will probably suffer intermittent radio disconnections. Inside the coverage area there can be *holes* (zones with no radio coverage or very poor radio coverage). If the mobile terminal makes a long lasting stay in one of these holes, it will suffer a radio disconnection. A total elimination of indoor holes requires allocating a lot of AP. But this increases the number of *roaming* processes (that also produces a temporal radio disconnection). A total elimination of outdoor holes requires allocating a lot of BS. But this increases the economic cost of the 3G network. Another source of service disruptions is the interference of collocated radios or concurrent access to shared radio channel (Zhu &Yang, 2007). The mobile telephone shown in Figure 2 accesses a VoD Server in Fixed Internet (circles inside the coverage area represent holes) while it is moving.

The physical disconnections of the mobile telephone imply it will perceive variable bandwidth at link level. But also an unfair MAC protocol also will affect the perceived bandwidth by the mobile telephone (we center our attention only in the wireless link). Laksono (2004) presented an interesting study about service disruptions at link level. On the one hand, he mathematically demonstrated that long lasting disconnections (seconds or few minutes) can produce buffer underrun. On the other hand, he argued that those disconnections must be controlled using buffers, monitoring the network and controlling the input video rate. We thought it was a good solution at link level, but insufficient for the end to end application level. This is the reason we explored application level mechanisms to mitigate the adverse effects of service disruptions.

Figure 2. Mobile telephone accessing Internet streaming server in an area of coverage where there are internal zones of no coverage and it is sometimes in a zone where there is no coverage

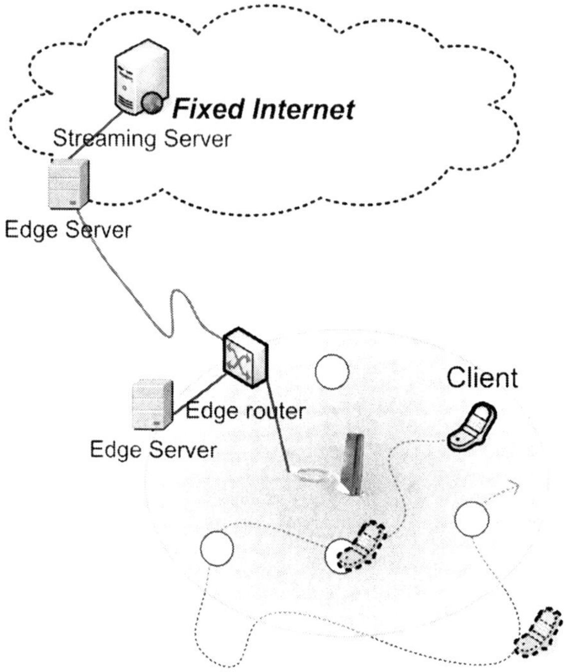

At Network level, we considered mobile telephones and servers with public static *Internet Protocol* (*IP*) address. In the last recent years some work has been done at Internet Transport level to mitigate the effects of Client unavailability. Daniel and Macedo (2008) presented a survey of problems that arise in the context of Internet main protocols for gadgets and mobile telephones. New transport protocols have been installed in the VoD Server to communicate video and signaling. The more important are *TFRC* (Handley & Floyd, 2003), *Datagram Congestion Control Protocol* (*DCCP*) (Kholer & Handley, 2006) and *Stream Control Transmission Protocol* (*SCTP*) (Stewart & Xie, 2000). Only TFRC is used in current VoD servers. We adapted the working philosophy of reliable sockets (Ansari & Sathyanath, 2007; Zandy & Miller, 2002) for UDP/*Real Time Protocol* (*RTP*) in order them not to close the connection for a long lasting service disruptions. But the implementation of original reliable sockets produced random failures and there was no way to control the abort of sockets in the Client and in the Server (Moreno, 2009).

In the current operating systems, the WNIC driver reports information about the wireless signal strength in the mobile terminal. Lee and McKillop (2006) integrated the wireless signal strength, a GPS, an accelerometer and a *ping* signaling message (at network level) to patent an outdoor user warning system. The main idea was to proactively warn the user before its mobile telephone will be out of coverage or the wireless signal strength will be poor. To the best of our knowledge, this patent has not been implemented in current mobile telephones. A mechanism to mitigate the adverse effects of service disruptions can integrate the ideas of this patented method with other application level parameters. For instance, let us suppose: First, the moving mobile telephone can read a coverage map of the geographical zone. Second, there is a minor temporal variation of wireless channel behavior. Third, it knows it will be in a zone with poor or no coverage after *m* minutes. Lastly, the buffer can store up to *s* seconds of video in a VoD session. If the mobile telephone requests the VoD Server to send it *s* seconds of video before the *m* minutes, it will support service disruptions up to *s* seconds. Let us note that this cannot be done in live sessions and service disruptions longer than *s* cannot be supported.

In Figure 3 we show a schema that represents the effects of two consecutive service disruptions. If a service disruption occurs, a new one will start from scratch. The visualization is paused for a while and the lost packets flood the wireless channel decreasing its effective bandwidth. For those reasons commercial video servers detect the availability of the Client and try to reconnect it till 2 minutes after. But open source servers like VideoLAN does not control it.

We reached the following conclusions to be taken into account for the design of an application level mechanism to mitigate the adverse effects of service disruptions:

- The mechanism must manage parameters of different levels of the network architecture. It must control the availability of the mobile telephone and the termination of the video streaming session. It must avoid the flooding of the wireless channel with video frames that will never be received by the mobile telephone. It must buffer the video frames to send them to the mobile telephone when it will be available after a service disruption. It must use standard Internet protocols.
- Buffering policy of the clients and servers is usually a well-kept secret. The mechanism the Server uses to control the availability of the Client is often a well-kept secret. The encapsulation of video on data packets is generally not revealed. The

Figure 3. A temporal data and signaling streaming communication in presence of service disruptions

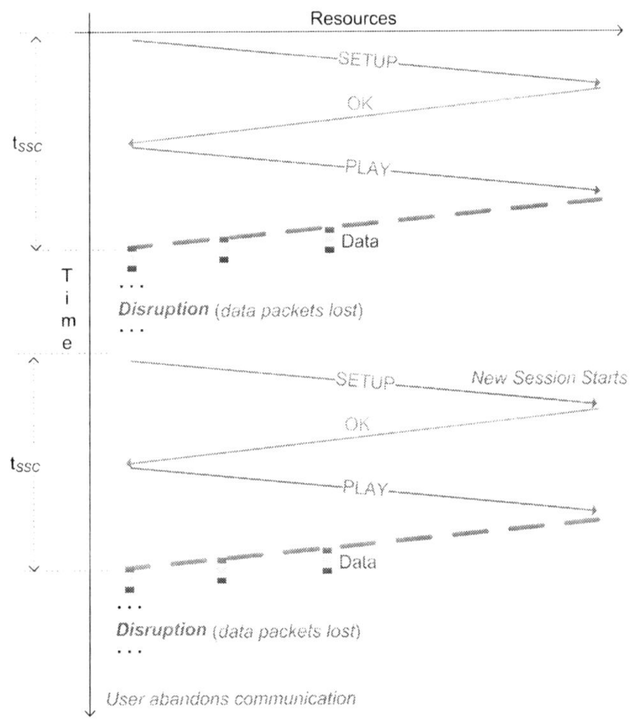

tested version of reliable sockets does not work well to control service disruptions.

- It is unrealistic to implement the above summed up actions making the modification to the code of commercial Client and Server. The change of the code of the open source Client and Server is a time consuming and error prone task. It is impossible to change the code of video streaming protocols because the understanding of the semantic of the messages is very difficult.

A PROXY AGENT BASED MULTI-PLATFORM FRAMEWORK TO MITIGATE THE ADVERSE EFFECTS OF SERVICE DISRUPTIONS

As already stated, the mechanism to mitigate the adverse effects of service disruptions must consider different levels of the network architecture. In each level one or more variables (parameters) must be taken into account to formulate an optimization problem. Three examples of parameters are the quality of the physical wireless channel, the size of the Client buffer and the bandwidth perceived by the Client for each flow. It is very difficult to find an exact and general mathematical solution to this optimization problem.

Cross layer techniques explore particular optimal solutions to the above optimization problem. Schaar (2007) presented a must-read introduction to cross-layer technique for multimedia applications. The basic idea of cross-layer technique is to optimize different parameters of the network architecture (at different levels) to obtain the optimal values of each of them to globally optimize an objective function. Cross layer is difficult to implement in existing mobile networks because operators block the access to critical parameters.

A more realistic solution is to use an application level proxy, because it manages free available parameters. A proxy traditionally controls the traffic flow between the Client and the Server taking into account particular parameters. For instance, Vandalore and Feng (2001) considered the state of wireless channel, Chakareski and Chou (2006) took into account previously stored resources and Gao and Zhang (2003) took into consideration the live video traffic. They did not test their solutions in mobile telephones. We implemented a proxy architecture in a commercial telephone N95 considering link, transport and application levels parameters (Suarez & Macias, 2008). We verified that the programming of the proxies is tedious, hard, time consuming and error prone.

An algorithmic solution to the above optimization problem can be specified programming the concurrent behavior of software agents (Shoham & Leyton-Brown, 2009). This solution has several important properties. Firstly, the behavior of the agents contains the rules to optimize a particular parameter. The programming of the behavior of an individual agent is dead easy. Secondly, the messaging, simple and easy to specify, declares the cooperation among agents. The correct specification of that cooperation represents the global optimization of the different parameters. Lastly, the solution scales well with the number of parameters because it is very easy to add a new agent and new messages. Bellavista and Corradi (2005) used a proprietary platform of proxy agents to control service disruptions caused by roaming. They also used proactive buffering scheme implemented as part of the agents' behavior to obtain good simulation results. To the best of our knowledge, those agents have not been implemented in commercial mobile telephones. In contrast, our objective in (Suárez & Menza, 2007) was to use an open source platform named JADE. We showed that JADE spectacularly improves the user experience using proactive buffering. But it did not introduce extra delay and jitter.

In this chapter we present our new solution inspired by the work in (Suárez & Menza, 2007) which was specially designed for portable computers. Direct migration of proxies' code to mobile telephones was not possible. We used JADE-LEAP because JADE was not originally designed for mobile telephones (LEAP replaces some of the JADE core functionality but it does not change the JADE's API). JADE-LEAP imposed several architectural software design changes to meet particularities of mobile telephones. The LEAP agents warn the Server to terminate the session for service disruptions. LEAP runs on mobile telephones compatibles with *Java 2 Standard Edition* (*J2SE*) using *Java Development Kit* 1.4 and later versions; furthermore it runs on mobile telephones compatibles with J2ME (using MIDP 1.0 and later versions). Moreover, J2SE is supported by all the main operating systems for current high-end mobile telephones. These facts prove that our mechanism can be implemented in a wide range of mobile telephones.

We also wanted to improve the time needed to design the software of our previous work (Suarez & Macias, 2008). This was possible due to JADE-LEAP ability to control the service disruptions. It raised appropriated events that were processed by the agents. We only specified the behavior of the agents and messaging among them.

The Description of Our Mechanism Supported by JADE-LEAP

In this section we discuss the kinds of proxies, the mapping of agents to devices and the main actions of our mechanism.

We use two kinds of proxies: *Proxy Client* and *Proxy Server*. A schema of their interconnection is shown in Figure 4.a. The network is between the proxies. An internal extra signaling channel between proxies is used to control the Client availability. The Proxy Server is in charge to buffer video frames while the Client is not available. When the Client will be available after a service disruption, the Proxy Server delivers the buffered video frames. In this way, the Client always recovers the lost video frames.

We used JADE-LEAP in *Split mode* because it achieved a better computational load balancing. In this mode the container of the Proxy Client was

Figure 4. a) General software architecture to solve the problems of wireless media streaming on 4G Networks. b) Mapping of agents to devices

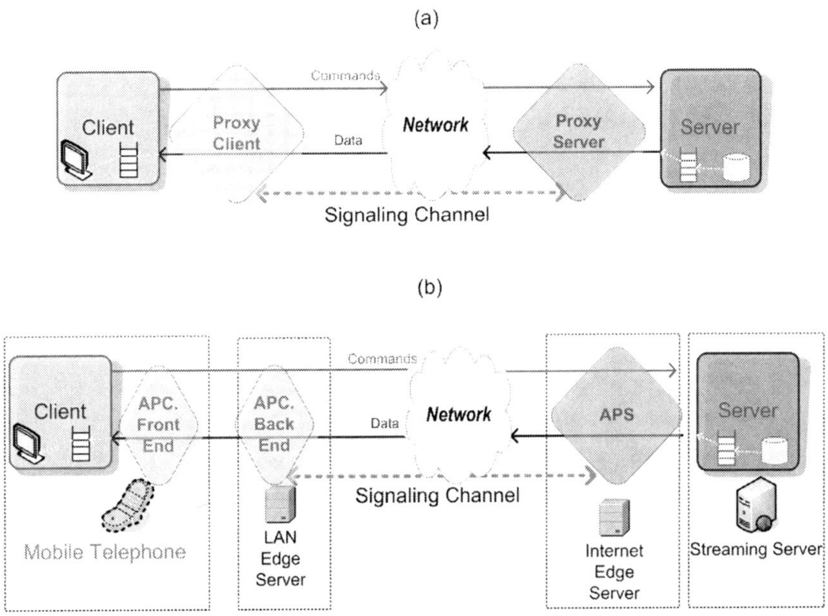

divided into two permanently connected components. One of them was directly associated with the mobile telephone (*FrontEnd*) and the other was associated with the Server (*BackEnd*). The Split mode also achieved an improved management of the wireless communication. Firstly, it automatically handled disconnections between the *FrontEnd* and the *BackEnd*. Secondly, the *BackEnd* automatically stored (in an internal buffer) all the messages it sent to the *FrontEnd*. Lastly, the *FrontEnd* periodically tried to reconnect automatically with the *BackEnd*. Let us note that these three actions eased the programming of our mechanism.

The mapping of agents to devices is shown in Figure 4.b. We justify that mapping as follows:

Server. In general it should be mapped to a fixed Internet machine. But we mapped it to a machine in a *Local Area Network* (*LAN*) because of we only treated the service disruptions in the wireless link. This is shown in the top row of Figure 5.

Client. It was mapped to the mobile telephone. Its behavior could change depending on the displaying policy of original and lost video frames.

Figure 5. The main steps (actions) of our mechanism

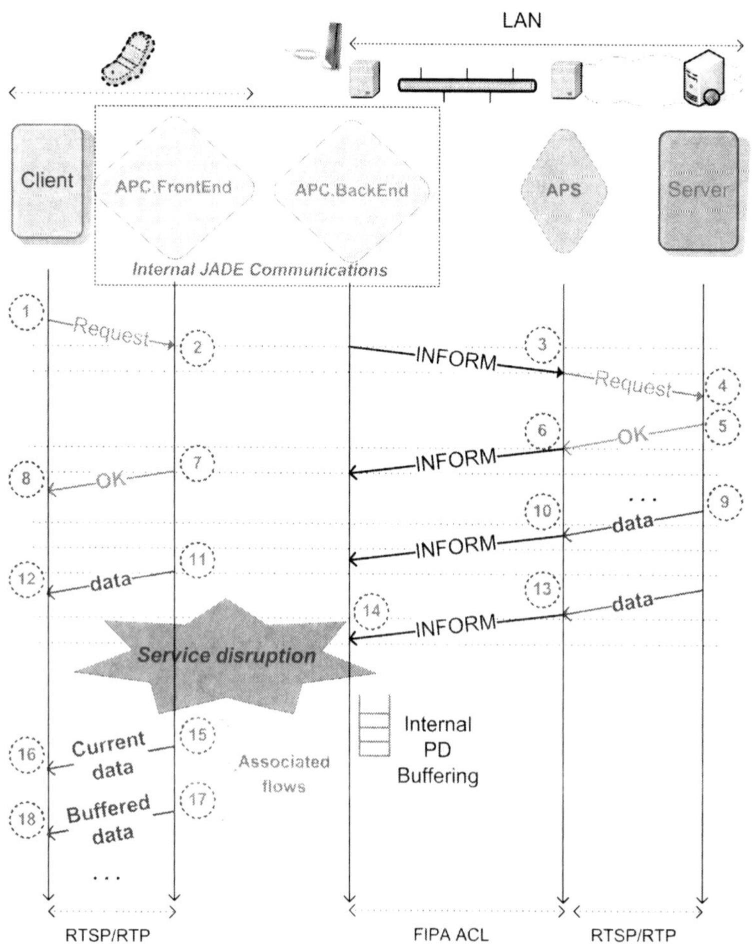

Proxy Server. We call it *Agent Proxy Server* (*APS*). In general it should be mapped to a fixed Internet edge router (Server). But we mapped it to a computer connected to the LAN in order to minimize (to zero) the service disruptions between it and the Server.

Proxy Client. We call it *Agent Proxy Client* (*APC*). As we mentioned above it was divided into two components: *APC.FrontEnd* and *APC.BackEnd*. The first was mapped to the mobile telephone. The second should be mapped to a fixed Internet edge router. But we mapped it to a computer in the LAN to minimize (to zero) the service disruptions between it and the APS. That computer connected with the APS (which is in another computer in the LAN) and with the *APC.FrontEnd*.

The *APC.FrontEnd* and *APC.BackEnd* communicated internally in the platform. We used *Agent Communication Language* (*ACL*) to specify the communication between APS and *APC.BackEnd*. We used *INFORM* ACL messages with a simple structure (it was not necessary to use an ontology). An INFORM message encapsulated a RTSP message. In theory, the actions of that encapsulation process must overlap with the actions of buffering in the *APC.BackEnd* and the communication between the APS and the Server.

In Figure 5 we show the main actions of our mechanism. We denote these actions as *Action #*, where # is a number inside a circle in the Figure 5.

The *actions 1* to *8* manage each RTSP command and its associated response:

Action 1. The Client sends a *RTSP request* to the Server (*RTSP "SETUP URL" message*, where *SETUP* is a RTSP command and *URL* is the Web address of the Server).

Action 2. The *RTSP request* is internally delivered by the JADE-LEAP, from the *APC.FrontEnd* to the *APC.BackEnd*. The *APC.BackEnd* encapsulates the *RTSP request* in an INFORM message.

Action 3. The INFORM message is received by the APS. The APS takes the encapsulated RTSP message and sends it to the Server.

Action 4. The Server receives the original *RTSP request* sent by the Client.

Actions 5 to 8. The Server responds a RTSP "*200 OK*" command and the Client will receive it.

The response to a *PLAY* is the only one that causes problems when a service disruption will occur. The agents only process the PLAY and its response (*actions 9* to *12*).

Actions 9 to 12. The Server sends the video frames; the APS receives them, makes the encapsulation and sends them into an INFORM message (*Action 10*). This message will arrive at the *APC.BackEnd* which relays it internally to the *APC.FrontEnd* (*Action 11*). Lastly, the *APC.FrontEnd* sends a RTP data *packet* that contains the video frames (*Action 12*).

The JADE-LEAP's *Persistent Delivery* (*PD*) service automatically controls the availability of the mobile telephone. This service must be manually installed and configured due to it is not provided in the standard JADE-LEAP distribution. The *APC.FrontEnd* and the *APC.BackEnd* coordinates the PD service. A buffering mechanism, implemented in the *APC.BackEnd*, is part of the PD service. The *APC.BackEnd* stores in the buffer video frames previously sent by the APS (*Action* 13). The buffering and the control of availability are implemented in *Action 14*. Let us note that the PD substitutes the original buffering in the APS (we maintain the APS for compatibility of our previous solutions in portable computers).

Actions 13 and 14. The APS sends the INFORM messages that contain the video frames (*Action 13*) in response to a successful PLAY. The PD service stores INFORM messages in the internal buffer and controls the mobile telephone availability (*Action 14*).

Let us now discuss the duration of a service disruption and the buffer size in order to define the agents' behavior. Let d be the duration of the service disruption (mobile telephone is disconnected). Let y be the size of the buffer for duration d (the amount of video frames stored during the service disruption). Let t be a time *threshold* intended to control d and y. The value of t can be fixed in the behavior of the agents (such as 2 minutes). But in general it must be adapted taking into account several parameters. For example, the kind of video, the specific wireless environmental conditions and a set of previous values. The combination of parameters d, y and t produces different alternatives to specify the agents' behavior. Two of them come in question. When will the Server terminate the session? How will the Client visualize lost video frames?

Let us suppose $d \geq t$. That is, the probability the mobile telephone will not be available after d is high. Let us suppose a VoD session and the mobile telephone's buffer can't store video frames necessary to support a service disruption of duration d. Let us also suppose a live session and the same mobile telephone's buffer size. It is impossible for the mobile telephone to store in its buffer enough video frames to support the service disruption of duration d. Under those assumptions the lost video frames will flood the wireless link. But they will not arrive at the mobile telephone. The efficient behavior of the APS should warn the Server to terminate the video streaming session. Also the *APC.FrontEnd* will indicate the user that the session will be closed: the user must reconnect manually later.

Let us now suppose $d<t$. That is, the mobile telephone is available before t. There are several alternatives to display the lost video frames (y) and the current video frames. The principal idea is to implement the best schedule of frames that should result in the best user experience. Let us note that all these frames will be stored in the internal buffer of PD service. The lost frames stay up to d units of time in the buffer and the current ones must be sent to the Client as soon as possible. An alternative is to schedule the video frames displaying sequentially. That is, the current video frames are displayed after the lost video frames. That should be appropriated for VoD. Another alternative is to display in parallel the lost video frames and the current ones. That should be appropriated for VoD and live session. This last alternative is implemented in *actions 15* to *18*.

Actions 15 to 18. The PD Service detected the re-connection of the mobile telephone. The *APC.FrontEnd* internally receives, from the *APC.BackEnd*, the current (*Action 15*) and the buffered (*Action 17*) video frames alternatively (one by one). The Client will display the current (*Action 16*) and the buffered video frames (*Action 18*) alternatively (one by one). The current video frames are displayed in a big window and the buffered video frames in a small window.

In theory, these two video streams from the *APC.BackEnd* to the Client must be supported by the available bandwidth of the wireless technology. Otherwise, *APC.BackEnd* must send a reduced resolution version of the lost video frames. On the other hand, it must be verified, in practice, the overlapping of *actions 15 to 18* with the other *actions* of our mechanism.

Let us now suppose that buffered video frames of a service disruption are being displayed. In that moment a new (or multiple) service disruption occurs. The lost video frames of the new service disruption are added to the rear of the buffer proactively (after the video frames of the first service disruption). The lost video frames of all service disruptions are displayed sequentially in the second window.

The Experimental Results

The implementation of our mechanism was intended to show JADE-LEAP did not increase latency and jitter, our mechanism overlapped its actions efficiently and the Client displayed two parallel video streams to improve the user experience.

We implemented a Client (*Player*) and a Server in order to control their internal actions and measure performance parameters that was not accessible in open source or commercial clients and servers (such as the amount of initial buffering needed in the Client and the delivery rate of the Server). Without loss of generality, the Server implements the actions associated to a basic set of RTSP commands: *SETUP, PLAY, PAUSE, TEARDOWN, DESCRIBE, SET_PARAMETER* and *GET_PARAMETER*. It also communicated video data frames encapsulated in RTP packets.

We implemented the Client and the *APC.FrontEnd* in a *N95* mobile telephone which incorporated a WiFi WNIC at 11 Mbps. The Server, the APS and the *APC.BackEnd* run in a portable computer with a 1.40 GHz Pentium Mobile *Central Processing Unit* (*CPU*), 1GB *Random Access Memory* (*RAM*), and a WiFi WNIC at 54 Mbps. It had a low computation power and RAM memory capacity. The WiFi technology did not provide *Quality of Service* (*QoS*). These conditions represented the most difficult case to test in putting our mechanism in practice.

Our mechanism is independent of the video frames compression format because the Server, the agents and the Client only communicate and display them. Moreover, the cadence of the INFORM packets (from *APS* to *APC.BackEnd*) is the same as the cadence of the RTP packets (from the Server to the *APS*). We used the *Motion Joint Photographic Expert Group* (*MJPEG*) which basically compressed each video frame as a JPEG image. Each frame was encapsulated in a RTP packet. The used RTP payload format was compliant with the one defined by Fenner and Frederick (1998). The resolution of JPEG images was adapted to the mobile telephone screen (*120x90* pixels). We used MJPEG because it represented the least functional *codec*. It did not take advantage of temporal compression. That means the Server must send approximately a similar amount of traffic in each RTP packet which should cause a high waste of bandwidth. However, MJPEG actually simplified the control of service disruptions.

In Figure 6 we show a photograph of the portable computer screen and the mobile telephone used for experimentation. This photograph shows the frames sent during a service disruption. The Server sent video frame number 517 as indicated by the sentence "*Sending Frame 517*" on a window of the portable computer screen. In others windows are shown the directory of agents in JADE and source code of agents. They delivered the video frames streams to the Client in INFORM messages that were coded with a conversation and protocol identifications. In this photograph the Client is displaying two video frames streams. For that reason, the screen of the mobile telephone is divided

Figure 6. Two images in a mobile telephone with the corrective actions in the player

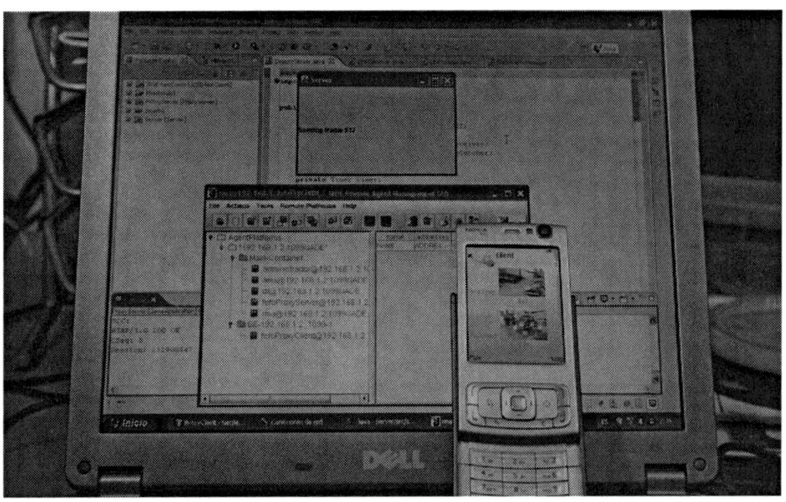

into two windows. The frame number 517 (current video frame stream) is shown on the top window and the bottom window shows frame number 221 (corresponding to one of the lost video frames stream). The scheduling of these two streams is one by one. That is, the *APC.BackEnd* perfectly alternates the delivery of the video frames of both streams. We improved the user experience, when using two video windows, because we minimized the adverse effects of chained T_{SSC} (Figure 3) while respecting the original duration of the videos, therefore, allowing the user to quickly compare the contents of the two windows, at glance.

We were not interested in the absolute performance evaluation of the JADE-LEAP. For that reason, we sought a reference for comparing purposes. The Client-Server system without proxies was the appropriated reference because we did not find other experimental works like this. On the one hand, we measured the jitter and delay of video frames using JADE-LEAP. On the other hand, we measured jitter and delay in the lone Client-Server system and compared both mea-

Figure 7. a) Delay of the communication without proxies. b) Delay of the communication with proxies

(a)

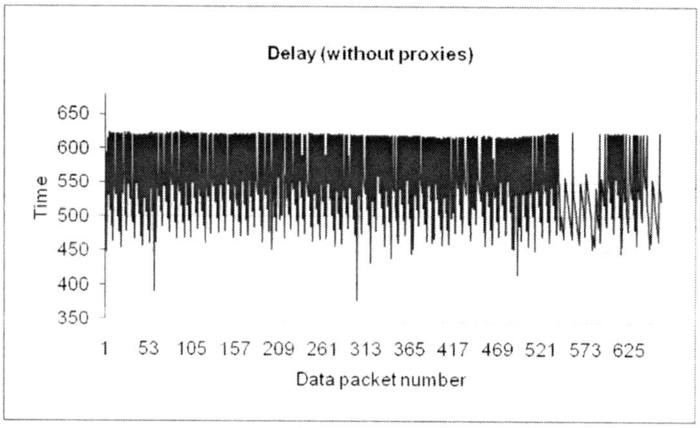

(b)

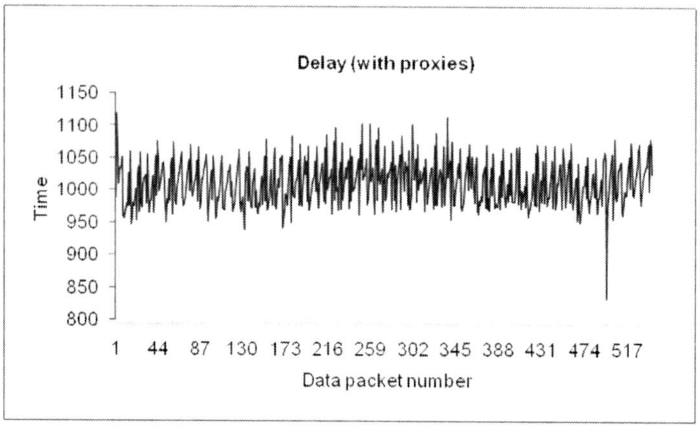

sures. In this simple way we observed the extra problems JADE-LEAP introduced in the WiFi Channel. Without loss of generality we set the sampling frequency to 2 *frames per second* (*fps*) due to we were interested in the comparison of the above systems not in the absolute performance of JADE-LEAP.

We made several experiments: In the first one the mobile telephone did not suffer any service disruption. The mobile telephone was in a fixed location and it did not move. In this scenario we measured the delay and jitter to test the lone Client Server system. Then we did the same but including the JADE-LEAP agents.

In Figure 7.a we show the delay, of the delivered packets to the Client, without proxies. It was clear the delay was maintained in the interval 450 and 630 ms approximately. That is, the mean delay was around 0.5 s. In Figure 7.b we show the delay using our mechanism (JADE-LEAP). The delay was doubled (1 s) because the JADE-LEAP implemented an internal buffering (*APC. BackEnd*) for all the video frames. But it was maintained in the interval 950 and 1050 approxi-

Figure 8. a) Jitter of the packets arrival to the Client without proxies. b) Jitter of the packets arrival to the Client with proxies

(a)

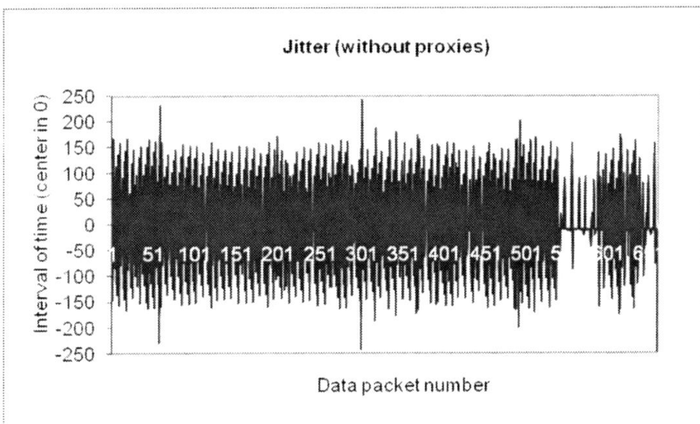

(b)

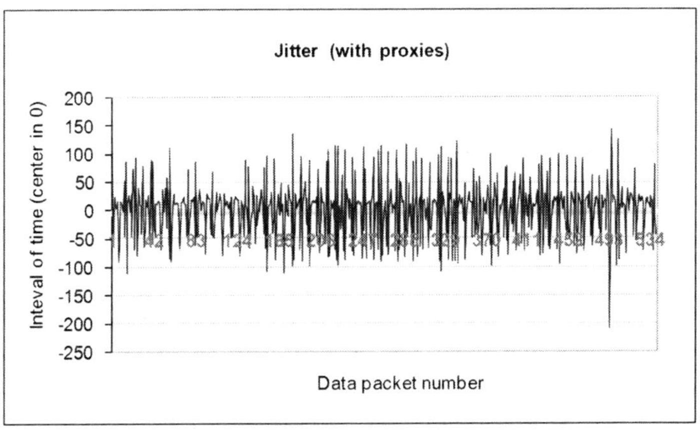

mately. It means the variation of the delay was less than in the case we did not use proxies. This is better observed in Figure 8 where the jitter in the case we did not use proxies was doubled by the case we used proxies. That good news was expected due to the mentioned internal buffering of JADE-LEAP. We can therefore conclude that the use of JADE-LEAP is beneficial because it smoothes the jitter. But it should be improved to achieve lower delay and increase the throughput. The code of the Client must be optimized to do that. But it is important to remark the problems introduced by the system software in the *N95* that did not allow us to freely optimize the Client code. The upcoming mobile telephones must improve this aspect in their SDK.

In the second experiment we compared the performance of our mechanism in presence of service disruptions. In the experiment we put a WiFi AP indoor, in a zone which physical characteristics were not previously known. We moved the mobile telephone in the surroundings of the AP and we randomly detected areas with very low level of radio signal strength (the mobile telephone did not receive video frames). In those areas a service disruption occurred. After some time we returned to the area of high level of radio signal strength. We repeated that several times.

The Client-Server system always produced video frames loss while the mobile telephone was unavailable and the user always had to start a new streaming session from scratch. On the contrary, the behavior of JADE-LEAP in presence of service disruptions was spectacular. The buffered packets for a service disruption were sent to the Client once the mobile telephone was available again. The Server did not terminate the session for a service disruption. We achieved an experiment in which the service disruption lasted 90 s in order to show that JADE-LEAP behaved spectacularly. Let us explain it (Figure 9.a):

The Initial period of time when the mobile telephone was available. The delay of packets was initially around 1 s from packet number 1 to packet number 69. It was observed as a straight line because the vertical axis range was high. From packet number 70 to packet number 77 the delay increased around 1.5 s. This is shown better in the small graphic. The reason of that delay was the mobile telephone was moving to an out of coverage area.

Disconnection (service disruption). The mobile telephone was unavailable while the communication of the packet number 79 was achieved (around packet number 79) and a service disruption of 90 s was produced. This service disruption provoked the buffering of packets between numbers 79 and 257 in the internal buffer of JADE-LEAP (*APC. BackEnd*). The reason the packet number 78 was delayed 120 s (120000 ms) was that JADE-LEAP before communicated others packets stored in its internal buffer. The delay of packets between numbers 79 and 257 was around 90 s (second straight line).

Final period of time when the mobile telephone was available again. In theory, the third straight line must be centered on 1 s (like the first one). But the packet number 78 accumulated a delay of 30 s (120 s minus 90 s of disconnection). For that reason, the delay was around 35 s from the packet number 258 and successive.

The variation of the jitter is shown in Figure 9.b. The bigger graphic contains the same three straight lines as the Figure 9.a. These lines are separated by extreme values of the jitter that mark the different periods of time above explained. In the smaller graphic is shown that it varied in the interval [-50, 50] (0.1 s). It is also shown that for packet 69 the jitter was 425 ms. This jitter was due to the mobile telephone leaved an area where the radio signal strength was good. This is shown as a point in the big graphic. Let us observe that other similar behaviors are found in that graphic: they explain the slight variations of the jitter.

In Figure 10 we show the overlapping of the both streams: the current video frames and the lost video frames. In Figure 10.a we show the

Figure 9. a) Delay of communication in presence of a service disruption. b) Jitter of communication in presence of a service disruption

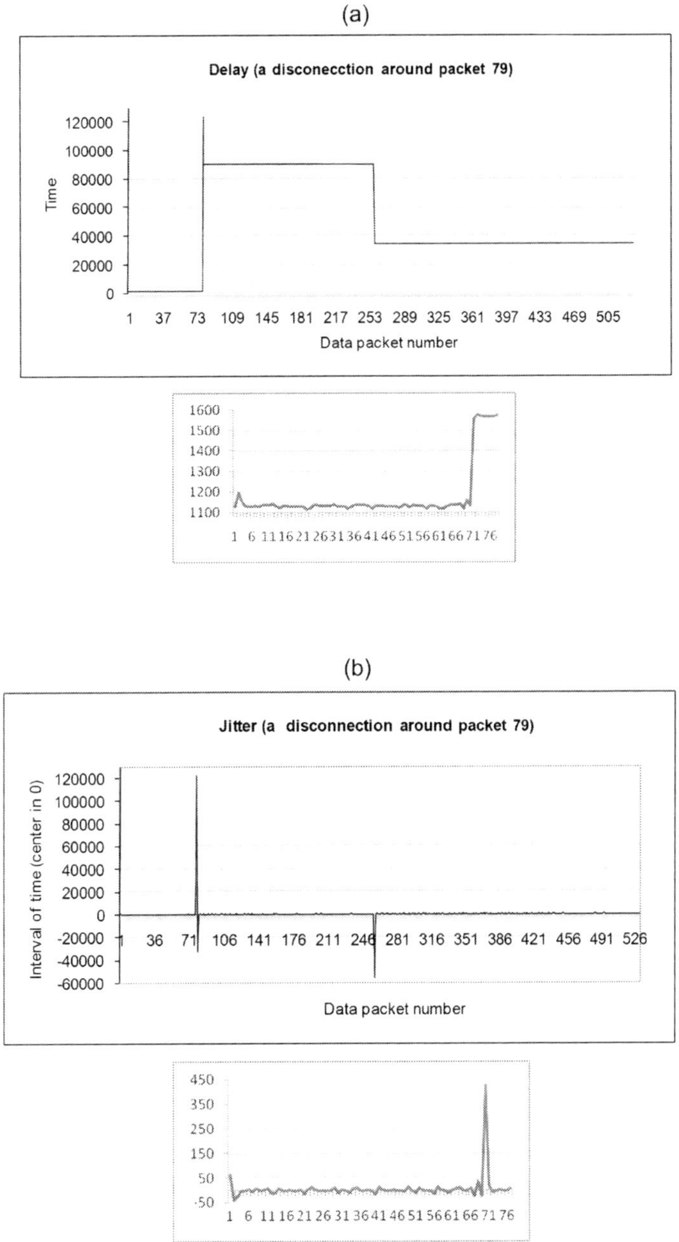

instant of time of arrival at the Client for each individual packet:

Initial period of time when the mobile telephone was available. The packets numbers 1 to 69 initially arrived at consecutive instants of time: the packet number 77 arrived at around 40 s (38890 ms).

Disconnection (*service disruption*). After 38890 time point in the horizontal axis there was a disconnection of 90 s (no packets arrives at the

Figure 10. a) Instant of time of arrival of each packet to the Client b) Interleaving of both streams

(a)

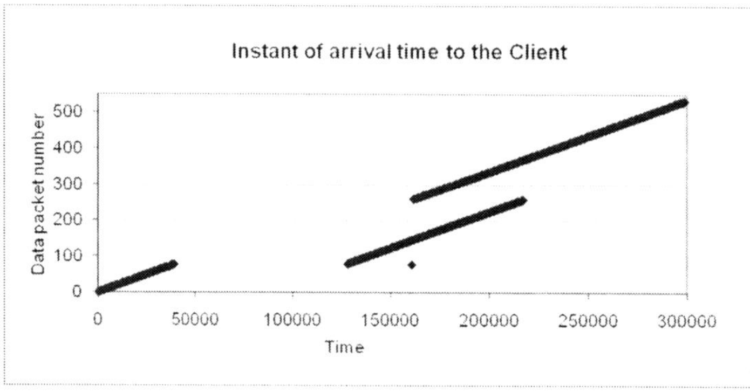

(b)

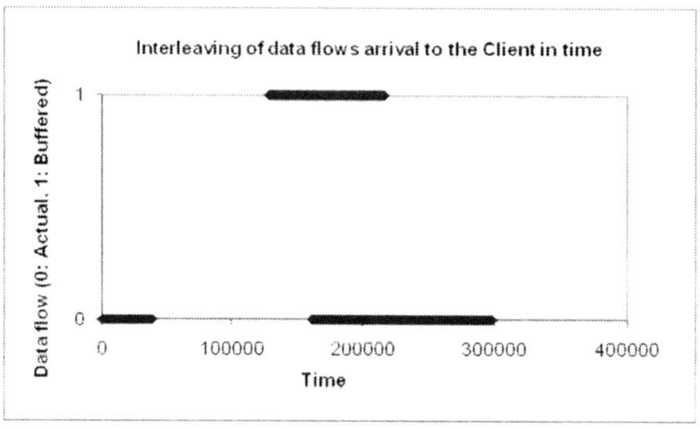

Client). During this time the Server continued sending packets to the APS. All of these packets were stored in the internal buffer of JADE-LEAP (*APC.BackEnd*).

Interleaving of both streams of video frames arrival to the Client. In theory, these two streams must arrive at the Client as indicated above (one by one). But it is shown that during 30 s the Client only received packets of one stream. After this, it received the packet number 78, and then it received both streams. Let us remark that the arrival of packets to the Client was not perfectly interleaved as the theory indicates. But considering the average time of arrival of packets of both streams to the Client, the interleaving was acceptable.

Final period of time when the mobile telephone was available after a service disruption. From packet 257 and successive, the Client only received one stream.

We have explained the interleaved arrival of packets to the Client but we have not shown exactly the identification of the streams. The horizontal axis of Figure 10.b is the time of arrival and the vertical axis has been divided into two values: 0 for the current video frames and 1 for the lost video frames.

FUTURE RESEARCH DIRECTIONS

Video streaming technique has a crucial impact in current and upcoming multimedia services. The

Internet of People will demand rich multimedia interfaces on mobile telephones. The *Internet of ideas* will give more importance to video and audio exposition of ideas. Lastly, the *Internet of things* will make the mobile telephones enrich the people communication using wireless sensors. The adverse effects of service disruptions could slow the impact of video streaming in upcoming mobile multimedia services.

In this chapter we reviewed some challenges. Firstly, a practical multi-platform SDK must be developed for any mobile telephone and operating system. Lastly, the mobile streaming software must be standardized.

Our mechanism admits further improvements:

Including more kinds of agents. For instance, one agent, in charge to observe the level of wireless signal strength, could proactively warn the *APC. FrontEnd* about the state of the mobile telephone (available or unavailable).

Rate distortion control. The APS could cooperate with an agent in charge to implement a rate distortion mechanism.

Efficient multicast control. An agent could manage multicast subscriptions and forced unsubscriptions due to service disruptions.

Adaptive threshold. An agent will control the adaptation of the threshold to specific scenarios.

Extending the multi-platform software. We must test the programmability of our mechanism in upcoming mobile telephones with new operating systems.

Exploring different strategies for displaying video frames. Other video formats could help us to tag the video in order to avoid the visualization of lost frames.

We have detected some possible improvements for JADE-LEAP:

Message Transport Protocol. It could be improved to make it slight in order to accelerate the communication of messages.

Internal buffering and transmission of INFORM messages. It must allow user policies for different buffering and internal transmission schemes.

We presented experimental results for N95 mobile telephone. But we also tested our software architecture in Windows Mobile (HP iPAQ 5555 using the *Jeode Java Virtual Machine*). We recently initiated the implementation of our mechanism in android (HTC Hero mobile telephone) and in an iPhone mobile telephone. More work is expected to be done for both android and iPhone platforms.

We presented our experimental results for MJPEG. Recently, we successfully used ".rm" video archives in the N95 mobile telephone. That means modern MPEG4 is supported by our mechanism. However, more work is needed in order to test our mechanism using two video windows in android, iPhone and MAEMO.

CONCLUSION

A lot of current and upcoming mobile multimedia services use video streaming technique to efficiently improve the user experience. Mobile telephones exhibit powerful hardware architecture, full operating system and a galaxy of applications. In spite of this dramatic evolution, there is still not a true multi-platform SDK. On the contrary, each traditional mobile telephone has a particular SDK, even though recent operating systems allow the usage of a SDK for different types of mobile telephones.

Non multi-platform SDK implies the existence of a jungle of different video streaming clients, both commercial and open source. These clients normally interoperate better with their counterpart Server and proprietary protocol. These restrictions make a complex task to modify current video streaming software. On the one hand, the modification of the commercial video streaming software is only possible for the enterprises that supply it. On the other hand, the modification or adaptation of open source software is very hard, time consuming and error prone.

We showed, with a simplistic mathematical model for the kernel of the streaming technique, the adverse effects of service disruptions. These service disruptions affect negatively the user experience. Moreover, it is impossible to predict when a service disruption will happen, due to the variability of the wireless channel conditions. The service disruptions can be characterized as a mathematical multi-parameter optimization problem. Multi-platform software agents can be used to provide an algorithmic solution to that optimization problem. We used JADE-LEAP, a Java based framework, for programming our agent based solution. We mitigated the adverse effects of service disruptions improving the user experience spectacularly. Once the mobile telephone was available (after a service disruption) the Client (Player) displayed the current and the buffered video frames (while the mobile telephone was unavailable). The newly modified JADE-LEAP agents were able to warn the Server to terminate the session in the case of a server disruption or to maintain it open for short disruptions. We use a static threshold to distinguish the duration of the service disruptions.

The experimental results, compared with the ones in the lone Client-Server, revealed the quality of our mechanism which smoothed the jitter and did not introduce excessive delay. Moreover, it produced an acceptable overlapping of buffering in the Client and in the Server, the communication of the video frames and the internal buffering in the JADE-LEAP. Lastly, an important additional fact was that our mechanism represents a minimal effort of programming.

REFERENCES

Allen, C., Arnold, W., Balkan, A., Cannasse, N., Grden, M., Gunesch, M., & Zupko, A. (2008). *The essential guide to open source flash development*. Berkeley, CA, USA: APRESS.

Ansari, F., & Sathyanath, A. (2007). STEM: Seamless transport endpoint mobility. *ACM Computing and Communications Review*, *11*(2), 1–13.

Apostolopoulos, J., Trott, M., Kalker, T., & Wai-Tian, T. (2005). Enterprise streaming: Different challenges from Internet streaming. In *IEEE International Conference on Multimedia and Expo*. (pp. 1386-1391). Amsterdam, The Netherlands: IEEE Press.

Belda, A. (2006). *Contribution to the multimedia communications in wireless networks for mobile devices* (in Spanish). Unpublished doctoral dissertation, Technical University of Valencia, Spain.

Bellavista, P., Corradi, A., & Giannelli, C. (2005). Mobile proxies for proactive buffering in wireless Internet multimedia streaming. In *IEEE International Conference on Distributed Computing Systems Workshop* (vol. 3, pp. 297-304). Columbus (OH), USA: IEEE Press.

Bellifemine, F., Caire, G., Poggi, A., & Rimassa, G. (2003). JADE, A white paper. *Journal of Telecom Italia Lab.*, *3*(3), 6–19.

Bolla, R., Mangialardi, S., Rapuzzi, R., & Repetto, M. (2009). Streaming multimedia contents to nomadic users in ubiquitous computing environments. In *Proceedings of the 28th Conference on Computer Communications (Mobile Video Delivery Workshop, INFOCOM '09)* (pp. 1-6). Washington, DC., USA: IEEE Press.

Caire, G., & Pieri, F. (2009). LEAP USER GUI. *Telecom Italia Lab*. Retrieved on January 30, 2009, from http://jade.tilab.com/ doc/ LEAPUserGuide.pdf.

Chakareski, J., & Chou, P. (2006). RaDiO edge: Rate-distortion optimized proxy-driven streaming from the network edge. *IEEE/ACM Transactions on Networking*, *14*(6), 1302–1312.

Chou, P. A. (2007). Streaming media on demand and live broadcast. In Van der Schaar, M., & Chou, P. A. (Eds.), *Multimedia over IP and wireless networks: Compression, networking, and systems* (pp. 453–502). London, UK: Academic Press.

Costa, C., Cunha, I., Borges, A., Ramos, C., Rocha, M., Almeida, J., & Ribeiro-Neto, B. (2004). Analyzing client interactivity in streaming media. In *ACM International Conference on Wide World Web Conference.* (pp. 534-543). Beijing, China: ACM Press.

Daniel, F., Macedo, A. L., & Pujolle, G. (2008). From TCP/IP to convergent networks: Challenges and taxonomy. *IEEE Communications Surveys & Tutorials*, *10*(4), 40–55.

Doblander, A., Zoufal, A. & Rinner B. (2009). A novel software framework for embedded multiprocessor smart cameras. *ACM Transactions on Embedded Computing Systems, 8*(3), 24:1-24:30.

Fenner, W., Frederick, R., Frederick, S. & Stewart, P. (1998). *RTP payload format for JPEG-compressed video,* request for comments: 2435, standard track, the Internet Engineering Task Force.

Frey, P. W., Hasler, A., & Alonso, G. (2009). Server-efficient high-definition media dissemination. In *Proceedings of the 19th International Workshop on Network and Operating Systems Support for Digital Audio and Video,* Williamsburg, Virginia (USA) (pp. 49-54). New York, NY, USA: ACM Press.

Gao, L., Zhang, Z., & Towsley, D. (2003). Proxy-assisted techniques for delivering continuous multimedia streams. *IEEE/ACM Transactions on Networking*, *11*(6), 884–894.

Gomez, G. (2009). *Modelling and evaluation of the streaming services (end to end) for networks with wireless access* (in Spanish). Unpublished doctoral dissertation, Technical University of Malaga, Spain.

Goyal, V. (2006). *Pro Java ME MMAPI: Mobile Media API for Java Micro Edition*. Berkeley, CA, USA: APRESS.

Gupta, V., Williams, M., Johnston, D., McCann, S., Barber, Ph., & Ohba, Y. (2006). *IEEE 802.21: Overview of Standard for Media Independent Handover Services (Tech. Rep.)*. New York, NY: IEEE Press.

Handley, M., Floyd, S., Padhye, J. & Widmer, J. (2003). *TCP Friendly Rate Control (TFRC): Protocol Specification.* Request for comments: 3448, standard track, the Internet Engineering Task Force.

Ippolito, L. J. (2008). *Satellite communications systems engineering: Atmospheric effects, satellite link design and system performance*. West Sussex, England: John Wiley & Sons, Ltd.

Kholer, E., Handley, M. & Floyd, S. (2006). *Datagram Congestion Control Protocol (DCCP)* Request for comments: 4340, standard track, the Internet Engineering Task Force.

Kumar, K. G., Lipscomb, J.S, Ramchandra, A., Chang, S. P., Gaddy, W. L., Leung, R. H.,...Menon, J. P. (2001). The HotMedia architecture: Progressive and interactive rich media for the Internet. *IEEE Transactions on Multimedia, 3*(2), 253:267.

Laksono, I. (2004). *Achieve end-to-end Qos for wireless video streaming*. Retrieved August 30, 2009, from http://www.eetindia.co.in/ ARTICLES/2004NOV/B/2004NOV16_RFT_QA_TA.pdf? SOURCES=DOWNLOAD.

Le Moal, D., Moralo, D., & Campello, J. (2009). Power efficient real-time disk scheduling. In *Proceedings of the 19th International Workshop on Network and Operating Systems Support for Digital Audio and Video,* Williamsburg, Virginia (USA) (pp. 55-60). New York, NY: ACM Press.

Lee, M. M., McKillop, C., & Schell, S. V. (2006). Wireless communication out of range indication. *United States patent application assigned to Apple Computer Corp.*

Ming, L. X. (2000). Streaming technique and its application in distance learning system, *IEEE International Conference on Signal Processing* (vol. 2, pp. 1329-1332). Washinton, DC., USA: IEEE Press.

Moreno, E. (2009). *Modification of the reliable sockets interconnection mechanism and its use in video streaming applications* (in Spanish). Unpublished Master doctoral dissertation, University of Las Palmas de G. C., Spain.

Pagani, M. (Ed.). (2009). *Encyclopedia of multimedia technology and networking.* Hershey, PA: IGI Global.

Pagani, M., & Schipani, D. (2005). Motivations and barriers to the adoption of 3G mobile multimedia services: An end user perspective in the Italian market. In Deans, P. (Ed.), *E-commerce and M-commerce Technologies* (pp. 80–95). Hershey, PA: IGI Global.

Pantos, R. (2009). *HTTP live streaming.* Internet-draft: Standard track, the Internet Engineering Task Force (expires November 2009).

Qadeer, M. A., Ahmad, R., Khan, M. S., & Ahmad, T. (2009). Real time video streaming over heterogeneous networks, *IEEE 11th International Conference on Advanced Communication Technology* (vol. 2, pp. 1117 - 1122). Washington, DC., USA: IEEE Press.

Sanders, W. B. (2008). *Learning flash media server 3.* Sebastopol, CA (USA): O' Reily Media, Inc.

Savo, G. G. (2006). *Advanced wireless networks: 4G technologies.* West Sussex, England: John Wiley & Sons, Ltd.

Schaar, M. (2007). Cross-layer wireless multimedia. In Van der Schaar, M., & Chou, P. A. (Eds.), *Multimedia over IP and wireless networks: Compression, networking, and systems* (pp. 122–138). London, UK: Academic Press.

Schulzrinne, H., Rao, A. & Lanphier, R. (1998). *Real Time Streaming Protocol (RTSP)* request for comments: 2326, standard track, the Internet Engineering Task Force.

Shoham, Y., & Leyton-Brown, K. (2009). *MULTIAGENT SYSTEMS algorithmic, game-theoretic, and logical foundations.* Cambridge, England: Cambridge University Press.

Stewart, R., Xie, Q., Morneault, K., Schwarzbauer, H., Taylor, T., Rytina, I.,...Paxson, V. (2000). *Stream control transmission protocol.* Request for cmments: 2960, standard track, the Internet Engineering Task Force.

Suárez, A., Macías, E., Martín, J., Gutierrez, Y., & Gil, M. (2008). Light protocol and buffer management for automatically recovering streaming sessions in WiFi mobile telephones. In *International Conference on Mobile Ubiquitous Computing, Systems, Services and Technologies* (pp. 70-76). Valencia, Spain: IEEE Press.

Suárez, A., Menza, M., Macías, E. M., & Sunderam, V. (2007). Automatic resumption of streaming sessions over WiFi using JADE. *IAENG International Journal of Computer Science, 33*(1), 92–100.

Tse, D., & Viswanath, P. (2005). *Fundamentals of wireless communication.* Cambridge, England: Cambridge University Press.

Vandalore, B., Feng, W., Jain, R., & Fahmy, S. (2001). A survey of application layer techniques for adaptive streaming of multimedia. *International Journal of Real-Time Imaging*, *7*(3), 221–235.

Weaver, J. L., Gao, W., Chin, S., & Iverson, D. (2009). *Pro JavaFX™ platform script, desktop and mobile RIA with Java™ technology*. Berkeley, CA, USA: APRESS.

Wu, S., Hsu, J., & Chen, C. (2007). Headlight prefetching for mobile media streaming. In *ACM Workshop on Data Engineering for Wireless and Mobile Access* (pp. 67-74). Beijing, China: ACM Press.

Zandy, V., & Miller, B. P. (2002). *Reliable sockets* (Computer Sciences Tech. Rep. 2002 LISA XVI November 3- 8). Philadelphia, PA (USA): University of Wisconsin.

Zhou, G. (2009). (To be published). Geo-referencing of video flow from small low-cost civilian UAV. *IEEE Transactions on Automation Science and Engineering*, *6*(3).

Zhu, C., Li, Y., & Niu, X. (Eds.). (2011). *Streaming media architectures, techniques, and applications: Recent advances*. Hershey, PA: IGI Global.

Zhu, J., Yang, X., & Guo, X. (2007). Disruptive CSMA with Credit Payback (CP) protocols for multi-radio network. In *IEEE 2nd International Conference on Cognitive Radio Oriented Wireless Networks and Communications* (pp. 206-215). Orlando (FL), USA: IEEE Press.

ADDITIONAL READING

Bennett, S., & Marsh, D. (2009). *Handbook of online education.* Great Britain: MPG Books Ltd (http://continuumbooks.com).

Burnett, E. (2009). *Hello, Android: Introducing Google's mobile development platform (2nd edition).* USA: Pragmatic Bookshelf (http://www.pragprog.com).

Chang-Gen, J., Ming-He, H., & Tang, J. (2009). Design and implementation of video conference system based on flex and J2EE. In *Proceedings of the International Conference on Image Analysis and Signal Processing,* Halifax, Canada (pp. 184-186). Washington, DC., USA: IEEE Press.

Dahr, S. (2005). MANET: Applications, issues, and challenges for the future. *International Journal of Business Data Communications and Networking*, *1*(2), 66–92.

Eichhorn, A. (2008). Middleware abstractions for cross-layer controlled media streaming. In *ACM Workshop on Middleware-Application Interaction* (pp. 13-18). Oslo, Norway: ACM Press.

Gou, L., Kim, J.-H., Chen, H.-H., Collins, J., Goodman, M., Zhang, X., & Giles, C. G. (2009). MobiSNA: A mobile video social network application. In *Proceedings 8th International ACM Workshop on Data Engineering for Wireless and Mobile Access* (pp. 53-56). New York, NY: ACM Press.

Hayun, R. B. (2009). *Java ME on Symbian OS: Inside the smartphone model*. West Sussex, United Kingdom: John Wiley & Sons.

Kovalick, A. (2009). *Video systems in an IT environment: The basics of networked media and file-based workflows* (*2nd Edition*). Burlington, MA (USA): Focal Press (Elsevier).

Lee, I. (Ed.). (2009). *Emergent strategies for e-business processes, services and implications: Advancing corporate frameworks*. Hershey, PA: IGI Global.

Lin, N.-H., Huang, T.-H., & Chen, B.-Y. (2007). 3D model streaming based on JPEG 2000. *IEEE Transactions on Consumer Electronics*, *53*(1), 182–190.

Maher, A. (2009). *iPhone SDK 3 programming: Advanced mobile development for Apple iPhone and iPod touch.* West Sussex, United Kingdom: John Wiley & Sons.

Margalef, F. P., López, A., & Fernández, G. (2009). Frame-accuracy synchronisation for mobile TV interactivity. In *IEEE International Symposium on Broadband Multimedia Systems and Broadcasting* (pp. 1 - 5). Bilbao, Spain: IEEE Press.

Neville-Neil, G. V. (2009). High bandwidth, low latency, and multihoming challenge the sockets API. *Communications of the ACM*, *52*(6), 51–55.

Perea, R. M. (2008). *Internet multimedia communications using SIP: A modern approach including Java practice*. Burglington, MA, USA: Morgan Kauffman Publishers.

Schütz, S., Eggert, L., Schmid, S., & Brunner, M. (2005). Protocol enhancements for intermittently connected hosts. *ACM* SIGCOMM. *Computer Communication Review*, *35*(2), 5–17.

Stauder, J., & Erbas, F. (2005). Mobile multimedia over wireless networks. In Rahman, S. M. (Ed.), *Multimedia networking: Technology, management and applications* (pp. 422–471). Hershey, PA: IRM Press.

Tan, W.-T., Cheung, G., Ortega, A., & Shen, B. (2009). Community streaming with interactive visual overlays: System and optimization. *IEEE Transactions on Multimedia*, *11*(5), 986–997.

Tatnall, A. (Ed.). (2009). *Web technologies: Concepts, methodologies, tools and applications*. Hershey, PA: IGI Global.

Tonev, G., Sunderam, V., Loader, R., & Pascoe, J. (2002). Location and network quality issues in local area wireless networks. In *International Workshop on Network and Operating Systems Support for Digital Audio and Video* (LNCS 2299, pp. 131-148). Kalsrüe, Germany: Springer-Verlag.

Tu, W., Steinbach, E., Muhammad, M., & Li, X. (2009). Proxy caching for video-on-demand using flexible starting point selection. *IEEE Transactions on Multimedia*, *11*(4), 716–729.

Wootton, R., Patil, N. G., Scott, R. E., & Ho, K. (Eds.). (2009). *Telehealth in the developing world.* Canada: Royal Society of Medicine Press/IDRC (http://www.idrca.ca).

Wu, W., Yang, Z., & Nahrstedt, K. (2009). Dynamic overlay multicast in 3D video collaborative systems. In *Proceedings of the 19th International Workshop on Network and Operating Systems Support for Digital Audio and Video,* Williamsburg, Virginia (USA) (pp. 1-6). New York, NY: ACM Press.

Xue, P., & Chandra, S. (2006). Revisiting multimedia streaming in mobile Ad hoc networks. In *Proceedings of the 19th International Workshop on Network and Operating Systems Support for Digital Audio and Video,* New Port, Rhode Island (USA). New York, NY: ACM Press.

Yao, P., & Durant, D. (2009). *Programming. NET Compact Framework 3.5* (2nd ed.). Boston, MA, USA: Addison-Wesley Professional.

Zammetti, F. (2009). *Practical Palm Pre WebOS Projects*. Berkeley, CA, USA: APRESS.

About the Contributors

Elsa Macias is anAssociate professor of Telecommunications at Las Palmas de Gran Canaria University, Department of Telematics Engineering, Spain. She received her PhD in Telecommunications (2001) from Las Palmas of Gran Canaria University for her work on Parallel Computing on a LAN-WLAN Cluster Controlling at Runtime the Variation of the Number of Proccesses. She received her MS in Telecomunications (1997) from the same University for her work on Parallelization of Diffuse IR Radiation System Simulation for Indoor Applications. Her research interests are in parallel and distributed computing and infrastructure wireless networks for collaborative computing. Her current research efforts have focused on the management of wireless channel disconnections to prevent abrupt endings of applications. She has published about 8 papers in refereed journals, 40 papers in refereed conferences, 1 paper in Spanish magazine, one educational book and co-editor of one book. She is member of Program & Organizing Committees & Chair sessions for several international and Spanish conferences. She has collaborated in several research projects. Professor Macías teaches telecommunications at the beginning, advanced, and graduate levels, and advises graduate theses in the area of wireless communications and parallel and distributed computing.

Alvaro Suarez is a Full Professor of Telecommunications at the University of Las Palmas de Gran, Canaria, Spain. He is Member of the Experts Commission of research of the University of Las Palmas de Gran Canaria from 1999-2002. He, in 1990, started working in systolic computing in the Technical University of Catalonia. Then he turned his attention to network computing and heterogeneous computing in 1994 when he returned to the University of Las Palmas de Gran Canaria where he founded the Concurrency and Architecture Group (GAC). His research interests are in parallel and heterogeneous distributed computing, infrastructure wireless networks for collaborative computing and collaborative frameworks. His current research efforts have focused on the management of wireless channel disconnections to prevent abrupt endings of applications. Professor Suarez teaches telecommunications at the beginning, advanced, and graduate levels, and advises graduate theses in the area of wireless communications and parallel and heterogeneous distributed computing. He also gives lectures at European universities. The *Multimedia and Ubiquitous Wireless Access Networks and Services* conference was lectured at University Ca' Foscari di Venezia (May 2005).

***

A. J. G. Abelém has a PhD in Computer Science by PUC/RJ (2003). Adjunct professor of the Federal University of Pará: Director of Technology of Information and Communication Center. He participates in master and doctorate Programs of the Department of Electrical Engineering and Master of the Department of Computer Science of UFPA. His research interests include: architectures of computer networks; multicast; wireless networks and mobile systems; optical networks; QoS and QoE. Nowadays, he coordinates partnership with the Government of the State of Pará/PRODEPA in order to offer solutions of Networks of access based in wireless mesh of low cost, as a of promoting digital inclusion, spreading and improving the coverage of the data network of the State in locations where the infrastructure of communication is poor or nonexistent. He is a Coordinator and/or participant in the Research Network in the State of Pará in Technologies of Information and Communications and at the Brazilian Institute of Research in WEB Sciences, both passed in publications of FAPESPA and CNPq foundations, respectively.

Patricia Acelas graduated in Telecommunications Engineering from Santo Tomas University (USTA), Bucaramanga, Colombia, in 2007 and got the MS degree in Telematics from the Universidad Politécnica de Valencia (UPV), Spain, in 2008. She currently works as a researcher at the Institute of Telecommunications and Multimedia Applications (iTEAM), Valencia. Her areas of interest are video encoding, QoS and wireless ad hoc networks.

Elisangela Aguiar is a PhD student in the post-graduation program of electrical engineering (PPGEE), with emphasis in Applied Computing, of the University of Pará (UFPA). Since January of 2004, she has been an adjunct professor I of the graduation course in Computer Science of the University of the Amazônia (UNAMA). She has served as Analyst of the Federal Service of Data Processing (SERPRO) since August of 2004, where currently she is the Regional Coordinator of Technology, and a member of the Group of Research and Development of SERPRO. Her interests of research are computer networks and telecommunication, mobile computing, networks in wireless mesh and ad hoc, quality of service and of experience, mobile communication systems, cloud computing, e-government, security and management of networks.

Laith Al-Jobouri received his BSc in 1995 from the University of Technology in Iraq and then worked in different wireless and microwave projects. He obtained his MSc from the University of Mosul in Iraq in 2001 then returned to work as a project manager and head of department for national and international companies in wireless and GSM. He is undertaking PhD research for a degree at the University of Essex, UK. His research interests are in video over wireless, forward error correction, and wireless technologies.

Salah S. Al-Majeed holds a BSc from Baghdad University, Iraq in 1992. He went on to take an MSc in Mobile and Satellite Communication at Westminster University, London, UK. His research interests are cross-layer design of control congestion, quality-of-service issues, scheduling algorithms, bandwidth allocation mechanisms and multi- connection techniques, especially in WiMAX broadband wireless. He is editor-in-chief of *Computer Science and Engineering: International Journal.*

Pau Arce received his Telecommunications Engineering degree and the MS in Telematics from the Universidad Politécnica de Valencia (UPV), Spain, in 2005 and 2007 respectively. Currently he works as a researcher at the Institute of Telecommunications and Multimedia Applications (iTEAM). His research interests include multimedia QoS, routing on wireless ad hoc networks and performance evaluation of computer systems.

Paulo Bezerra graduated as a Technologist in Data Processing from the University of Amazonia UNAMA (1997). He has a specialization in Computer Networks completed (1999) at the same University. He received a Master in Electrical Engineering from Federal University of Pará UFPA (2010) and PhD student of the Graduate Program in Electrical Engineering from the same university. He is a Professor Federal Institute of Education, Science and Technology - IFPA and Professor and Coordinator of Teaching Computer Science, University of Amazonia - UNAMA, and Delegate Institutional Brazilian Computer Society - SBC.

Wilder E. Castellanos received his Electronic Engineering degree from Industrial University of Santander, Bucaramanga, Colombia, in 2002 and the MSc degree in Telematics Engineering from the Universidad Politécnica de Valencia (UPV), Spain, in 2009. Currently, he is working towards the PhD degree in the Institute of Telecommunications and Multimedia Applications (iTEAM) at UPV. His current research interests include QoS-aware architectures and multimedia streaming in mesh/ad hoc networks.

Cesare Celozzi received the MSc degree in computer engineering from Politecnico di Torino, Torino, Italy, in 2004. He is currently a PhD student with the Dipartimento di Automatica e Informatica, Politecnico di Torino. His research interest includes virtual and augmented reality environments, human-machine 3D interfaces and segmentation of medical images.

Eduardo Cerqueira, CNPq fellowship PQ2, received his PhD and post-doc in Informatics Engineering from the University of Coimbra (UC) in 2008 and 2009, respectively. He is an Associate Professor at Faculty of Computer Engineering of the University of Para in Brazil. His research interests include Quality of Service, Quality of Experience, Sensor Wireless and Networks, Mobility and Multimedia. His publication includes four books, four patents and over than 50 papers in national and international refereed journal/conferences. He is founder of a series of workshops on Future Multimedia Networking (FMN). He is coordinating or participating several research national and international projects such as UbiquiMesh, MultiMesh, OTIMA, Universal and COMAH.

Chun-Cheng Chiang received his BS degree from National Tsing Hua University and MS degree from National Taiwan University in 2007 and 2009, respectively, both in Electrical Engineering. His research interests include wireless multimedia transmissions and cross layer optimization in wireless networks.

Alan Costa was a graduate at Computer Network Technology from College of the Amazon (2009) with a specializations of MBA in Corporate Finance from UGF (2010), in Computer Network Security from UGF (2010), in Network Administration from IESAN GNU Linux (2011) and a Master's in Computer Science from UFPA. His areas of interest include Quality of Experience (QoE), QoS management, Multimedia, Computer Networks and Management of Next Generation Computer Networks. His lines of research are Network Management, Distributed Systems, Computer Network Security, Access Control, 4G Networks, Multicast, Resource Reservation, Mobility, Multimedia, Quality of Service (QoS) and Quality of Experience (QoE).

João C. W. A. Costa has a degree in Electronic Engineering from the Federal University of Pará (1981). He has a Master of Electrical Engineering and Telecommunications from Pontifical Catholic University of Rio de Janeiro (1989) and PhD in Electrical Engineering and Telecommunications from Universidade Estadual de Campinas (1994). He was a Telecommunications engineer in DENTEL (FCC) and Funtelpa (1981-1986). Since 1994, he has been a professor of UFPA, currently an associate professor in the School of Computer Engineering Institute of Technology. He was a Research Director and coordinator of the UFPA PIBIC (abr/1999-jul/2001 and jul/2005 to Dec/2006). From 2007-2010, he was Deputy Secretary for Development, Science and Technology (NAVEGAPARÁ coordinator) and he was a Vice-president of the Brazilian Society of Microwaves and Optoelectronics (2010-2012) as well as a member of the Brazilian Society of Telecommunications and CNPq Researcher (since 1994). He is a Researcher at the Graduate Program in Computer Science and Graduate Program in Electrical Engineering (communications networks, electromagnetism and applied computer science). His research interests include modeling devices that include high-frequency performance evaluation of mobile communication networks and optical access networks and communication solutions for surveillance systems. He is a Coordinator and participant in funded projects (CNPq, FINEP, CAPES, SECTAM) and by public and private companies (Ericsson, Brasilsat, RedeCelpa, Eletronorte, TBE).

Adam Dou is a PhD candidate at the University of California at Riverside. His research area is distributed and real-time systems. He is currently working on distributed caching for improved Web access latencies. He has previously worked with MapReduce systems on smart-phones and scheduling for flash equipped sensor devices. He received his BASc from the Univeristy of Toronto in 2005.

José Espino is a PhD student of Information Technology. He was held at the *Innovation Center for Information Society* (*CICEI*) from the *University of Las Palmas de Gran Canaria* (*ULPGC*) since 2006. He also graduated in Computer Science from the *Facultad de Informatica* at ULPGC. Among his research areas of interest include telematics, multimedia content broadcasting and mobile devices. He has practiced professionally as a researcher in the ULPGC and is currently writing his doctoral thesis entitled *Using intelligent agents to automatically retrieve video-streaming sessions on mobile phones* to be presented in 2010.

Martin Fleury is currently employed as a Senior Lecturer at the University of Essex, UK, where he is a Director of Graduates (Research). Martin has authored or co-authored over one hundred and ninety articles on topics such as document and image compression algorithms, performance prediction of parallel systems, software engineering, reconfigurable hardware and vision systems. He has also published a book on high-performance computing for low-level image- and signal-processing algorithms, published by Wiley and Sons, NY, 2001. His current research interests are video communication over MANS, WLANs, PANs, BANs, MANETs, and VANETs.

Richard Fox received a BS in Computer Science from the University of Missouri Rolla in 1986, and an MS and PhD in Computer and Information Sciences from The Ohio State University in 1988 and 1992 respectively. Dr. Fox is currently an associate professor in the Department of Computer Science at Northern Kentucky University (since 2001). Prior to this, he taught at the University of Texas – Pan American between 1992 and 2001, where in 2000 he received the Outstanding Faculty Award for Teaching Excellence. Dr. Fox has authored or co-authored over 40 peer reviewed publications, primarily in the area of Artificial Intelligence.

Francisco Fraile was born in Murcia, Spain in 1979. He obtained a degree in Telecommunication Engineering from the Universidad Politécnica de Valencia (UPV) and a MSc Degree in Microwave Engineering from the University of Gävle in 2004. Since then, he has worked as a research engineer for the Swedish company Interactive TV Arena. In 2006, Francisco joined the Multimedia Communications research group (COMM) of the Institute of Telecommunications and Multimedia Applications at UPV, to proceed with his doctoral studies as an industrial PhD student, interested in networked electronic media.

Carlos. R. L. Frances, CNPq fellowship PQ2, graduated with a Bachelor in Computer Science from Federal University of Pará (1995), a Master of Computer Science and Computational Mathematics at the University of São Paulo (1998) and a PhD in Computer Science and Computational Mathematics at the University of São Paulo (2001). He is currently associate professor of School of Computer Engineering, Federal University of Pará, was president of Data Processing Company of the State of Pará(UFPA) from December 2007 to December 2010 and member of the advisory committee of the National Institute of Computing Studies and Research (INEP-MEC). He has experience in computer science, with emphasis on Performance Evaluation, acting on the following topics: Digital TV, Access Technologies, Markov Models and performance simulation, correlation techniques (Bayesian networks) and optimization.

Luis Angel Galindo is anExecutive MBA from IESE Business School. He has a Master in Services and Security in IP networks from Technical University of Madrid, a Master in Production Management, and an MSc degree in telecommunications from the UPM. From his position, he has managed various processes in the area of M&A. In the last, from the adquisition of Tuenti, he has led the development of activities necessary to provide MVNO capabilities. Luis is a 'creovator' (creative+innovator) bringing innovation to the user. Luis is the founder and leader of various initiatives (WIMS 2.0, Innovation 2.0, etc). He is a regular writer in journals on business and strategy issues and has authored several books. He conducts research to define a new sustainable business model in today's society. He performs in-company courses to outside companies addressing issues from organizational behavior to socio-organizational aspects in the transformation of a traditional company to a '2.0' company. He is also a freelance consultant of business and strategy.

Mohammed Ghanbari is the co-author of "Principles of Performance Engineering", a book published by IET press in 1997, the author of "Video Coding: An Introduction to Standard Codecs", a book also published by IET press in 1999, which received the year 2000 best book award by the IEE, and the author of "Standard Codecs: Image Compression to Advanced Video Coding" also published by the IET press in 2003. Prof. Ghanbari has authored or co-authored about 500 journal and conference papers, many of which have had a fundamental influence in this field.

Juan C. Guerri was born in Valencia, Spain in 1969. He received his MS and PhD (Dr. Ing.) degrees, both in Telecommunication Engineering, from the Polytechnic University of Valencia, in 1993 and 1997, respectively. He is a professor in the E.T.S. Telecommunications Engineering at the Polytechnic University of Valencia, and he leads Multimedia Communications research group (COMM) of the iTEAM (Institute of Telecommunications and Multimedia Applications). He is currently in research and development projects about multimedia QoS/QoE, wireless ad hoc networks and video performance evaluation.

Dimitrios Gunopulos is a Professor at the University of Athens. His research topics include data mining and knowledge discovery in databases, design and analysis of algorithms, and computational geometry. He is currently focusing his research on algorithms for data mining applications using geospatial data, and database indexing. He has held a Professor position at UCR, a Research Associate position in IBM Almaden, and a Postdoctoral position in the Max-Plank-Institut for Informatik. He received his MA and a PhD from Princeton University, in 1992 and 1995 respectively, and his BS from the University of Patras , Greece, in 1990.

Hung-Yun Hsieh received the BS and MS degrees in Electrical Engineering from National Taiwan University, Taipei, Taiwan, ROC, and the PhD degree in Electrical and Computer Engineering from Georgia Institute of Technology, Atlanta, Georgia, USA. He joined the Department of Electrical Engineering and the Graduate Institute of Communication Engineering at National Taiwan University in 2004, where he is currently an Associate Professor. His research interests are in the areas of wireless ad hoc networks, wireless access networks, and cognitive radio networks, with focuses on multimedia support in heterogeneous wireless communication systems, network protocols for vehicular ad hoc networks, and multiple access technologies in cognitive radio networks. He received the ACM MobiCom Best Student Paper Award in 2003.

Vana Kalogeraki is an Assistant Professor at the Athens University of Economics and Business. Her research interests include distributed and real-time systems, sensors and sensor systems, peer-to-peer systems, and resource management. She received her PhD from the University of California, Santa Barbara in 2000. She held a Research Scientist position at Hewlett-Packard Labs in Palo Alto, CA and was a Professor at the University of California, Riverside.

Fabrizio Lamberti graduated with a degree in computer engineering in 2000, and received the PhD degree in computer engineering in 2005 from Politecnico di Torino, Italy. Since 2006 he has been an assistant professor at Politecnico di Torino. He has published a number of technical papers in international journal and conferences in the areas of mobile and distributed computing, computer graphics, image processing, and visualization. He has served as a reviewer and program or organization committee member for several conferences. He is a member of the Editorial Advisory Board of international journals. He is member of the IEEE and the IEEE Computer Society.

Yen-Chi Lee received his BS degree in Electrical Engineering from National Taiwan University in 2008. Now he is studying for his PhD degree in Graduate Institute of Communication Engineering, National Taiwan University. His research interests are in the areas of wireless communications, including adaptive multimedia codec switching over wireless channels, optimal relay functionality in multiple relay networks, and physical layer approaches for wireless security.

Ulrike Lucke is Head of the Chair for Complex Multimedia Application Systems and Chief Information Officer (CIO) at the University of Potsdam, Germany. After graduation in Computer Science she worked at the University of Rostock from 1999 to 2010. She coordinated a number of research projects in the areas of eLearning, distributed architectures and mobile/pervasive computing, and was Co-Organizer and Program Co-Chair of several scientific workshops and conferences. She finished her PhD in 2006 with a work on architectures and tools for structured document processing. In 2010 she completed her habilitation thesis on innovative IT infrastructures for future Universities, and was offered a chair by the University of Potsdam. Ulrike Lucke is the speaker of the SIG eLearning of the German Informatics Society. She is married and has three children.

Taneli Mielikainen is working on large-scale data processing, mining and visualizations with Disco in Data Insight team at Nokia Research Center Palo Alto. He is also an adjunct professor in Computer Science at University of Helsinki, Finland. Prior to joining Nokia, he was doing research at HIIT BRU and FDK on various topics, including computational biology, data mining and combinatorial optimization. He defended his PhD Thesis on summarization techniques for pattern collections in data mining in May 2005. For his MSc thesis, he studied the computational complexity of determining relative orientations of cryo-electron microscope images of macromolecular complexes and implemented a software system Orient Express for experimenting with different heuristic search techniques.

Rosinei Oliveira graduated in Information Systems from the Federal University of Pará (UFPA) in 2008. In 2009, Oliveira earned a Masters in Electrical Engineering, also from the same institution. She has experience in Computer Science with emphasis in Computer Networking and Educational Technologies and also forms part of the research group of the Laboratory of High Performance Networks (LPRAD) acting on the following topics: measuring and modelling to evaluate performance of access technologies for digital inclusion, multimedia applications, Quality of Service and Quality of Experience.

Antonio Ou received his BS and MS degree both in Electrical Engineering from National Taiwan University in 2007 and 2009, respectively. His research interests include wireless multimedia transmissions and cross layer optimization in wireless networks.

Gianluca Paravati received the BSc and MSc degrees in electronic engineering from Politecnico di Torino, Torino, Italy, in 2005 and 2007, respectively. He is currently a Research Assistant with the Dipartimento di Automatica e Informatica, Politecnico di Torino. His research interest include real time image processing, target tracking in infrared imagery, and collaborative virtual environments. He has served as a reviewer for several international journals.

Nadia N. Qadri has more than four years of teaching and research experience at the renowned universities of Pakistan viz. Mehran University of Engineering & Technology, Fatima Jinnah Womens University and COMSATS Institute of Information Technology. Her research interests include video streaming for mobile ad hoc networks and vehicular ad hoc networks, along with P2P streaming.

Joaquín Salvachúa is a PhD on Telecommunications Engineering (1994) by the School of Telecommunications at the Polytechnic University of Madrid. He is a Telecommunications Engineer (1989) by ETSIT UPM. He is associated professor of the Department of Telematic Systems Engineering since 1995. He has worked in the validation and verification of protocols and distributed systems using formal languages, especially LOTOS, distributed multimedia applications for collaboration, distance learning and telepresence, and systems both for sharing P2P video stream to the performance database NoSQL high-performance data. Currently his research, within the group of next generation networks, focuses on distributed identity in social networks, analysis and recommendations based on science of complex networks, cloud computing oriented collaboration (Collaboration as a Service) and mixed & augmented reality. He has made numerous publications in magazines and books.

Andrea Sanna graduated in electronic engineering in 1993, and received the PhD degree in computer engineering in 1997, both from Politecnico di Torino, Italy. Currently, he has an associate professor position at the First Faculty of Architecture of Politecnico di Torino, Italy. He has authored and coauthored several papers in the areas of computer graphics, virtual reality, distributed computing, and computational geometry. He is a senior member of ACM.

Ádamo L. Santana received the BS degree in Computer Science from the University of the Amazon in 2002, master's (2005) and PhD (2008 in Applied Computing for Electric Engineering from the Federal University of Pará. Has experience in Computer Science, acting on the following subjects: computational intelligence, pattern recognition, data mining and time series.

Gilbert Seilheimer is currently working as a research assistant at the University of Applied Science Offenburg, Germany. After his apprenticeship in Information and Communication Technology he worked as a freelancer in various Web projects from 1998 to 2003. Gilbert Seilheimer studied Business Administration with the focus on Information Technology at the University of Applied Sciences Offenburg, Germany. Some of the core areas of his studies were marketing studies, human resources developments and poverty reduction strategies. Since living abroad in Buenos Aires (working experience), Gilbert Seilheimer has been working at the Department of Production Systems Engineering and Management since 2007. His research interests cover multimedia services, database and Web applications as well as integration of multimedia services into E-Commerce applications.

Tibor Szkaliczki received the MS and PhD degrees in electrical engineering from the Technical University of Budapest, Budapest, Hungary in 1992 and 1998, respectively. His PhD thesis discusses some algorithmic problems in routing of *Very-Large-Scale Integration* (*VLSI*) circuits. He joined the Computer and Automation Research Institute of the Hungarian Academy of Sciences, Hungary in 1997 and is currently a research fellow at the eLearning Department. He received Gyula Farkas Award from János Bolyai Mathematical Society in Hungary for the applications of mathematics. He received research scholarship at Institute of Information Technology of Klagenfurt University for five months in 2005 where he started studying algorithmic problems arising at multimedia systems. His research interests include distributed multimedia adaptation, multimedia adaptation in constrained and streaming environments, VLSI routing, and combinatorial optimization.

Ville H. Tuulos is a Member of Research Staff in Data Insight at the Nokia Research Center Palo Alto. Before joining NRC Palo Alto in 2007, he worked with the Complex Systems Computation (CoSCo) research group at the University of Helsinki, Department of Computer Science / HIIT. His main research topics were, and still are, scalable systems, cluster computing, statistical information retrieval (search engines), and general data mining and modeling.

Ngozi V. Uti received a BS summa cum laude in Computer Science from Northern Kentucky University (NKU) in 2002, and an MS in Computer Science from Indiana University (IU) in 2004. She is currently working towards the completion of her PhD in Computer Science and Engineering at the University of Cincinnati. She was an Associate Instructor in the Department of Computer Science at IU from 2002-2004 and an Instructor in the Department of Computer Science at NKU from 2008-2009. She was awarded the Outstanding Senior in Computer Science Award at NKU in 2003. She is an NSF Graduate Research Fellow and a Ford Pre-Doctoral Fellow for her 2007 awards of the National Science Foundation Graduate Research Fellowship and the Ford Foundation Pre-Doctoral Fellowship, both for her doctoral research. Her research interests include Mobile Computing, Artificial Intelligence, and Real-Time Mobile Video Compression, Analysis, and Streaming over Cell Phone Networks.

Zhung-Han Wu received his BS and MS degrees both in Electrical Engineering from National Taiwan University in 2008 and 2010, respectively. His research interests include wireless multimedia transmissions and cross layer optimization in wireless networks.

Ping-Cheng Yeh received his BS degree in Mathematics and MS degree in Electrical Engineering from the National Taiwan University, in 1996 and 1998, respectively. In 2005, he received his PhD degree in Electrical Engineering and Computer Science from the University of Michigan, Ann Arbor. He joined the Department of Electrical Engineering and the Graduate Institute of Communication Engineering at the National Taiwan University in August 2005. His research interests include wireless multimedia transmissions, physical layer security, cooperative communications, and cross-layer design in wireless networks. Dr. Yeh has received various awards in the past, including EECS Outstanding GSI Award (2002), University of Michigan Outstanding GSI Award (2003), NTU Excellent in Teaching Award (2008, 2009), and NTU Distinguished Teaching Award (2010).

Martin Zimmermann is professor in the Department of Production Systems Engineering at the University of Applied Sciences Offenburg, Germany. From 2002 to 2008, he managed the department as dean. He is the coordinator of the Lean Manufacturing lab. His research interests cover networked multimedia services with the vision of "seamless multimedia communications", i.e. media delivery platforms, streaming architectures, streaming to mobile devices, quality of service (e.g. network engineering), content distribution networks for streaming, and context aware communications. He studied Computer Science with the focus on Telecommunications at the Karlsruhe University of Technology, Germany, where he received the MSc (Dipl.-Inf.) degree in 1988. Working as a scientific assistant, he received his PhD in 1993. From 1993 until 1996 he worked as a senior consultant at the IBM European Networking Center in Heidelberg and Deutsche Bank Group in Frankfurt, Germany. He published over 50 papers in journals and conferences and received a best paper award at the *7th Advanced International Conference on Telecommunications AICT 2010.*

Index

Symbols

A

B

C

D

E

F

G

H

I

J

L

M

N

O

P

Q

R

S

T

U

V

CPSIA information can be obtained at www.ICGtesting.com
Printed in the USA
BVOW062146290911

272463BV00004B/6/P

9 781613 501443